ADVANCES IN PHOTONIC CRYSTALS

Edited by **Vittorio M. N. Passaro**

Advances in Photonic Crystals
http://dx.doi.org/10.5772/47821
Edited by Vittorio M. N. Passaro

Contributors

Luca Marseglia, Robinson Savarimuthu, Igor Guryev, Juan Ricardo Cabrera Esteves, Jose Amparo Andrade Lucio, Igor A. Sukhoivanov, Oscar Ibarra Manzano, Everardo Vargas Rodriguez, Natalia Gurieva, Wenfu Zhang, Wei Zhao, Marcin Koba, Volodymyr Fesenko, Sergiy Shulga, Tiziana Stomeo, Antonio Qualtieri, Ferruccio Pisanello, Luigi Martiradonna, Pier Paolo Pompa, Marco Grande, Antonella D'Orazio, Massimo De Vittorio, Boris Malomed, Thawtachai Mayteevarunyoo, Shunji Nojima, Zhiyuan Li, Lavrinenko, Vittorio M. N. Passaro

Published by InTech

Janeza Trdine 9, 51000 Rijeka, Croatia

Publishing Process Manager Ana Pantar
Technical Editor InTech DTP team
Cover InTech Design team

Additional hard copies can be obtained from orders@intechopen.com

Advances in Photonic Crystals, Edited by Vittorio M. N. Passaro
p. cm.
ISBN 978-953-51-0954-9

For EU product safety concerns:
IN TECH d.o.o., Prolaz Marije Krucifikse Kozulić 3, 51000 Rijeka, Croatia,
info@intechopen.com or visit our website at intechopen.com.

Meet the editor

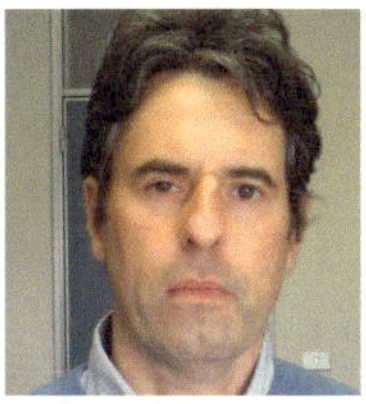

Vittorio M. N. Passaro received his Laurea degree in Electronic Engineering (cum laude) from University of Bari, Italy, in 1988, and his Ph.D. degree from Politecnico di Bari in 1992. In 2000 he joined Politecnico di Bari as an associate professor of Electronic Technologies and Photonics. Since 1988 his research interests have been focused on several theoretical and experimental aspects of integrated photonic technologies, in particular in silicon Photonics. In 2004 he formed the Photonics Research Group (URL: http://dee.poliba.it/photonicsgroup). He is the author or co-author of more than 270 papers in international journals and conference proceedings, the editor of three books and coauthor of two international patents, with more than 1200 citations. Prof. Passaro is IEEE Senior Member and OSA Senior Member.

Contents

Preface

After 1987 Yablonovitch's milestone paper, photonic crystals have been the topic of a huge number of papers. For many years, photonic crystals have been investigated both theoretically and experimentally because of their peculiar and intriguing properties for nanophotonic applications, such as laser generation, optical sensing, beam filtering, anisotropic property control, high field confinement, and so on. In particular, after the first demonstration of two-dimensional photonic crystal at optical wavelengths, planar slabs have been investigated to efficiently fabricate two-dimensional photonic crystals by etching the hosting slab or by forming pillars over the slab. To this aim, several technologies have been applied to derive photonic crystal properties in hosting materials, such as semiconductor slabs (III/V alloy compounds, silicon and compounds), metamaterials and others, as well as in photonic crystal fibers. Nowadays, many international research groups are still very active in this topic, since many theoretical aspects in modeling and design of photonic crystals, as well as in fabrication aspects, are not yet well standardized.

This book presents some advances of the international research in the field, collecting many chapters relevant to different theoretical and experimental aspects of photonic crystals, mainly two-dimensional, for Nanophotonics applications. Chapters are written by some important international research groups. The book is divided in two parts, a theoretical section followed by a section devoted to experiments and applications. First part includes chapters developing several numerical methods for analysis and design of photonic crystal devices, such as 2D ring resonators for filters, single and coupled nanobeam cavities, birefringence in photonic crystal cavities, propagation in anisotropic photonic crystals, threshold analysis in photonic crystal lasers, gap solitons in photonic crystals, novel photonic atolls, dynamic characteristics of linear and non linear photonic crystal filters. Second part includes four chapters focusing on many aspects of photonic crystals fabrication and applications, such as nitrogen defect technology in diamond, silicon nitride free standing membranes, silicon photonic crystals structures, applications of photonic crystals for optical sensing.

I would like to acknowledge the efforts of all the contributing authors for the best quality of the chapters collected in this book. Moreover, I would like to thank Ms. Mirna Cvijic, Ms. Sandra Bakic and Ms. Ana Pantar, who have subsequently followed the book publishing process, for their great help in the preparation of this book, in particular the tasks of chapter proposal collection and manuscript editing and correction.

Vittorio M. N. Passaro
Associate Professor
Politecnico di Bari
Bari, Italy

Section 1

Theory

Chapter 1

Photonic Crystal Ring Resonator Based Optical Filters

S. Robinson and R. Nakkeeran

Additional information is available at the end of the chapter

http://dx.doi.org/10.5772/54533

1. Introduction

Photonic Crystals are periodic nanostructures that are designed to affect the motion of photons in the same way as the periodic potential in a semiconductor crystal affects the electron motion by defining allowed and forbidden electronic energy bands [1, 2]. Generally, PCs are composed of periodic dielectric, metello-dielectric nanostructures, which have alternative lower and higher dielectric constant materials in one, two and/or three dimensions to affect the propagation of electromagnetic waves inside the structure. As a result of this periodicity, the transmission of light is absolutely zero in certain frequency ranges which is called as Photonic Band Gap (PBG).

By introducing the defects (point defects or line defects or both) in these periodic structures, the periodicity and thus the completeness of the PBG are entirely broken which allows to control and manipulate the light [1, 2]. It ensures the localization of light in the PBG region which leads to the design of the PC based optical devices.

2. History of photonic crystals

Electromagnetic wave propagation in periodic media is first studied by Lord Rayleigh in 1888. These structures are One Dimensional (1D) Photonic Crystals (1DPCs) which have a PBG that prohibits the light propagation through the planes. Although PCs have been studied in one form or another since 1887, the term "Photonic Crystal" is first used over 100 years later, after Yablonovitch and John published two milestone papers on PCs in 1988. Before that Lord Rayleigh started his study in 1888, by showing that such systems have a 1D PBG, a spectral range of large reflectivity, known as a stop-band. Further, 1DPCs in the form of periodic multi-layers dielectric stacks (such as the Bragg mirror) are studied extensively. Today, such structures are used in a diverse range of applications such as reflective coatings for enhancing the efficiency of Light Emitting Diodes (LEDs) and highly reflective mirrors in certain laser cavities.

In 1987, Yablonovitch and John have proposed 2DPCs and 3DPCs, which have a periodic dielectric structure in two dimensions and three dimensions, respectively. The periodic dielectric structures exhibit a PBG. Both of their proposals are concerned with higher dimensional (2D or 3D) periodic optical structures. Yablonovitch's main motivation is to engineer the photonic density of states, in order to control the spontaneous emission of materials that are embedded within the PC. In the similar way, John's idea is to affect the localization and control of light inside the periodic PC structure. Both of these works addresses the engineering of a structured material exhibiting ranges of frequencies at which the propagation of electromagnetic waves is not allowed, so called PBGs - a range of frequencies at which light cannot propagate through the structure in any direction.

After 1987, the number of research papers concerning PCs has begun to grow exponentially. However, owing to the fabrication difficulties of these structures at optical scales, early studies are either theoretical or in the microwave and optical regime, where PCs can be built on the far more readily accessible nanometer scale. By 1991, Yablonovitch has demonstrated the first 3D PBG in the microwave regime.

In 1996, Thomas Krauss made the first demonstration of a 2DPC at optical wavelengths. This opened up the modern way of fabricating PCs in semiconductor materials by the methods used in the semiconductor industry. Although such techniques are still to mature into commercial applications, 2DPCs have found commercial use in the form of Photonic Crystal Fibers (PCFs) and optical components. Since 1998, the 2DPCs based optical components such as optical filters [3,4], multiplexers [5], demultiplexers [6], switches [7], directional couplers [8], power dividers/splitters [9], sensors [10,11] etc., are designed for commercial applications.

3. Types of photonic cyrstals

PCs are classified mainly into three categories according to its nature of structure periodicity, that is, One Dimensional (1D), Two Dimensional (2D), and Three Dimensional (3D) PCs. The geometrical shape of 1DPCs, 2DPCs and 3DPCs are shown in Figure 1 where the different colors represent material with different dielectric constants. The defining structure of a PC is the periodicity of dielectric material along one or more axis. The schematic illustrations of 1DPCs, 2DPCs and 3DPCs are depicted in Figures 2(a), 2(b) and 2(c), respectively.

3.1. One dimensional PCs

In 1DPCs, the periodic modulation of the refractive index occurs in one direction only, while the refractive index variations are uniform for other two directions of the structure. The PBG appears in the direction of periodicity for any value of refractive index contrast i.e., difference between the dielectric constant of the materials. In other words, there is no threshold for dielectric contrast for the appearance of a PBG. For smaller values of index contrast, the width

http://dx.doi.org/10.5772/54533

of the PBG appears very small and vice versa. However, the PBGs open up as soon as the refractive index contrast is greater than one ($n_1/n_2 > 1$), where n_1 and n_2 are the refractive index of the dielectric materials. A defect can be introduced in a 1DPCs, by making one of the layers to have a slightly different refractive index or width than the rest. The defect mode is then localized in one direction however it is extended into other two directions. An example for such a 1DPC is the well known dielectric Bragg mirror consisting of alternating layers with low and high refractive indices, as shown in Figure 2(a).

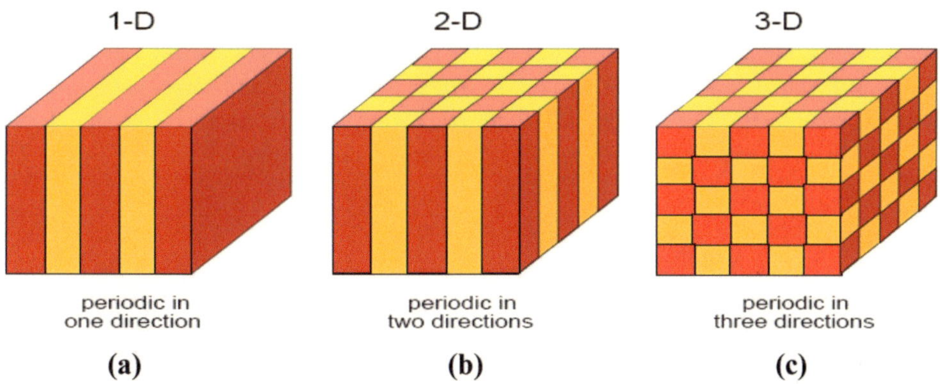

Figure 1. Geometrical shapes of photonic crystals (a) 1D (b) 2D and (c) 3D

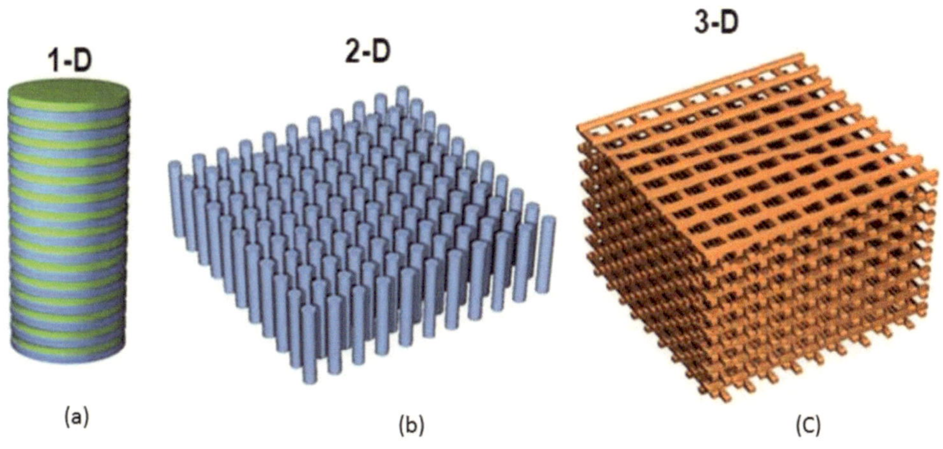

Figure 2. Schematic illustrations of photonic crystals (a) 1D (b) 2D and (c) 3D

The wavelength selection and reflection properties in 1DPCs are used in a wide range of applications including high efficiency mirrors [12,13], optical filters [14,15, 16], waveguides [17], and lasers [18]. Also, such structures are widely used as anti-reflecting coatings which dramatically decrease the reflectance from the surface and used to improve the quality of the lenses, prisms and other optical components.

3.2. Two dimensional PCs

PC structure(s) that are periodic in two different directions and homogeneous in third direction are called 2DPC which is shown in Figure 1.(b) and 2(b). In most of the 2DPCs, the PBG occurs when the lattice has sufficiently larger index contrast. If the refractive index contrast between the cylinders (rods) and the background (air) is sufficiently large, 2D PBG can occur for propagation in the plane of periodicity perpendicular to the rod axis.

Generally, 2DPCs consist of dielectric rods in air host (high dielectric pillars embedded in a low dielectric medium) or air holes in a dielectric region (low dielectric rods in a connected higher dielectric lattice) as shown in Figures 3(a) and 3(b). The dielectric rods in air host give PBG for the Transverse Magnetic (TM) mode where the E field is polarized perpendicular to the plane of periodicity. The air holes in a dielectric region give (Transverse Electric) TE modes where H field is polarized perpendicular to the plane of periodicity.

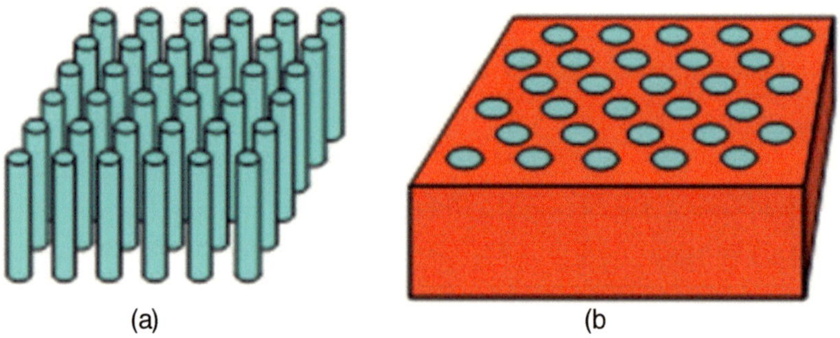

Figure 3. Structure of (a) dielectric rods in air and (b) air holes in dielectric region

Based on the value of vertical index contrast the structures can have, they are categories into the following four geometries:

- Membrane Holes : Hole type PCs with a high vertical index contrast
- Membrane Pillars : Pillar based PCs with a high vertical index contrast
- Deeply etched Holes : Hole type PCs with a low vertical index contrast
- Deeply etched Pillars : Pillar based PCs with a low vertical index contrast

Above all, the membrane holes and pillars with high vertical index contrast received a crucial role for device realization.

3.3. Three Dimensional PCs

A 3DPCs is a dielectric structure which has periodic permittivity modulation along three different axes, provided that the conditions of sufficiently high dielectric contrast and suitable periodicity are met, a PBG appears in all directions. Such 3D PBGs, unlike the 1D and 2D ones, can reflect light incident from any direction. In other words, a 3D PBG material behaves as an omnidirectional high reflector. As an example, Figure 4 depicts the 3D woodpile structure.

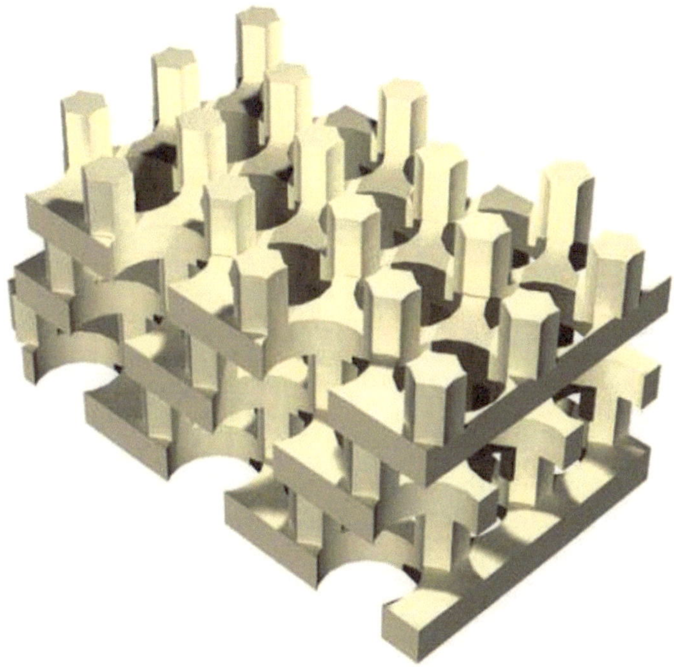

Figure 4. Structure of 3D woodpile photonic crystals

Due to the challenges involved in fabricating high-quality structures for the scale of optical wavelengths, early PCs are performed at microwave and mid-infrared frequencies [19, 20]. With the improvement of fabrication and materials processing methods, smaller structures have become feasible, and in 1999 the first 3DPC with a PBG at telecommunications frequencies is reported [21, 22]. Since then, various lattice geometries have been reported for operation at similar frequencies [23, 24]. Waveguide and the introduction of intentional defects in 3DPCs has not progressed as rapidly as in 2DPCs, due to the fabrication difficulties and the more complex geometry required to achieve 3D PBGs.

4. Numerical analysis

There are many methods available to analyze the dispersion behavior and transmission spectra of PCs such as Transfer Matrix Method (TMM) [25], FDTD method [26], PWE method [27], Finite Element Method [28] (FEM) etc.,. Each method has its own pros and cons. Among these, PWE and FDTD methods are dominating with respect to their performance and also meeting the demand required to analyze the PC based optical devices.

The PWE method is initially used for theoretical analysis of PC structures, which makes use of the fact that Eigen modes in periodic structures can be expressed as a superposition of a set of plane waves. Although this method can obtain an accurate solution for the dispersion properties (propagation modes and PBG) of a PC structure, it has still some limitations. i.e., transmission spectra, field distribution and back reflections cannot be extracted as it considers only propagating modes. An alternative approach which has been widely adopted to calculate both transmission spectra and field distribution is based on numerical solutions of Maxwell's equations using FDTD method. Typically, the PWE method is used to calculate the PBG and propagation modes of the PC structure and FDTD is used to calculate the spectrum of the power transmission.

5. Applications of 2DPCs

The ability to control and manipulate the spontaneous emission by introducing defects in PCs, and related formation of defect state within PBG has been used for designing the optical devices for different applications that are directed towards the integration of photonic devices. 2DPCs is the choice of great interest for both fundamental and applied research, and also it is beginning to find commercial applications. K. Inoue et al 2004 have summarized the use PCs in various applications as shown in Figure 5.

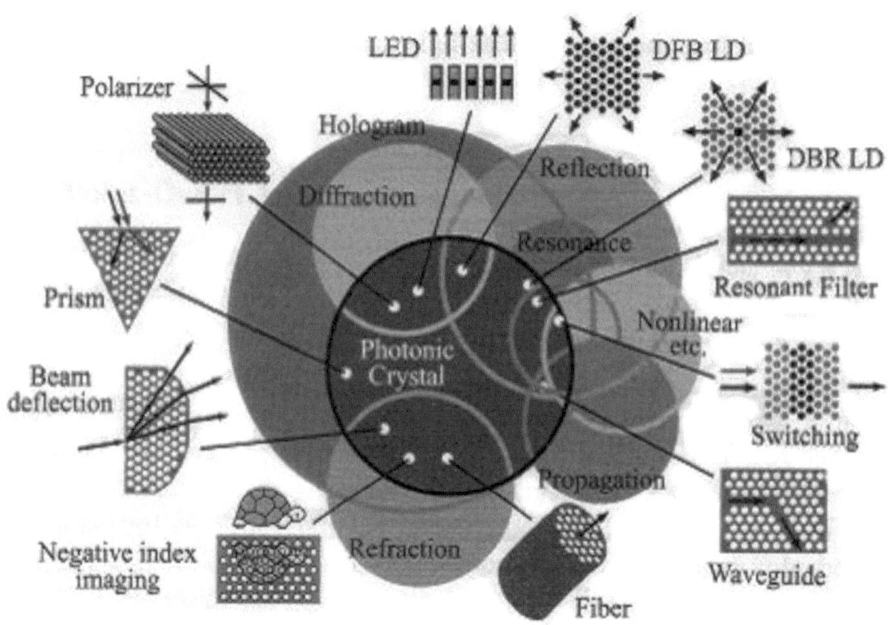

Figure 5. Applications of photonic crystals

The majority of PC applications utilize the phenomenon of PBG that opens the new road to design optical components in micrometer (μm) range. Waveguides that confine light via PBGs are a new development. Generally, the waveguide is intended to transport waves of a particular frequency from one place to another place through a curved path. Using this waveguide many optical components are reported in the literature such as power splitter/ power divider [29] which divides the power in an input waveguide equally between output waveguides, Y splitter [30], and directional couplers [31] and so on.

It is also possible to design a cavity, formed by the absence of a single rod or group of rods (point defects), which is positioned between two waveguides each of which is formed by the absence of a row of rods (line defects). Various geometries of the micro cavities have been explored over the years with a goal of increasing the Q factor of a cavity, while reducing the cavity size. Two main cavity geometries can be distinguished as those are based on point defect based cavity [32] and line/point defect (PCRR) based cavities [33]. Such a cavity is useful for optical filters [34], lasers [35], multiplexers and demultiplexers [36] etc,.

6. Optical ring resonator

An optical ring resonator is positioned between two optical waveguides to provide an ideal structure of the ring resonator based ADF. At resonant condition, the light (signal) is dropped from the bus (top) waveguide and it is sent to the dropping (bottom) waveguide through ring resonator. The schematic structure of the ring resonator based ADF is shown in Figure 6, which consists of a bus waveguide and dropping waveguide, and ring resonator. The ring resonator acts as a coupling element between the waveguides. Also, it has four ports, ports 1 and 2 are the input terminal and transmission output terminals whereas ports 3 and 4 are forward and backward dropping terminals, respectively.

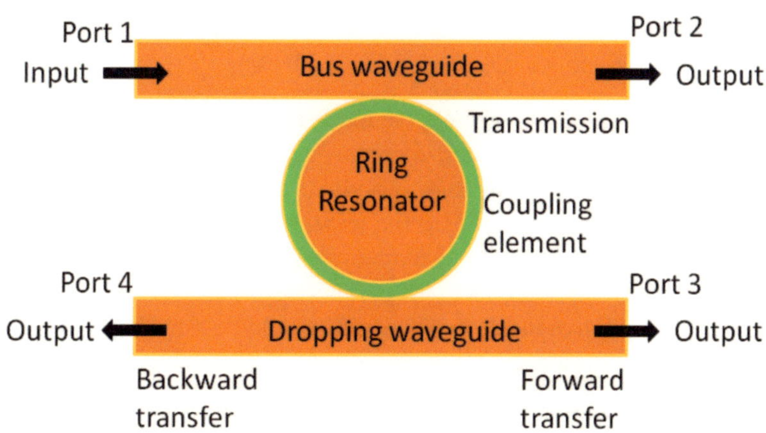

Figure 6. Schematic structure of the ring resonator based ADF

In PC structures, there are two ways to design optical resonator as follows,

i. Line defect or point defect based resonators - changing the size or dielectric constant of rods in the structure

ii. Ring Resonators (RRs) - removing some rods in order to have a ring shape

In RR based devices, the choice of the ring size is determined by the desired resonant wavelength and the tradeoff between the cavity Q and the modal volume V [34]. Compared to point defect or line defect PC cavities, Photonic Crystal Ring Resonators (PCRRs) offer scalability in size, flexibility in mode design due to their multi mode nature [37], easy integration with other devices and adaptability in structure design.

6.1. Operating principle

If the ring resonator supports only one resonant mode, it will decay through both waveguides along the forward and backward directions which introduces the reflection. Hence, in order for complete transfer to happen, at least two modes are needed for the decaying amplitudes to cancel either the backward direction or the forward direction of the bus waveguide (Fan et al 1998).

Two mirror planes can be considered for this structure, one is perpendicular to the waveguides and another is parallel to the waveguides. In order to cancel the reflected signal, a structure with a mirror plane symmetry perpendicular to both waveguides is considered. Assume that there exist two localized modes that have different symmetries with respect to the mirror plane: one has even symmetry and another has odd symmetry. The even mode decays with the same phase into the forward and backward directions as shown in Figure 7(a), however the odd mode decays into the forward direction, out of phase with the decaying amplitude along the backward direction as shown in Figure 7(b). When the two tunneling processes come together, the decaying amplitudes into the backward direction of both waveguides are canceled, which clearly depicts in Figure 7(c). It should be noted that, in order for cancellation to occur, the line shapes of the two resonances should overlap. It means both resonances must have significantly the same resonant wavelength and the same bandwidth [32].

Also, due to the occurrence of degeneracy, the incoming wave interferes destructively with the decaying amplitude into the forward direction of the bus waveguides, causing all the power traveling in the bus waveguide to be cancelled. The symmetry of the resonant modes with respect to the mirror plane parallel to the waveguides determines the direction of the transfer wave in the ADF. For instance, as it apparent from Figures 8(a), 8(b) and 8(c), when both of the modes are even with regard to the parallel mirror plane, the decaying amplitudes along the backward direction of the drop waveguide would be canceled, letting all the power be transferred into the forward direction of the drop waveguide. On the other hand, the even mode could be odd with respect to the mirror plane parallel to the waveguides. When the accidental degeneracy between the states occurs, the decaying amplitudes cancel in the forward direction of the drop waveguide (Figures 8(a), 8(b) and 8(c)). Entire power is transferred into the backward direction of the drop waveguide [32].

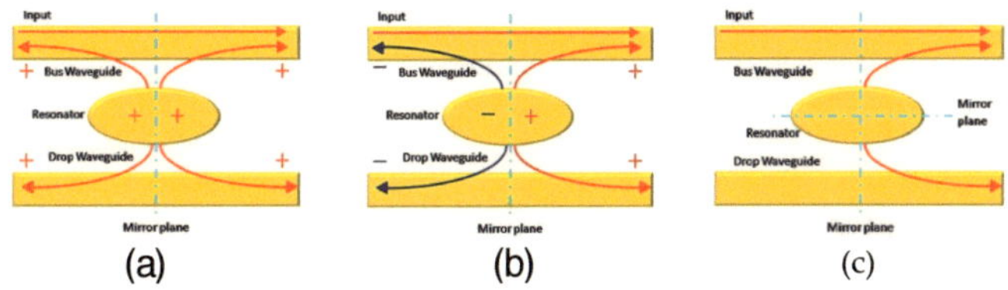

Figure 7. Channel drop tunneling process for a resonator system that supports forward transfer of signal

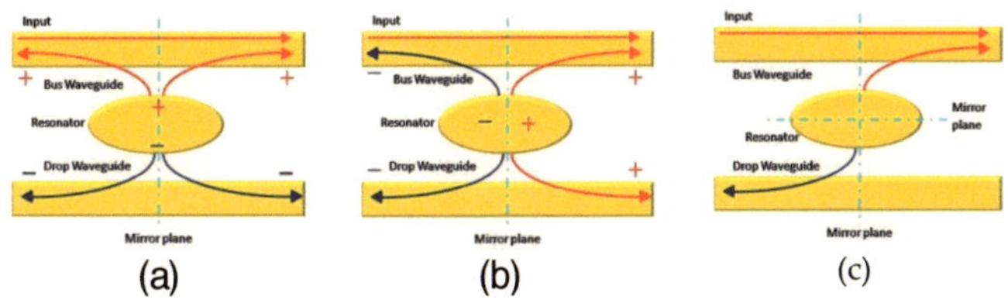

Figure 8. Channel drop tunneling process for a resonator system that supports backward transfer of signal

The PCRR resonant coupling occurs due to the frequency and phase matching between the propagating waveguide mode and the PCRR resonant cavity mode. The coupling direction is mainly determined by the modal symmetry and the relative coupling between the PCRRs. The direction is the same for the propagating wave in the waveguide and the coupled wave inside PCRR. However, the direction may be the same or reverse for the coupling between PCRRs, depending upon the coupling strength and the modal symmetry [32]. Both forward dropping and backward dropping can be obtained depending upon the mode symmetry properties with respect to the coupling configurations.

6.2. Requirements of the ADF

The filter performance is determined by the transfer efficiency between the two waveguides. Perfect efficiency corresponds to complete transfer of the selected channel in either forward or backward direction in the dropping waveguide without forward transmission or backward reflection in the bus waveguide. All other channels remain unaffected by the presence of optical resonators.

To achieve complete transfer of the signal at resonance, the PCRR based ADF must satisfy the following three conditions:

i. The resonator must possess at least two resonant modes, each of them must be even and odd, with respect to the mirror plane of symmetry perpendicular to the waveguides

ii. The modes must degenerate

iii. The modes must have equal Q

All three conditions are necessary to achieve complete transfer of the signal from the bus waveguide to PCRR and PCRR to drop waveguides.

7. Photonic crystal ring resonator based ADF

The PCRR based ADF is designed using two dimensional pillar type PC with circular rods and consists of an array of rods in square lattice, as shown in Figure 9(a). The number of rods in 'X' and 'Z' directions is 21. The distance between the two adjacent rods is 540 nm, which is termed as lattice constant, 'a'. The Si rod with refractive index 3.47 is embedded in the air. The radius of the rods is 0.1 μm and the overall size of the device comes around 11.4 μm × 11.4 μm. The band diagram in Figure 9(b) gives the propagation modes and PBG of the PC structure, which has TM PBG ranging from 0.295 a/λ to 0.435 a/λ whose corresponding wavelength lies between 1241 nm and 1830 nm. It covers the entire wavelength range of third optical communication window. The guided modes (even and odd) inside PBG region resulting due to line and point defects (21×21 PC) are shown in Figure 9(c) which supports the complete channel transfer in turn higher output efficiency at resonance. The structure is surrounded by Perfect Matched Layer (PML) as absorbing boundary conditions to truncate the computational regions and to avoid the back reflections from the boundary [38].

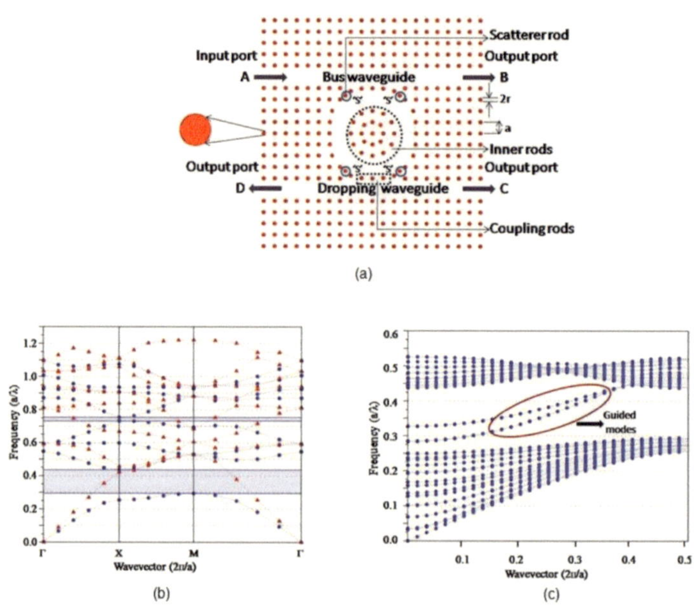

Figure 9. a) Schematic structure of circular PCRR based ADF (b) band diagram of 1 × 1 PC (unit cell) and (c) band diagram of 21 ×2 1 PC (super cell) structure after the introduction of line and point defects

The normalized transmission spectra of the circular PCRR based ADF is obtained using 2D Finite Difference Time Domain (FDTD) method. Although the real SOI structure, would, in practice, require 3D analysis, our 2D approach gives a general indication of the expected 3D behavior. 2D analysis carried out here allows us to identify qualitatively many of the issues in the cavity design (e.g. mode control, cavity Q and the placement of the scatterers in our quasi-square ring cavity) and the coupling scheme design. This can offer us the design trade-offs and guidelines before the real structure design based on a completely 3D FDTD technique, which is typically computational time and memory consuming.

The circular PCRR based ADF (in Figure 9 (a)) consists of two waveguides in horizontal (r-x) direction and a circular PCRR is positioned between them. The top waveguide is called as bus waveguide whereas the bottom waveguide is known as dropping waveguide. The input signal port is marked 'A' with an arrow on the left side of bus waveguide. The ports 'C' and 'D' of drop waveguide is the drop terminals and denoted as forward dropping and backward dropping, respectively, while the port 'B' on the right side of bus waveguide is designated as forward transmission terminal.

The bus and the dropping waveguides are formed by introducing line defects whereas the circular PCRR is shaped by creating point defects (i.e. by removing the columns of rods to make a circular shape). The circular PCRR is constructed by varying the position of inner rods and outer rods from their original position towards the center of the origin (r). The inner rods are built by varying the position of adjacent rods on the four sides, from their center, by 25%, on the other hand the outer rods are constructed by varying the position of the second rod on the four sides, from their center, by 25% in both 'X' and 'Z' directions. The number of rings that are formed by the ring is three. In order to improve the coupling efficiency, dropping efficiency and spectral selectivity by suppressing the counter propagation modes, the scatterer rods (labeled as 's') are placed at each corner of the four sides with half lattice constant. The material properties and dimension of the scatterer rods are similar to the other rods. The rods which are located inside the circular PCRR are called inner rods whereas the coupling rods are placed between circular PCRR and waveguides. At resonance, the wavelength is coupled from the bus waveguide into the dropping waveguide and exits through one of the output ports. The coupling and dropping efficiencies are detected by monitoring the power at ports 'B' and, 'C' and 'D', respectively.

A Gaussian input signal is launched into the input port. The normalized transmission spectra at ports 'B', 'C' & 'D' are obtained by conducting Fast Fourier Transform (FFT) of the fields that are calculated by 2D-FDTD method. The input and output signal power is recorded through power monitors by placing them at appropriate ports. The normalized transmission is calculated through the following formula:

$$T(f) = \frac{1/2\int real(p(f)^{monitor})dS}{Source\ Powe}$$

where T(f) is normalized transmission which is a function of frequency, p(f) is poynting vector and dS is the surface normal. The normalization at the output side does not affect the result because of source power normalization. Finally, the T(f) is converted as a function of wavelength.

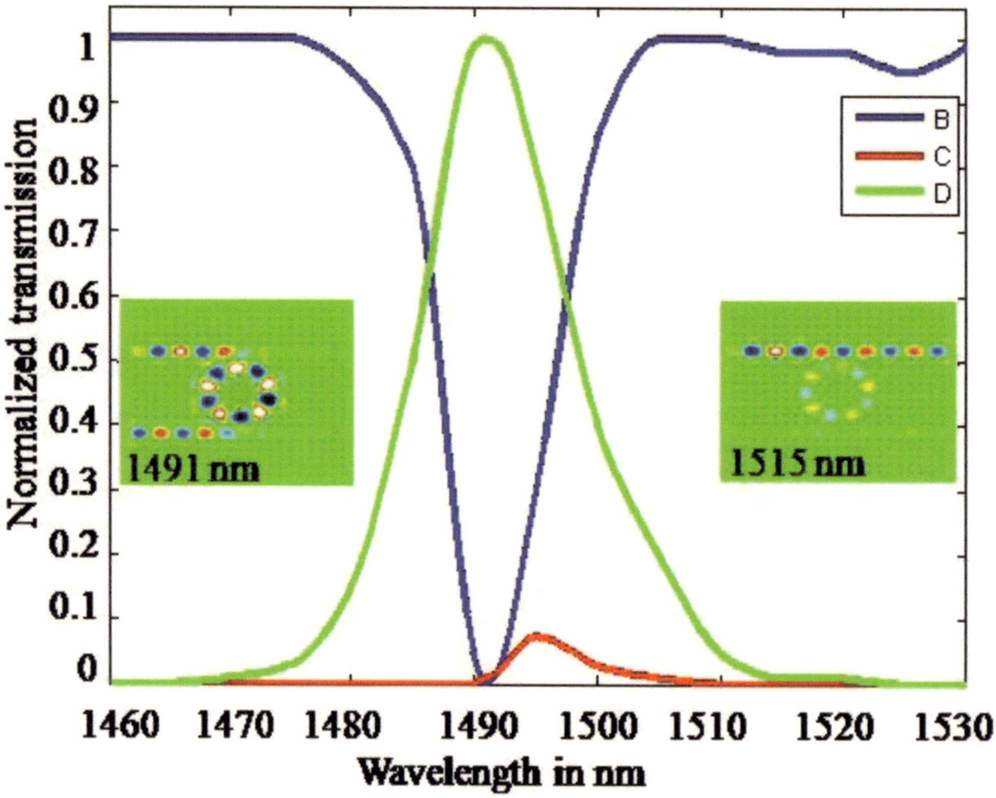

Figure 10. Normalized transmission spectra of circular PCRR based ADF

Figure 10 shows the normalized transmission spectra of circular PCRR based ADF. The resonant wavelength of the ADF is observed at 1491 nm. The simulation shows 100% coupling and dropping efficiencies and its passband width is 13 nm. The Q factor, which is calculated as $\lambda/\Delta\lambda$ (resonant wavelength/full width half maximum), equals to almost 114.69. The obtained results meet the requirements of ITU-T G 694.2 CWDM systems. The inset in Figure 10 depicts the electric field pattern of pass and stop regions at 1491 nm and 1515 nm, respectively. At a resonant wavelength, λ=1491 nm the electric field of the bus waveguide is fully coupled with the ring and reached into its output port D. In this condition there is no signal flow in port B. Similarly, at off resonance, λ=1515 nm the signal directly reaches the transmission terminal (the signal is not coupled into the ring). Figure 11 clearly illustrates the three dimensional view of PCRR based ADF. It shows the arrangement of Si rods in the structure and the overall dimensions of the device would come around 11.4 μm (length) × 11.4 μm (width). The effect of point to point network after incorporating the PCRR based ADF is discussed in the following sections.

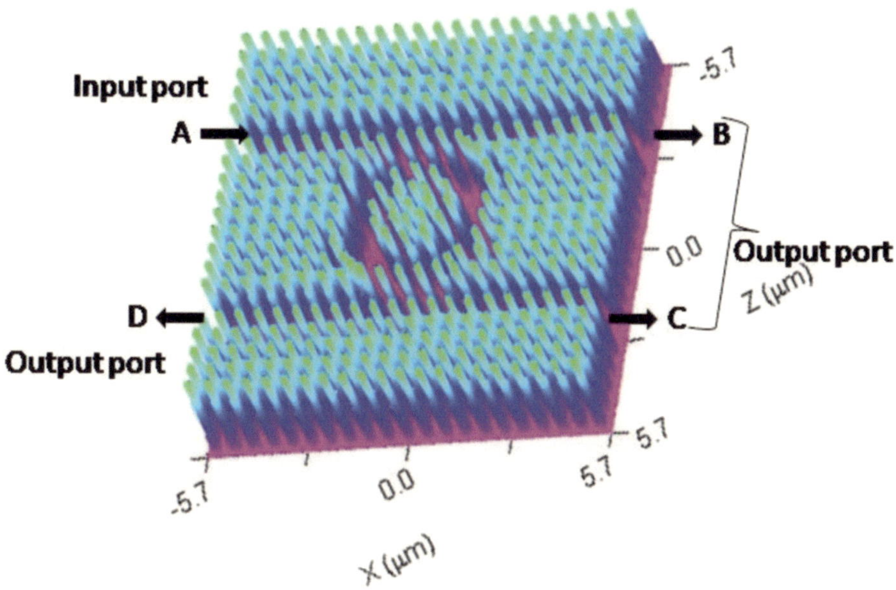

Figure 11. Three dimensional view of circular PCRR based ADF

7.1. Tuning of Resonant Wavelength

Although the PCRR based ADFs have a fixed operating wavelength, the application area will become much broader if the operating wavelength can be tuned dynamically and externally. This would greatly improve the utilization of PC based optical devices for real time and on demand applications. Generally, the resonant wavelength tuning of PCRR based ADF can be done by altering the structural parameters such as refractive index (dielectric constant), lattice constant and radius of the rods in the structure. Among these, the most efficient way to tune the resonant (operating) wavelength of the ADF is changing the refractive index of the material since it is not resulting in degradation of filter performance. Recent year, the exploration of tunablity for 2D PC based optical devices is mainly being carried out with respect to the refractive index [39, 40], lattice constant [41] and radius of the rod [42]. There are several tuning mechanisms such as thermal tuning [39], mechanical tuning [42, 43], MEMS actuator [44] etc., are reported to change the structural parameters. Here, the changes in refractive index, the radius of the rod and lattice constant are considered to examine the possibility of resonant wavelength tuning.

The normalized transmission spectra with respect to the refractive index difference, radius of the rod and lattice constant are shown in Figures 12 (a), (b) and (c), respectively. All the three cases, while varying the structural parameters the coupling and dropping efficiencies are not changing however there is a trivial change in passband width in turn Q factor.

It is observed that, while increasing (decreasing) the value of refractive index, lattice constant and radius of the rod, the resonant wavelength of the filter shifts into the longer wavelength (shorter wavelength). However, the other filter parameters such as coupling efficiency, dropping efficiency and Q factor are not affected while changing the refractive index and radius of the rod. There is a significant change is observed while varying the lattice constant.

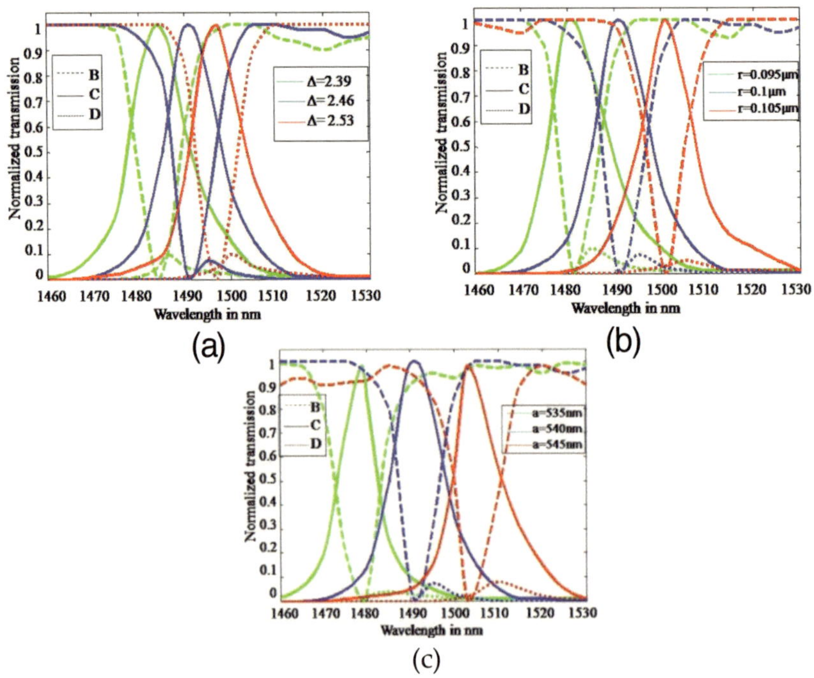

Figure 12. The effect of normalized transmission spectra of the circular PCRR based ADF for varying : (a) refractive index difference (b) radius of the rod and (c) lattice constant

Further, to investigate the impact of resonance for small variation in structural parameters, the simulation is carried out with very small step value. The accounted step value for refractive index difference, radius of the rod and lattice constant is 0.01, 0.001 μm and 1 nm, respectively whose corresponding resonant wavelength shift is shown in Figure 13(a). While considering the change in refractive index, other two parameters are kept constant and vice versa. The shift in resonant wavelength for an infinitesimal change in the refractive index, radius of the rods and lattice constant is given below:

$\Delta\lambda / \Delta n = 1$ nm / 0.01 (for refractive index difference)

$\Delta\lambda / \Delta r = 2$ nm / 0.001 μm (for radius of the rods)

$\Delta\lambda / \Delta a = 2.2$ nm / 1 nm (for lattice constant)

where $\Delta\lambda$ is the shift in resonant wavelength, Δn is the change in refractive index difference, Δr is the change in radius of the rod and Δa is the change in lattice constant. It means that there is 1 nm shift in resonant wavelength for every change in 0.01 values of the refractive index difference.

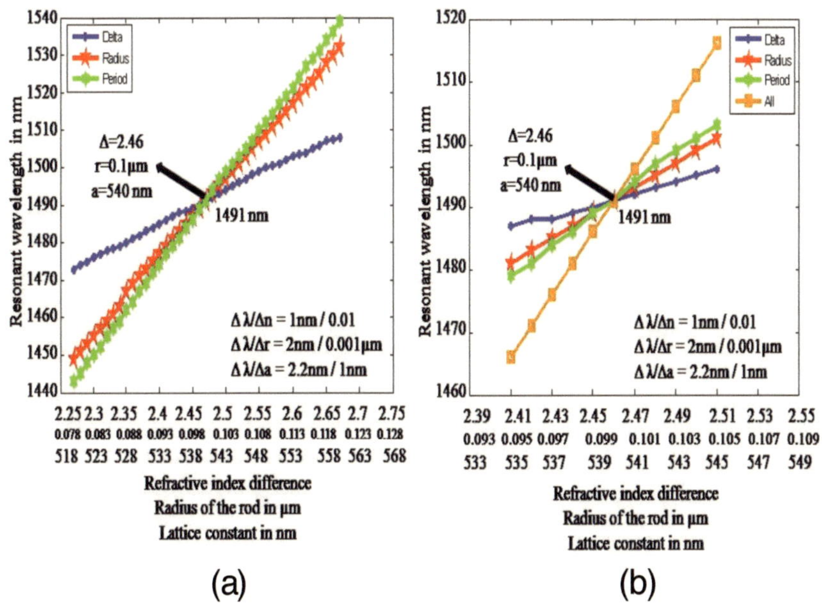

Figure 13. Effect of resonant wavelength shift with respect to refractive index difference, radius of the rod and lattice constant (a) individually and (b) combinedly

The wide tuning range (1471 nm to 1611 nm) is possible by altering any one of the structural parameters. If we considered only one parameter to arrive wide tuning range, the required change in parameter is large which affects the filter parameters. It can be figured out by varying all the structural parameters simultaneously instead of changing any one of the parameters. As expected, there is 5.2 nm resonance shift observed while simultaneously changing the refractive index difference, radius of the rod and lattice constant by 0.01, 0.001 μm and 1 nm, respectively, from the reference value. As discussed earlier, for every change in 0.01 refractive index, 0.001μm radius of the rod and 1 nm lattice constant, there is 1 nm, 2 nm and 2.2 nm resonant wavelength shift is observed. If there is a uniform step change in all the parameters, the cumulative individual resonance shift of (1nm+2nm+2.2nm) 5.2 nm is noted, which is shown in Figure 13(b).

8. BPF using quasi-waveguides

In Wavelength Division Multiplexing (WDM) systems, the number of incoming channels are departed into an optical fiber with designated wavelengths. Hence, optical filters are necessary to select a required channel(s) at any destination. The BPF is a right device to select either a single or multiple channels from the multiplexed signals. In the literature, PC based BPF has been designed by introducing point defects and/or line defects [44, 45], using bi-periodic structures [46] and using liquid crystal photonic bandgap fibers [47]. Moreover, no other methods are reported to design PC based BPF. The circular PCRR based BPF is designed by exploiting the coupling between the quasi-waveguides and circular PCRR and its simulation results are presented.

8.1. Design of the structure

The structural parameters such as radius of the rod (0.1μm), lattice constant (540 nm), and refractive index (3.46) are chosen to be similar to the previous one. However, the total number of rods in the structure in 'X' and 'Z' directions is 21 and 19, respectively. As the basic structure (rods in air) and its parameters are similar to previous one, therefore, the PBG ranges are also similar.

Figure 14 sketches the schematic structure of the circular PCRR based BPF. The BPF consists of two quasi waveguides in horizontal (Γ-x) direction and a circular PCRR between them. The Gaussian signal is applied to the port marked 'A' (arrow in the left side of top quasi waveguide) and the output is detected using power monitor which is positioned at the output port marked 'B' (arrow left side of the bottom quasi waveguide). The coupling rod is placed between circular PCRR and quasi waveguides, marked as 'c'. The reflectors, demarcated in a rectangular box, placed above and below the right side of circular PCRR are shown in Figure 14, which are used to improve the output efficiency of the BPF by reducing the counter propagation modes. In order to enhance the output efficiency and maintain the structure in symmetric nature, the number of periods (Si rods) in the reflector is kept constant, 9.

The structural parameters such as radius of the rod (0.1μm), lattice constant (540 nm), and refractive index (3.46) are chosen to be similar to the previous one. However, the total number of rods in the structure in 'X' and 'Z' directions is 21 and 19, respectively. As the basic structure (rods in air) and its parameters are similar to previous one, therefore, the PBG ranges are also similar.

Figure 14 sketches the schematic structure of the circular PCRR based BPF. The BPF consists of two quasi waveguides in horizontal (Γ-x) direction and a circular PCRR between them. The Gaussian signal is applied to the port marked 'A' (arrow in the left side of top quasi waveguide) and the output is detected using power monitor which is positioned at the output port marked 'B' (arrow left side of the bottom quasi waveguide). The coupling rod is placed between circular PCRR and quasi waveguides, marked as 'c'. The reflectors, demarcated in rectangular box, placed above and below the right side of circular PCRR are shown in Figure 14, which are used to improve the output efficiency of the BPF by reducing the counter propagation modes. In order to enhance the output efficiency and maintain the structure in symmetric nature, the number of periods (Si rods) in the reflector is kept constant, 9.

http://dx.doi.org/10.5772/54533

8.2. Simulation results and discussion

The normalized transmission spectra of PCRR based BPF are shown in Figure 15(a). The observed output efficiency is approximately 85% at 1420 nm and close to 100% over the range of wavelengths 1504 nm to 1521 nm whose corresponding bands are denoted as Band I and Band II, respectively. The center wavelength and FWHM bandwidth of these bands are 1420 nm and 1512.5 nm, and 20 nm and 35 nm, respectively. Also, the calculated Q factor of Band I and Band II is 71 and 50.41, respectively. As it is witnessed, the number of passbands depends on the number of inner rings that are formed by the inner rods. Here, the two inner rings considered results in two passbands. The size and shape of the ring resonator determines the resonant wavelength. The bandwidth and channel spacing are decided by the other structural parameters namely, radius of the rod, period and dielectric constant (refractive index) of the material.

The Figure 15(b) illustrates the relation between the output efficiency and wavelength shift for different dielectric constant of the structure. It can be seen clearly that the center wavelength of the bands shifts into the lower wavelength region when the dielectric constant of structure is decreased, and similarly the center wavelength of the bands shifts into the higher wavelength region when the dielectric constant of the structure is increased. It is also noticed that the output efficiency is not significantly changed while varying the dielectric constant of the structure. The magnitude of the wavelength shift is around 9 nm for every 0.5 change in dielectric constant value of the structure. However, the bandwidth is almost not affected by the variation of dielectric constant.

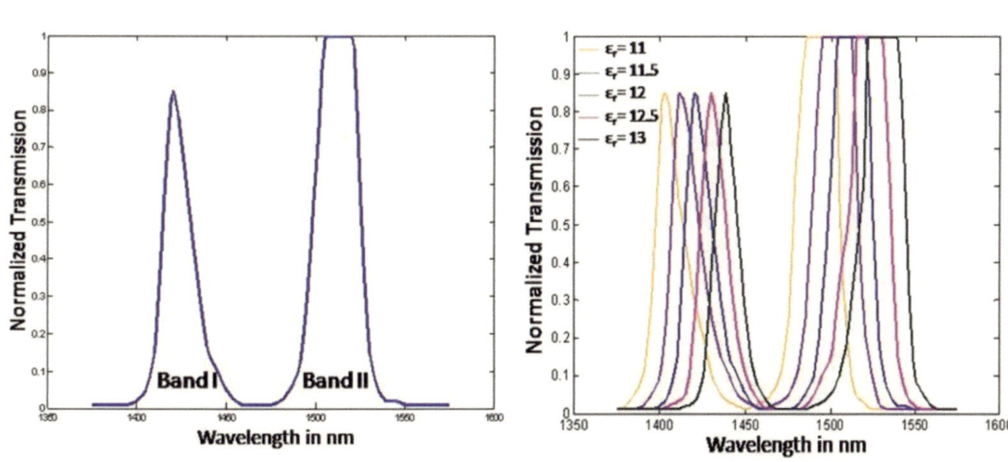

(a) (b)

Figure 15. Normalized transmission spectra of (a) circular PCRR based BPF and (b) for different values of dielectric constant

9. PCRR based BSF

Essentially, BSF is one of the prominent components to suppress (remove) either single or multiple unwanted channels from the multiplexed output channels, or also it passes most of the frequency range unaltered, however it attenuates/stops a specific range. In literature, the PC based BSF have been designed by introducing point and line defects [49], and using square and rectangular resonant cavity [50]. As the cavity size is small in the defects based BSFs, it does not provide the wide stopband width even it has higher stopband efficiency. Though, the square and rectangular cavities based BSFs offers a wide stopband width, it reduces the stopband efficiency owing to scattering at corners in resonance condition as it has proper corner. The proposed circular PCRR has gradual changes at corner and subtle in nature which is considered here for designing BSF.

9.1. Design of the structure

The proposed BSF is designed using 2D square lattice PCs with circular PCRR, which is shown in Figure 16. The number of rods in 'X' and 'Z' directions (21), lattice constant (540 nm), radius of the rod (0.1 μm) and refractive index of the rods (3.46) are similar as the filters discussed in the previous chapters. The PCRR based BSF, consists of a waveguide in horizontal (Γ-x) direction and a circular PCRR below the waveguide. The waveguide is called as bus waveguide and the ring resonator has 4 rings of Si rods in the inner rods (cavity). The bus waveguide is formed by introducing line defects and the circular PCRR is shaped by point defects.

The circular PCRR consists of four rings in the inner cavity, which is constructed by varying the position of both inner and outer rods from their original position towards center of the origin (Γ). In the four rings inner cavities, the center rod in the structure is considered as the first ring and the second ring is placed around the first ring and then third ring followed by the fourth ring. The inner rods are built by varying the position of adjacent rods in the four sides, from their center, by 25%, whereas the outer rods are constructed by varying the position of the second rod on four sides, from its center, by 25% both in 'X' and 'Z' directions where 'X' is the horizontal direction and 'Z' is the vertical direction. The position of the rods is varied by varying the lattice constant.

At resonance, the signal is coupled into the PCRR from bus waveguide and reflected back to the input port, hence the signal is not reached into the output at that resonant condition. This behavior is used to stop single or multiple channels from the multiplexed input/output channels. The stopband efficiency is obtained by monitoring the power at port 'B' at resonant condition.

Figure 17 shows the normalized transmission spectra of PCRR based BSF. The stopband efficiency of the BSF is approximately 98% and the width of the stopband is 11 nm. Here, the stopband width is calculated at FWHM point and the stopband efficiency is computed by subtracting the detected output power from the normalized transmission value and multiplying by 100.

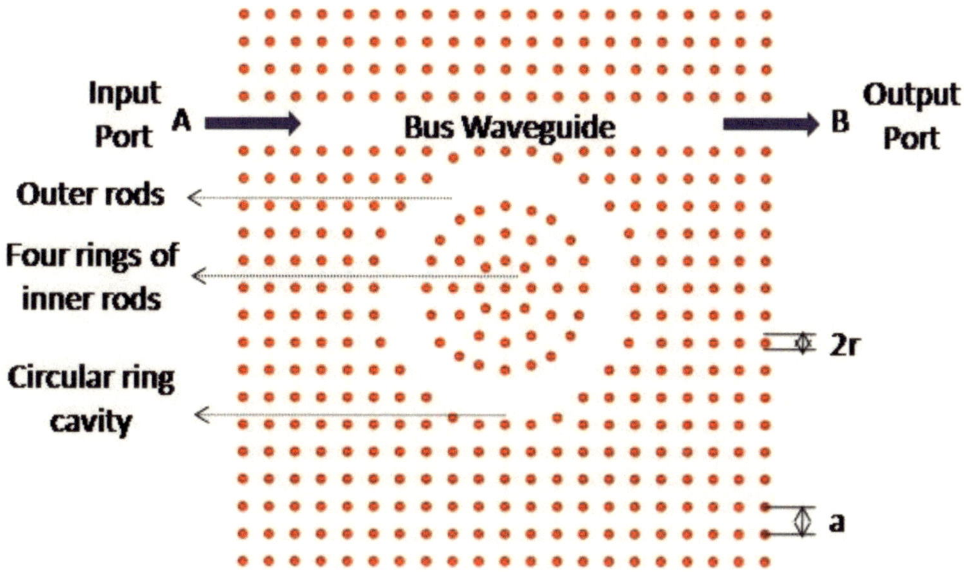

Figure 16. Schematic structure of circular PCRR based BSF

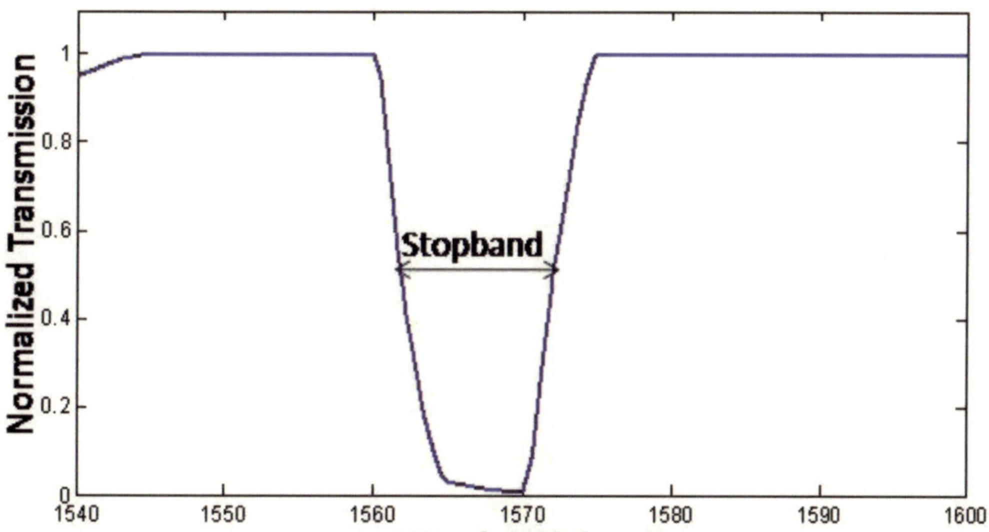

Figure 17. Normalized transmission spectra of circular PCRR based BSF

The Figures 18(a) and 18(b) depict the typical electric field pattern for pass and stop bands at 1550 nm and 1570 nm, respectively. At resonant wavelength, λ=1550 nm the electric field of the bus waveguide is fully transferred to the output port (OFF resonance), and hence, the maximum transfer efficiency is obtained, whereas at 'ON' resonance, λ=1570 nm the signal is coupled into the resonant cavity from the bus waveguide and reflected back to the input. Hence, the signal does not reach the output port which reduces the output power.

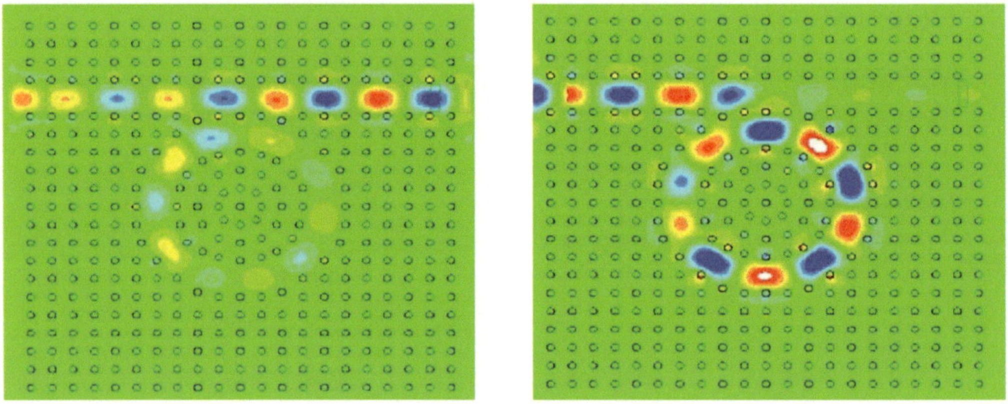

(a) (b)

Figure 18. Electric field pattern of the circular PCRR based BSF at: (a) 1550 nm and (b) 1570nm

10. Conclusion

In this Chapter, we have reviewed the progress of photonic crystal ring resonators and ring resonator devices. Emphasis has been on the principles and applications of ultra-compact photonic crystal ring resonators. We proved that circular PCRR based optical filters provide better performance than others. These findings make the PCRRs an alternative to current microring resonators for ultra-compact WDM components and applications in high-density photonic integration.

Author details

S. Robinson* and R. Nakkeeran

*Address all correspondence to: mail2robinson@pec.edu

Department of Electronics, School of Engineering and Technology, Pondicherry University, Puducherry, India

References

[1] Yablonovitch, E. (1987). Inhibited spontaneous emission on solid-state physics and electronics", Physics Review Letters, , 58(20), 2059-2062.

[2] John, S. (1987). Strong localization of photons in certain disordered dielectric superlattices", Physics Review Letters, , 58(23), 2486-2489.

[3] Ma, Z, & Ogusu, K. (2011). Channel drop filters using photonic crystal fabry-perot resonators", Optics Communications, , 284(5), 1192-1196.

[4] Mohmoud, M. Y, Bassou, Z. M, Taalbi, A, & Chekroun, Z. M. (2012). Optical channel drop filters based on photonic crystal ring resonators", Optics Communications, , 285(1), 368-372.

[5] Rawal, S, & Sinha, R. K. (2009). Design, analysis and optimization of silicon-on-insulator photonic crystal dual band wavelength demultiplexer", Optics Communications, , 282(19), 3889-3894.

[6] Benisty, H, Cambournac, C, Laere, F. V, & Thourhout, D. V. (2010). Photonic crystal demultiplexer with improved crosstalk by second-order cavity filtering", Journal of Lightwave Technology, , 28(8), 1201-1208.

[7] Wang, Q, Cui, Y, Zhang, H, Yan, C, & Zhang, L. (2010). The position independence of heterostructure coupled waveguides in photonic-crystal switch", Optik Optics, , 121(8), 684-688.

[8] Moghaddam, M. K, Attari, A. R, & Mirsalehi, M. M. (2010). Improved photonic crystal directional coupler with short length", Photonics and Nanostructures-Fundamentals and Applications, , 8(1), 47-53.

[9] Gannat, G. A, Pinto, D, & Obayya, S. S. A. (2009). New configuration for optical waveguide power splitters", IET Optoelectroncis, , 3(2), 105-111.

[10] Abdel Malek F ((2011). Design of a novel left-handed photonic crystal sensor operating in aqueous environment", IEEE Photonics Technology Letters, , 23(3), 188-190.

[11] Olyaee, S, & Dehghani, A. A. (2012). High resolution and wide dynamic range pressure sensor based on two-dimensional photonic crystal", Photonic Sensors, , 2(1), 92-96.

[12] Bruyant, A, Lerondel, G, Reece, P. J, & Gal, M. (2003). All silicon omnidirectional mirrors based on one-dimensional photonic crystals", Applied Physics Letters, , 82(19), 3227-3229.

[13] Li, Y, Xiang, Y, Wen, S, Yong, J, & Fan, D. (2011). Tunable terahertz mirror and multi channel terahertz filter based on one dimensional photonic crystals containing semiconductors", Journal of Applied Physics, , 110(7), 073111-073111.

[14] Chen, M. C, Luan, P. G, & Lee, C. T. (2003). Novel design of organic one-dimensional photonic crystal filter", in the Proceedings of the 5th IEEE International Conference on Lasers and Electro-Optics, , 2, 1-4.

[15] Nemec, H, Duvillaret, L, Garet, F, Kuzel, P, Xavier, P, Richard, J, & Rauly, D. (2004). Thermally tunable filter for terahertz range based on a one-dimensional photonic crystal with a defect", Journal of Applied Physics, , 96(8), 4072-4075.

[16] Lee, H. Y, Cho, S. J, Nam, G. Y, Lee, W. H, Baba, T, Makino, H, Cho, M. W, & Yao, T. (2005). Multiple wavelength transmission filters based on Si-SiO_2 one dimensional photonic crystals", Journal of Applied Physics, , 97(10), 103111-103111.

[17] Taniyama, H. (2002). Waveguide structures using one-dimensional photonic crystal", Journal of Applied Physics, , 91(6), 3511-3515.

[18] Lu, T. W, Chiu, L. H, Lin, P. T, & Lee, P. T. (2011). One-dimensional photonic crystal nanobeam lasers on a flexible substrate", Applied Physics Letters, , 99(7), 071101-071101.

[19] Ogawa, S, Tomoda, K, & Noda, S. (2002). Effects of structural fluctuations on three dimensional photonic crystals operating at near-infrared wavelengths", Journal of Applied Physics, , 91(1), 513-515.

[20] Yang, Y. L, Hou, F. J, Wu, S. C, Huang, W. H, Lai, M. C, & Huang, Y. T. (2009). Fabrication and characterization of three dimensional all metallic photonic crystals for near infrared applications", Applied Physics Letters, , 94(4), 041122-041122.

[21] Baohua, J, Buso, D, Jiafang, L, & Min, G. (2009). Active three dimensional photonic crystals with high third order nonlinearity in telecommunication", in the Proceedings of the IEEE International Conference on Lasers and Electro Optics, , 1-2.

[22] Deubel, M, Von Freymann, G, Wegener, M, Pereira, S, Busch, K, & Soukoulis, C. M. (2004). Direct laser writing of three dimensional photonic crystal templates for photonic bandgaps at telecommunication wavelengths", in the Proceedings of the IEEE International Conference on Lasers and Electro Optics, , 1-3.

[23] Ohkubo, H, Ohtera, Y, Kawakami, S, & Chiba, T. (2004). Transmission wavelength shift of +36 nm observed with Ta_2O_5-SiO_2 multichannel wavelength filters consisting of three dimensional photonic crystals", IEEE Photonic Technology Letters, , 16(5), 1322-1324.

[24] Liu, R. J, Li, Z. Y, Feng, Z. F, Cheng, B. Y, & Zhang, D. Z. (2008). Channel drop filters in three dimensional woodpile photonic crystals", Journal of Applied Physics, , 103(9), 094514-094514.

[25] Pendry, J. B. and MacKinnon A ((1992). Calculation of photon dispersion relation", Physics Review Letters, , 69(19), 2772-2775.

[26] Taflove, A. (2005). Computational Electrodynamics: The Finite-Difference Time Domain Method", Artech House, Boston, London.

[27] Pendry, J. B. (1996). Calculating photonic band structure", Journal of Physics: Condensed Matter, , 8(9), 1085-1108.

[28] Pelosi, G, Coccioli, R, & Selleri, S. (1997). Quick Finite Elements for Electromagnetic waves", Artech House, Boston, London.

[29] Djavid, M, Ghaffari, A, Monifi, F, & Abrishamian, M. S. (2008). Photonic crystal power dividers using L-shaped bend based on ring resonators", Journal of Optical Society of America B, , 25(8), 1231-1235.

[30] Borel, P. I, Frandsen, L. H, Harpoth, A, Kristensen, M, Jensen, J. S, & Sigmund, O. (2005). Topology optimized broadband photonic crystal Y-splitter", Electronics Letters, , 41(2), 69-71.

[31] Moghaddam, M. K, Attari, A. R, & Mirsalehi, M. M. (2010). Improved photonic crystal directional coupler with short length", Photonics and Nanostructures-Fundamentals and Applications, , 8(1), 47-53.

[32] Fan, S, Villeneuve, P. R, Joannopoulos, J. D, & Haus, H. A. (1998). Channel drop tunneling through localized states", Physics Review Letters, , 80(5), 950-963.

[33] Mai, T. T, Hsiao, F. L, Lee, C, Xiang, W. F, & Chen, C. C. (2011). Optimization and comparison of photonic crystal resonators for silicon microcantilever sensors", Sensors and Actuators A: Physical, , 165(1), 16-25.

[34] Qiang, Z, Zhou, W, & Soref, R. A. (2007). Optical add-drop filters based on photonic crystal ring resonators", Optics Express, , 15(4), 1823-1831.

[35] Siriani, D. F, & Choquette, K. D. (2010). In-phase, coherent photonic crystal vertical cavity surface emitting laser arrays with low divergence", Electronics Letters, , 46(10), 712-714.

[36] Shih, T. T, Wu, Y. D, & Lee, J. J. (2009). Proposal for compact optical triplexer filter using 2-D Photonic crystals", IEEE Photonics Technology Letters, , 21(1), 18-21.

[37] Manolatou, C, Khan, M. J, Fan, S, Villeneuve, P. R, Haus, H. A, & Joannopoulos, J. D. (1999). Coupling of modes analysis of resonant channel add drop filters", IEEE Journal of Quantum Electronics, , 35(9), 1322-1331.

[38] Berenger, J. P. (1994). A perfectly matched layer for the absorption of electromagnetic waves", Journal of Computational Physics, , 114(2), 185-200.

[39] Figotin, A, Godin, Y. A, & Vitebsky, I. (1998). Two-dimensional tunable photonic crystal", Physical Review B, , 57(5), 2841-2848.

[40] Levy, O, Steinberg, B. Z, Boag, A, Krylov, S, & Goldfarb, I. (2007). Mechanical tuning of two-dimensional photonic crystal cavity by micro electro mechanical flexures", Sensors and Actuators A, , 139(1-2), 47-52.

[41] Hadzialic, S, Kim, S, Sarioglu, A. F, Sudbo, A. S, & Solgaard, O. (2010). Displacement sensing with a mechanically tunable photonic crystal", IEEE Photonics Technology Letters, , 22(12), 1196-1198.

[42] Asano, T, Kunishi, W, Nakamura, N, Song, B. S, & Noda, S. (2005). Dynamic wavelength tuning of channel drop device in two dimensional photonic crystal slab", Electronics Letters, , 41(1), 37-38.

[43] Chew, X, Zhou, G, Chaum, F. S, Dens, J, Tang, X, & Loke, Y. C. (2010). Dynamic tuning of an optical resonator through MEMS driven coupled photonic crystal nano cavities", Optics Letters, , 35(15), 2517-2519.

[44] Costa, R, Melloni, A, & Martinelli, M. (2003). Band-pass resonant filters in photonic crystal waveguides", IEEE Photonics Technology Letters, , 15(3), 401-403.

[45] Chao, C, Li, X, Li, H, Xu, K, Wu, J, & Lin, J. (2007). Bandpass filters based on phase-shifted photonic crystal waveguide gratings", Optics Express, , 15(18), 11278-11284.

[46] Djavid, M, Ghaffari, A, Monifi, F, & Abrishamian, M. S. (2008). Photonic crystal narrow band filters using biperiodic structures", Journal of Applied Science, , 8(10), 1891-1897.

[47] Wei, L, Alkeskjold, T. T, & Bjarklev, A. (2010). Electrically Tunable Bandpass Filter on Liquid Crystal Photonic bandgap fibers", in the Proceedings of the international Conference on OFC/NFOEC, , 1-3.

[48] Oswald, J. A, Wu, B. I, Mcintosh, K. A, & Verghese, S. (2000). Metallo-dielectric photonic crystals for infrared applications", in the Proceedings of Quantum Electronics and Laser Science conference, , 42-43.

[49] Djavid, M, Ghaffari, A, Monifi, F, & Abrishamian, M. S. (2008). Photonic crystal narrow band filters using biperiodic structures", Journal of Applied Science, , 8(10), 1891-1897.

[50] Monifi, F, Djavid, M, Ghaffari, A, & Abrishamian, M. S. (2008). A new bandstop filter based on photonic crystals", in the Proceedings of Progress in the Electromagnetic Research, Cambridge, USA, , 674-677.

Chapter 2

Single and Coupled Nanobeam Cavities

Aliaksandra M. Ivinskaya, Andrei V. Lavrinenko,

Dzmitry M. Shyroki and Andrey A. Sukhorukov

Additional information is available at the end of the chapter

http://dx.doi.org/10.5772/55096

1. Introduction

In the coming decade in physics great effort will probably be devoted, among other things, to improving quantum storage and the development of quantum computer. To make use of quantum processes one should avoid decoherence influence of surroundings, or use specifically designed environment to modify the process considered. This is the case when an atom or a quantum dot — nanosized emitter in an active material — is located inside a medium exhibiting modified density of electromagnetic states, e.g., a photonic crystal. In fact, prospects to modify the density of states gave the major motivation to investigate photonic crystals back in the years of their inception. Still they generate large interest from the fundamental cavity quantum electrodynamics perspectives [1–3]. Photonic crystals based structures — beam splitters, cavities, slow light and logic devices — allow for a lot of diverse operations with light. Main advantages of dielectric photonic crystal components over, for instance, their plasmonic analogues are low-loss operation and low-cost production.

Photonic crystals (PhC's) are currently considered as a perspective platform to host low mode volume cavities with high quality factors. A defect can be formed in the photonic crystal lattice by breaking a perfect symmetry of the structure either by removing or shifting basic constitutive units or by local modification of the refractive index. For a quantum dot placed inside a defect in a photonic crystal, the radiation rate is directly connected with the ratio Q/V, where Q is the quality factor of the microresonator and V is the mode volume.

Basically, for a photonic crystal featuring a full three-dimensional (3D) band gap, a defect in it should give the highest possible Q-factor. However, fabrication of photonic crystals with full 3D band gaps, e.g., inverted opals or woodpile structures and defects in them is quite complicated. That is why a standard way to create a cavity is to use a 2D photonic crystal platform, mostly silicon or GaAs slabs with perforation. Position of holes in such slab is manually defined, giving thus flexibility in the design and optimization of cavities and other

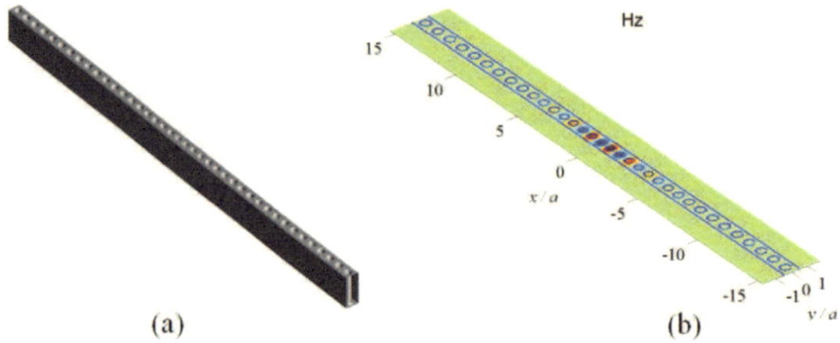

Figure 1. (a) The geometry of nanobeam cavity. (b) Magnetic field in resonance.

photonic components. A variety of high-Q, low-V cavity designs were proposed based on structural modifications in photonic crystal matrices [4–9].

The main channel for loosing energy from a free-standing membrane cavity is through the coupling to radiative leaky modes. In the plane of the slab the 2D photonic crystal acts as a distributed Bragg mirror, thus in-plane ($||$) leakage of radiation from the photonic crystal cavity is typically small. In the out-of-plane ($\perp$) direction light is primarily confined by total internal reflection, thus the magnitude of $\mathbf{k}_\perp$ vector should be as small as possible to reduce losses. There exists connection between in-plane and out-of-plane wave vector components stipulated by coupling to radiation modes. Increase in $\mathbf{k}_\perp$ originate from $\mathbf{k}_{||}$-vectors lying close to the light cone and the usual approach employed for optimization of cavities is through some guess for the design that would give $\mathbf{k}_{||}$ lying far enough from the light cone. Also in-plane mirror imperfections can lead to parasitic coupling to vacuum modes for some structures.

If some design is to be optimized to achieve high Q – mode, this should be done gently without abrupt changes in the structure geometry or refractive index, because otherwise undesirable leakage of radiation can appear. In this sense the best designs for optimization are waveguide-like ones [10] and nanobeam cavities [11, 12], Fig. 1, having simple arrangement of field maxima and minima along the straight line. Such 1D arrangement allows for application of the mode-matching rule [13], when the hole sizes and PhC pattern change gradually going from the cavity center towards the mirror part [14, 15]. For modes with more complicated symmetries, for instance, a hexapole mode in a one-hole-missing membrane [16], this approach is not readily applicable since the mode by itself can easily vanish due to a moderate geometry modification.

A photonic crystal nanobeam cavity created by perforating a photonic wire waveguide (nanoridge waveguide) with a row of holes, Fig. 1, reaches a Q-factor comparable to that of a photonic crystal membrane resonator while being much more compact and easy in fabrication. Even for a nanobeam cavity in a low refractive index material like SiO_2, fairly high Q-factors of several thousands were measured experimentally [17]. Besides a high Q-factor, a nanobeam cavity exhibits a set of other desirable characteristics: low mode volume V (less than the cubic wavelength of light) and the smallest footprint size among other high-Q cavities. This stimulates intensive investigations of nanobeam-related acousto-optic and optomechanic interactions [18, 19]. Recently all-optical logical switching [20] and quantum dot laser [11] have been demonstrated in nanobeam cavities. Tiny size of nanobeam cavities makes them also very promising for densely integrated photonic circuits.

Of particular interest are ensembles of cavities [21] with quantum dots placed inside. Full three-dimensional description of such systems is not yet a routine task, but it is very important for fundamental investigations of light-matter interactions [22–24]. Side-coupled nanobeams [14, 25, 26] offer new possibilities for shaping optical fields at nanoscale, which is potentially beneficial for various applications including trapping and manipulation of particles [27], sensing and optical switching through optomechanical interactions with suspended nanobeams [18, 19].

Strong and controllable coupling [28] is also required to create low-threshold lasers [29], observe Fano line shapes [30], design field concentrators for detection of molecules [26], create flat slow light passbands [31, 32], holographic storage [33], and enhance nonlinearities [34]. Formally, consideration of coupled cavities can be made in direct analogue with the molecular mode hybridization, that is why coupled resonators are often called 'photonic molecules' [35, 36].

On the other hand, for some applications reduction of coupling strength between the resonators is the key. Indeed, interaction between optical components can shift operation wavelength of the device. Avoiding of parasitic coupling of components is crucial for photonic integrated circuits and in optic network design. Realization of flexible control over the modes in arrays of nanocavities by their rearrangement contributes to the development of on-chip quantum-optical interferometers [23] and quantum computers [24].

2. The finite-difference frequency-domain method

Optics and photonics are rapidly developing fields building their success largely on use of more and more elaborated artificially nanostructured materials. To further advance our understanding of light-matter interactions in these complicated artificial media, numerical modeling is often indispensable.

One of the most challenging computational tasks is evaluation of the Q-factor of a resonator. The traditional way here is to use the finite-difference frequency-domain (FDTD) method to simulate these spatially extended structures with the subsequent extraction of Q by analyzing the ring-down of electromagnetic field components. Such time-domain simulations can take considerable time up to several days per single run for a high-Q 3D resonator.

If several modes are traced in the time domain within a single run, the accuracy of the Q-factor determination may degrade. The extraction of the separate mode field profiles requires the Fourier transformation of field evolution stored for some space volume and time interval. If two modes are degenerate, separating them with the FDTD method is even less trivial, especially if at the degeneracy point the coupled structure does not have a plane of symmetry allowing to split the modes by the appropriate domain reduction.

On the contrary, the frequency domain techniques grant an opportunity to get straightforwardly in one run maps of several modes, their eigenvalues and quality factors. When modes in the coupled cavities are degenerate, we can get an idea how they may look like — though the picture becomes now ambiguous.

As an competitive alternative to the time domain modeling we employ here the 3D finite-difference frequency-domain (FDFD) method. Details of the method are published elsewhere [37]. The eigenmode equation in the FDFD method is obtained through combining Maxwell's equations into the second-order differential equation for $\widetilde{\mathbf{H}} \equiv \sqrt{\mu}\,\mathbf{H}$:

$$\sqrt{\mu^{-1}}\,\nabla \times \epsilon^{-1} \nabla \times \sqrt{\mu^{-1}}\,\widetilde{\mathbf{H}} = \omega^2 \widetilde{\mathbf{H}}. \tag{1}$$

Then, Q-factor is straightforwardly found as $Q = Re(\omega)/2Im(\omega)$ after solution of the eigenproblem for complex ω. No other elaborated post-processing is needed. Thus transition to the frequency domain for cavity eigenmode analysis is very natural as it greatly reduces the computation time.

To impose boundary conditions we use perfectly matched layers (PMLs). The PMLs were originally designed to absorb propagating electromagnetic waves while evanescent fields can be even intensified in them. High-Q cavities have extremely intensive fields around them. To keep the domain size reasonably small and at the same moment to avoid evanescent tales to fall into the PMLs we use the free space squeezing procedure to set a buffer layer. For all simulations in this chapter we use one-lattice-constant-thick buffer layers. Two squeezing functions are applied to project infinite open space to this buffer layer: inverse hyperbolic tangent (*arctanh*) function and steeper $x/(1-x)$ function. To make our simulations efficient we also use the solution-adapted continuous grid density variation of lower resolution in the extended photonic crystal mirror part. If a smooth analytic function is used to create a non-equidistant mesh, it assures the impedance matched transformation leading to the absence of reflection in the transition region to the finer mesh. The symmetries of resonators are exploited to reduce the computational domain. For further insight in the FDFD simulations and free space squeezing we refer to [37].

3. Single nanobeam

3.1. Modeling in 2D: High-Q design

At the beginning we tailor the nanobeam cavity design in 2D to get a high-Q TE-mode (electric field in the $x-y$ plane). Fig. 2 shows a basic nanobeam sketch used to consider various cavity designs: a nanowire of refractive index 3.4 is suspended in air and has 20 holes in its half. Perforation consists of two regions: the chirped mode matched defect region and long periodic part acting as a Bragg reflector. In the reflecting part the lattice constant is a, hole diameter is d and total width of a nanobeam $t = 1.0a$. In the defect region the modified hole diameters and lattice constants are d_n and a_n, respectively, where n numbers a segment in the defect part of the cavity. For 2D simulations we put $\Delta x = \Delta y$. Along y-direction $1a$-wide buffer layers are squeezed with the hyperbolic arctangent function covered by PMLs on $1/3$. No air buffer is used along x-direction, just PMLs comprising 3 grid cells.

Intuitive variation of the defect region parameters – holes radii d_n and lattice constant a_n – in order to maximize the Q-factor led us to the following conclusions. First of all, when both parameters are constant (a_n=const, d_n=const) but differ from those in the reflecting part, the Q-factor can approach 10^5. Second, if one of the parameters slowly decreases in the defect region towards the center (for example, a_n=const and d_n is varied, or vice versa), Q rises to $10^6 \div 10^7$. Third, only if both a_n and b_n are gradually decreased from the periphery to the center of the cavity, Q reaches the highest value around $10^8 \div 10^9$ in 2D. In 3D it is usually one to two orders of magnitude less.

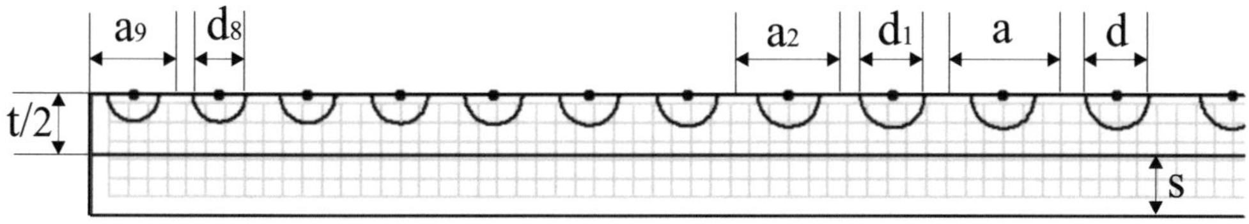

Figure 2. Nanobeam quarter, x-direction is pointing along the nanobeam perforation, y-direction is along the nanobeam width, s shows the buffer around the nanobeam.

	Design 1	Design 2	Design 3
Mirror: hole diameters	$d = 0.54a$	$d = 0.6a$	$d = 0.55a$
Defect: number of holes	$n = 1 \ldots 9$	$n = 1 \ldots 9$	$n = 1 \ldots 9$
Defect: hole diameter	$d_n = \frac{d}{\sqrt[10]{n}}$	$d_n = 0.6a_n$	$d_n = d - 0.012na$
Defect: lattice constant	$a_n = \frac{d_n}{0.6}$ $a_1 = 0.8a$	$r_1 = \frac{1}{0.843a}, r_2 = \frac{1}{a}$ $a_n = \frac{1}{r_1 + \frac{r_2 - r_1}{9} n}$	$a_n = \frac{d_n}{0.6}$

Table 1. Different designs of the nanobeam cavity sketched in Fig. 2.

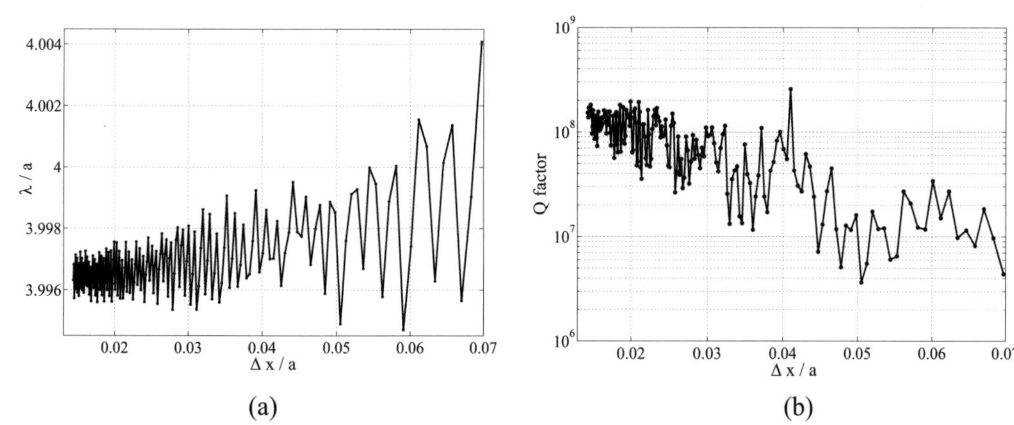

Figure 3. 2D (a) Q and (b) λ convergence for the third design from Table 1. Δx=Δy, y-buffer is $1a$-wide ($arctanh$ squeezed) with $1/3$ covered by the y-PMLs, x-PMLs comprise 3 grid cells.

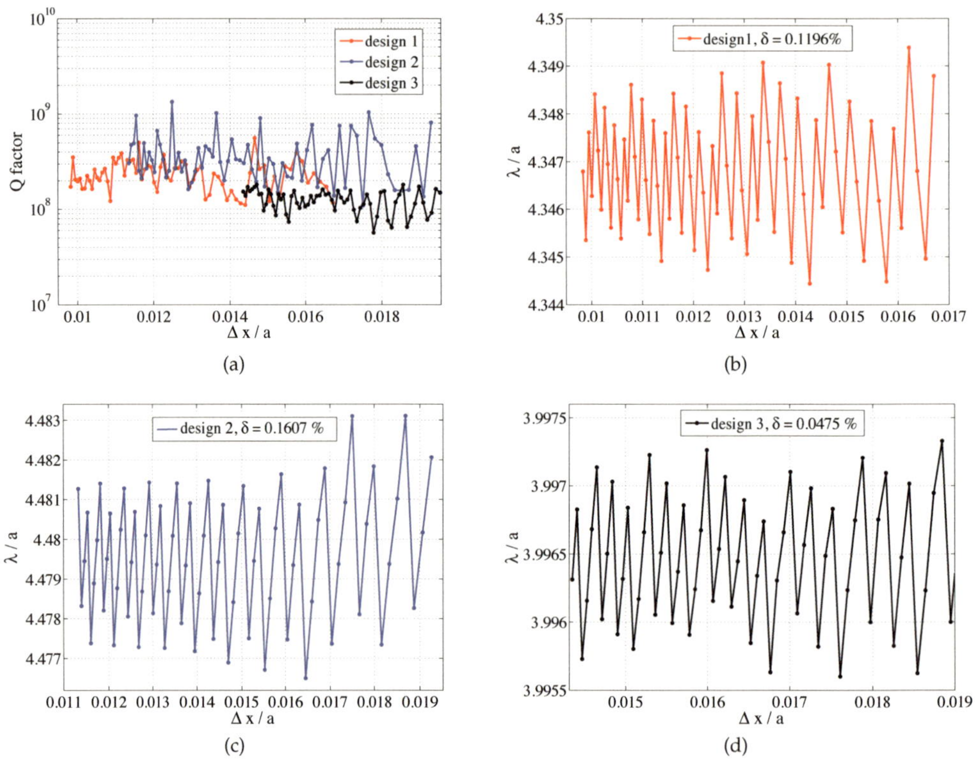

Figure 4. (a) Q-factor and (b)-(d) λ convergence for the three designs in the region of fine resolutions with estimation of spread δ shown in the legend. 2D simulations with $\Delta x = \Delta y$, $1a$-wide buffer layers ($arctanh$ squeezed) along the y-direction are covered on $1/3$ by y-PMLs, x-PMLs – 3 grid cells.

Having established that a_n and d_n should vary we investigated ways to do that. Several laws to tailor the nanobeam design have been compared. Among them are $1/\sqrt[10]{n}$ multiplier to decrease both a_n and d_n (design 1); cavity formation similar to [12] when the hole diameter and lattice constant vary linearly in the reciprocal space (design 2); and linear decrement of a_n and d_n towards the middle of the nanobeam (design 3) [15]. With all mentioned designs we were able to rise the Q-factor to the order of 10^8 simply by playing with parameters. Table 1 summarize details of different nanobeam cavity designs. For the first and third designs we start by defining modified hole diameter d_n, and modified segment size a_n is calculated afterwards. For the design 2 calculation of modified lattice constant a_n precedes evaluation of the defect hole diameters.

In Fig. 3 an example of the Q and λ convergence curves for the design three is plotted starting from a quite coarse resolution, while Fig. 4a–d allows to do more detailed comparison between different designs in the region of fine resolutions. All of the designs from Table 1 have similar Q-factor values, Fig. 4a, the design three revealing faster convergence than others. In Fig. 4b–d the eigenwavelength convergence is plotted for the three designs in the same Δx range as in Fig. 4a. To estimate the convergence rate, relative spread $\Delta\lambda$ of convergence curves around a central wavelength λ_0 can be introduced: $\delta = \frac{\Delta\lambda}{\lambda_0}100\%$. The design three has δ one order less than the designs one and two even at rougher resolutions.

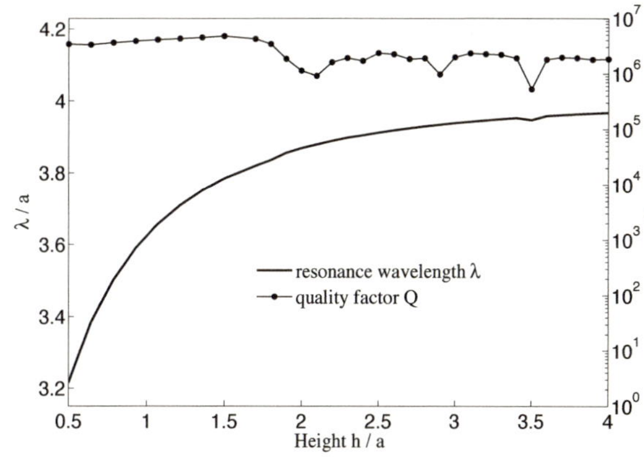

Figure 5. (a) Field patterns of the TE mode in the nanobeam cavity of the height $0.7a$. Only the quatre of the nanobeam is shown. (b) Resonance wavelength λ and Q-factor dependence on the nanobeam height h. 3D simulation is made with cubic grid cell of $0.07a - 0.1a$ size, $1a$-buffer along both y and z-directions is *arctanh*-squeezed, $a/3$ distance goes for PMLs, x-PMLs are 3 cells wide.

Thus, we ended with the sizes modification according to a linear law as better converging numerically. Design 3 is used in all further nanobeam cavity simulations. We think that the better numerical stability can lead also to the better stability with respect to the fabrication imperfections.

We pay special attention to this analysis, because the Q-convergence curve for the nanobeam cavity is worst than that for the PhC membrane resonator, compare with the results in [37]. A one-dimensional PhC basis for the nanobeam cavity is somewhat less reflecting than a 2D stop band utilized in the membrane resonators. Thus, if a small imperfection in a 1D mirror is present (due to either some perturbations during fabrication or inaccuracy due to finite difference description), radiation can easily escape through sidewalls of the 1D mirror, while in the case of a membrane leaking radiation can be captured in all in-plane directions in the surrounding 2D-PhC mirror. As discussed in Introduction, undesirable coupling to vacuum modes is directly connected to k_x (in 1D $k_{||} = k_x$) distribution around the light cone, and thus even a tiny variation in k_x might lead to the strong variation in the Q-factor, while the total wave vector is changed only slightly by perturbations and λ-curve preserves good convergence.

3.2. Modeling in 3D

In the 3D modeling a defect region in the nanobeam cavity is perforated according to design 3 (see Table 1). In 3D again $1a$-wide buffer layers along y and z-direction are covered by PMLs on $1/3$. No air buffer is used along x-direction, just PMLs comprising 3 grid cells. We use either equidistant or non-equidistant meshes, but always set $\Delta x = \Delta y = \Delta z$ in the center of the nanobeam (the defect region). Then for non-equidistant meshes Δx can be stretched up to three times towards the periphery of the nanobeam.

Fig. 5 shows dropping of the resonance wavelength of the TE-mode in the nanobeam cavity with the reduction of its height h along the z-direction. Decreasing of the nanobeam height

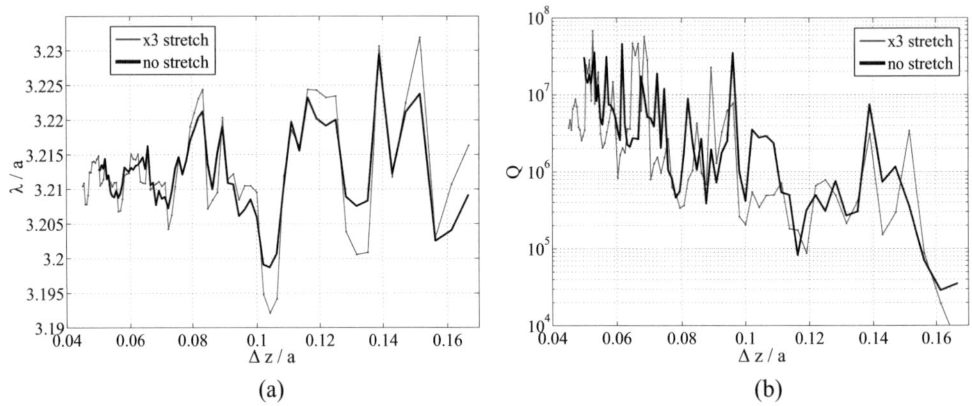

Figure 6. (a) The resonance wavelength and (b) Q-factor versus grid step for uniform (Δx=Δy=Δz) and nonuniform x-mesh with three times stretching in the mirror part (Δx=Δy=Δz in the center of the nanobeam). Buffer size along the y and z directions is $1a$ ($x/(1-x)$ squeezed) with 1/3 occupied by the PMLs, the PMLs width in the x-direction being 3 grid cells.

also greatly minimizes the mode volume, for example, for height $h = 0.5a$ it is equal to 0.86 $(\lambda/n)^3$. We keep this nanobeam height for the modeling of the coupled cavity structures.

Figure 6 demonstrates the Q- and λ-convergence for this height. The λ calculation has uncertainty less than 1% at sufficient resolutions while the convergence of Q is more unstable than in 2D. We can remind here that low accuracy of the Q-factor evaluation reflects high sensitivity of the Q-factors of 1D PhC-based structures to imperfections in their finite-difference description. And in 3D model we have an additional channel (z-direction) for coupling of a genuine cavity mode to leaky modes compared to the 2D case, what causes degradation of convergence. It is interesting to note that instabilities do not show up in Fig. 5. With variation of the height we do not change the resolution along the 1D mirror and thus k_x distribution is completely the same for all height values.

4. Coupled nanobeams

4.1. Two coupled nanobeams

When two identical cavities are positioned parallel to each other, their modes undergo hybridization. Supermodes possess symmetric or anti-symmetric profiles [25] and shift in frequency up and down from the former level. We will refer to this splitting as frequency detuning. The frequency detuning between the supermodes normally increases as the cavities are brought closer, and such sensitivity to the separation can lead to pronounced optomechanical phenomena. This may have various applications including the mechanically-induced frequency conversion for optical waves [38]. Analogous effects of modes splitting occur in coupled periodic waveguides, where several channels can enrich the band structure of a single mode waveguide in a controllable way, e.g. in slow light modes positioning at the band edge on demand [39].

4.1.1. Analysis of field profiles in 2D

First, we analyze two side-coupled nanobeam cavities as in Fig. 7a, where the right half of the structure is shown (the left half is symmetric). The individual identical nanobeams have

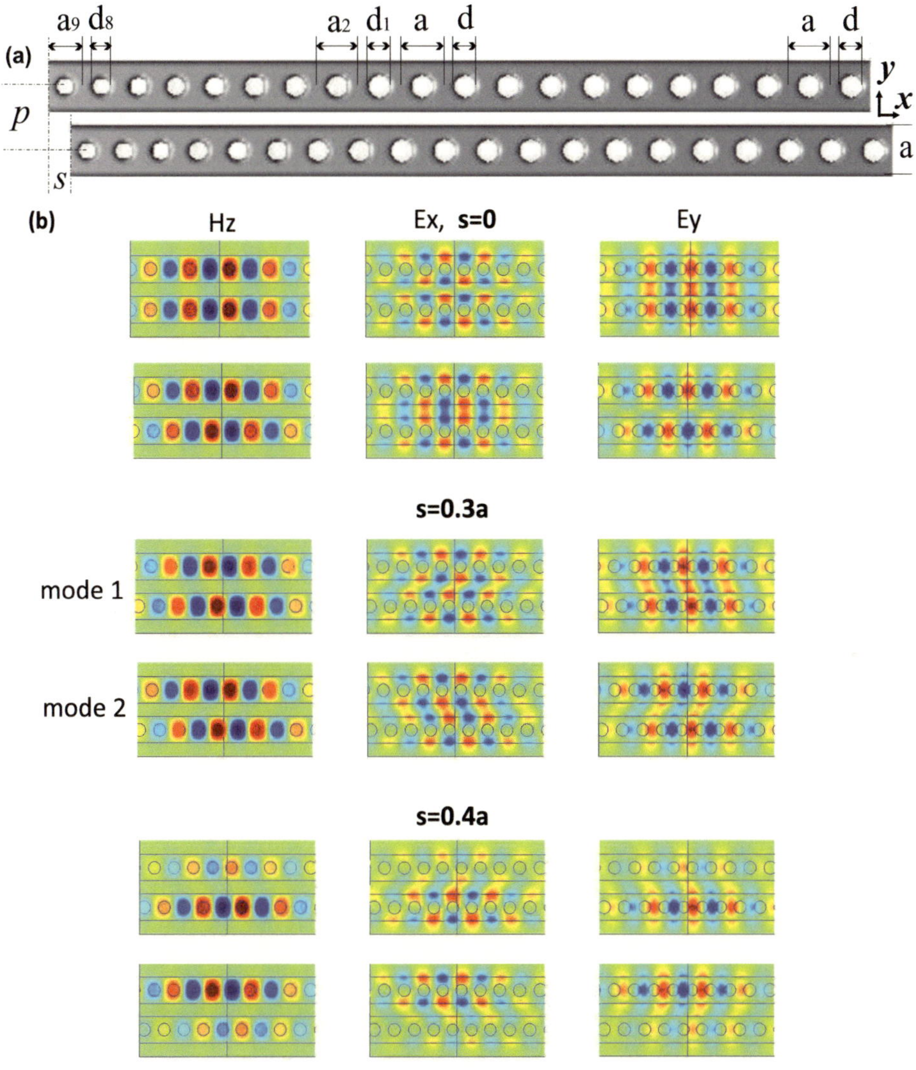

Figure 7. (a) Two nanobeams, each of width $t = a$, refractive index $n = 3.4$ and drilled with 20 holes in each half, separated one from another by p and shifted longitudinally by s. (b) 2D field profiles for modes (labeled mode 1 and 2) in coupled cavities with zero, $s = 0.3a$ and $s = 0.4a$ longitudinal shifts when separation $p = 1.5a$, for each shift mode one being positioned at the top while mode two takes the bottom position.

design 3 from Table 1. This gives a linearly chirped array of elements (holes), while other designs are also possible; the general mode properties are usually similar for different chirp functions. The resonance wavelength of a single nanobeam in 2D is $\lambda = 3.9964a$, $Q = 1.5{\cdot}10^8$, Fig 3. Two parameters describe the position of the second nanobeam cavity relative to the first one: transverse separation p and longitudinal shift s, Fig. 7a.

To make the computational work efficient all modeling is done at the beginning for 2D nanobeam geometries with the main emphasis on the field patterns redistribution as the coupled resonators are rearranged. As we are interested in coupling effects between two nanobeam resonators when they are shifted longitudinally and transversally with respect

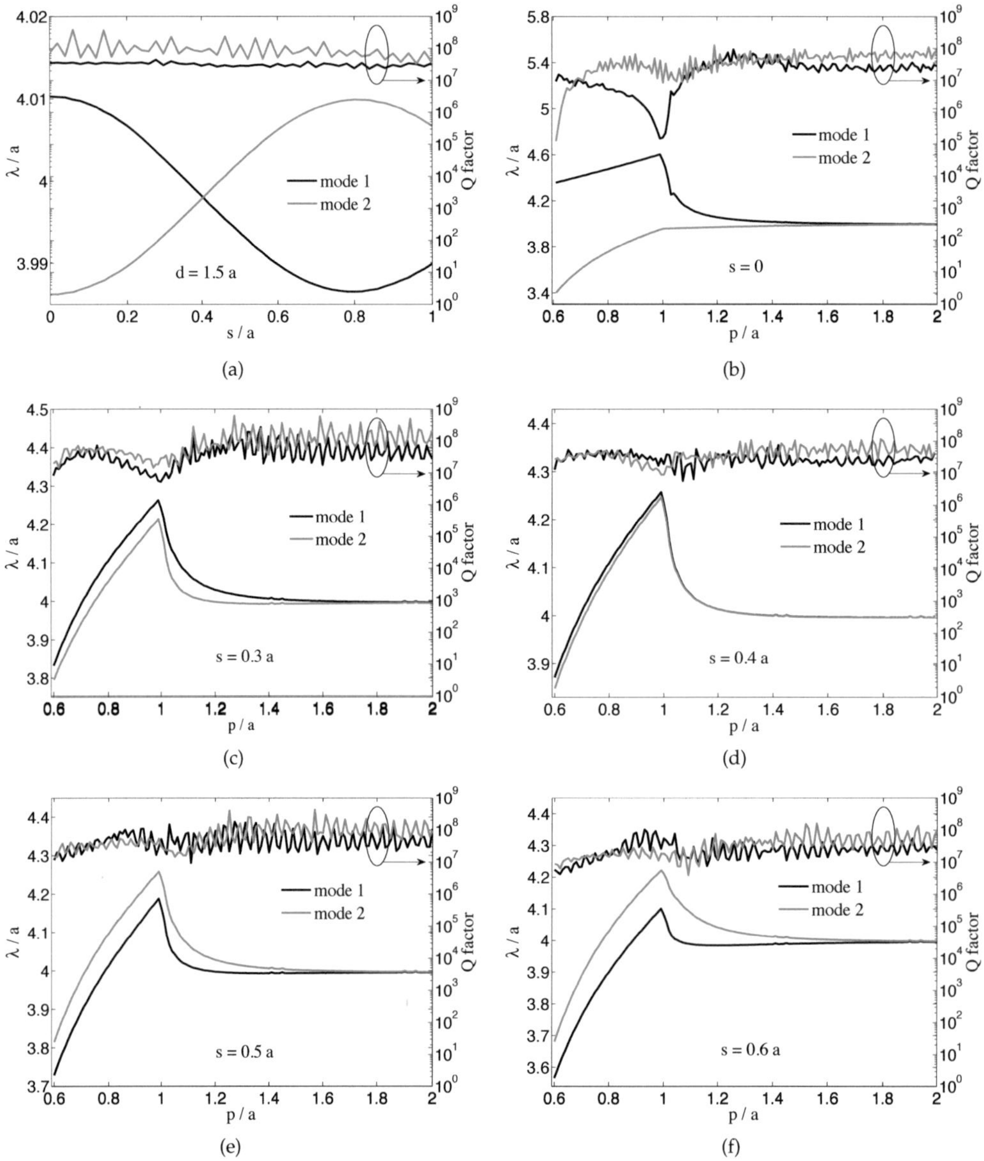

Figure 8. (a) Detuning of the mode one and mode two vs the longitudinal shift. Transverse separation between nanobeam axes is $p = 1.5a$. (b)-(f) Resonance wavelengths and Q-factors of eigenmodes in side-coupled nanobeam cavities vs. their transverse separation p. Results are presented for four different shifts: (b) $s = 0.0a$; (c) $0.3a$; (d) $0.4a$; (e) $0.5a$, (f) $0.6a$ as indicated by labels. Separation of $1.0a$ and less corresponds to a single dielectric beam with two rows of holes. 2D simulations are made with $\Delta x = \Delta y \simeq 0.004a$. $1a$-wide y-buffer is *arctanh* squeezed with half of it covered by PMLs. x-PMLs comprise 3 grid cells.

to each other, we need to perform a new computational cycle each time the structure is modified by a small displacement. In the 2D case the execution time is several minutes even for the huge resolution such as $n_{tot} = 6 \cdot 10^5$ grid nodes available with a 8Gb station. For 2D simulations we put $\Delta x = \Delta y \simeq 0.004a$. Along y-direction $1a$-wide buffer layers are squeezed with the inverse hyperbolic tangent function and covered by PMLs on 1/2. No air buffer is used along x-direction, just PMLs comprising 3 grid cells.

Mode profiles in Fig. 7b show formation of the symmetric and antisymmetric cavity modes when two nanobeam resonators are brought together without shifting, $s = 0$. Magnetic field hot spots coincide with the location of the holes. With separation $p = 1.5a$ we see only electric field in the air gap between the nanobeams. When one of the nanobeams is subjected to a longitudinal shift the system loses its symmetry and modes cannot be specified anymore as symmetric and antisymmetric. We will refer to notation 'mode 1' and 'mode 2' to call transformations of even and odd modes respectively with gradual shift starting from zero.

At zero longitudinal shift $s = 0$ E_x component of mode 1 and E_y component of mode 2 have a node plane passing through the middle of the air gap between the nanobeams ($y = 0$). For applications requiring high field intensities it would be preferable to avoid these zero-valued fields. It turns out that as the nanobeams are gradually shifted from $s = 0$ the node planes for both of these modes components are substituted by a plane with high field intensities, see Fig. 7b for $s = 0.3a$. At the same moment other electric field components (E_y for mode 1 and E_x for mode 2) still preserve quite high field values. Thus, a small longitudinal shift helps in removing areas of zero fields in the air gap and makes the electric field intensity more uniform across the gap between the two nanobeams. The field uniformity in the shifted nanobeams can be further improved by moving the nanobeams closer in the transverse direction.

From Fig. 8a it is evident that the modes experience degeneracy at around $0.4a$ shift. With larger shifts the eigenwavelength difference grows up again forming a periodic dependence of the frequency detuning on shift s. We also trace the effect of transverse cavity separation p on the resonant wavelengths and Q-factors for different longitudinal shifts s. Results are presented in Figs. 8b–f. Almost exact degeneracy is observed at $s = 0.4a$ for all transverse separations, Fig. 8d. The two principle eigenmodes are resolved in the FDFD numerical simulations with their frequency detuning being much smaller than for the other shifts.

Away from the degeneracy point each mode profile should support the 180° symmetry of the photonic structure around the central point ($x = y = 0$) between two cavities, for example for a magnetic field component it might be written: $H_z(x; y; z) = m\, H_z(-x; -y; z)$, where $m = +1$ or $m = -1$ for the two fundamental modes of the couple cavities [40]. These symmetries are visible for mode profiles shown in Fig. 7b and Fig. 9c,d. However, we note that exactly at the degeneracy point, the field profiles of the eigenmodes are defined with the certain ambiguity and must not satisfy the rotational symmetry, since any linear combination of eigenmodes is an eigenmode as well. As shift starts approaching $0.4a$ we see that field intensity in one of the nanobeams falls down, Fig. 7b for $s = 0.4a$. The connection between the modes weakens. As the result the modes settle mostly in one or another nanobeam, Fig. 11b, bottom panel. In more complicated structures, where cavities are tuned by infiltration, similar effects of anticrossing were registered experimentally [35] (there is always some perturbation present, which, strictly speaking, removes the degeneracy, thus in fact both terms — mode degeneracy and anticrossing — can mean the same here). In Fig. 9a the

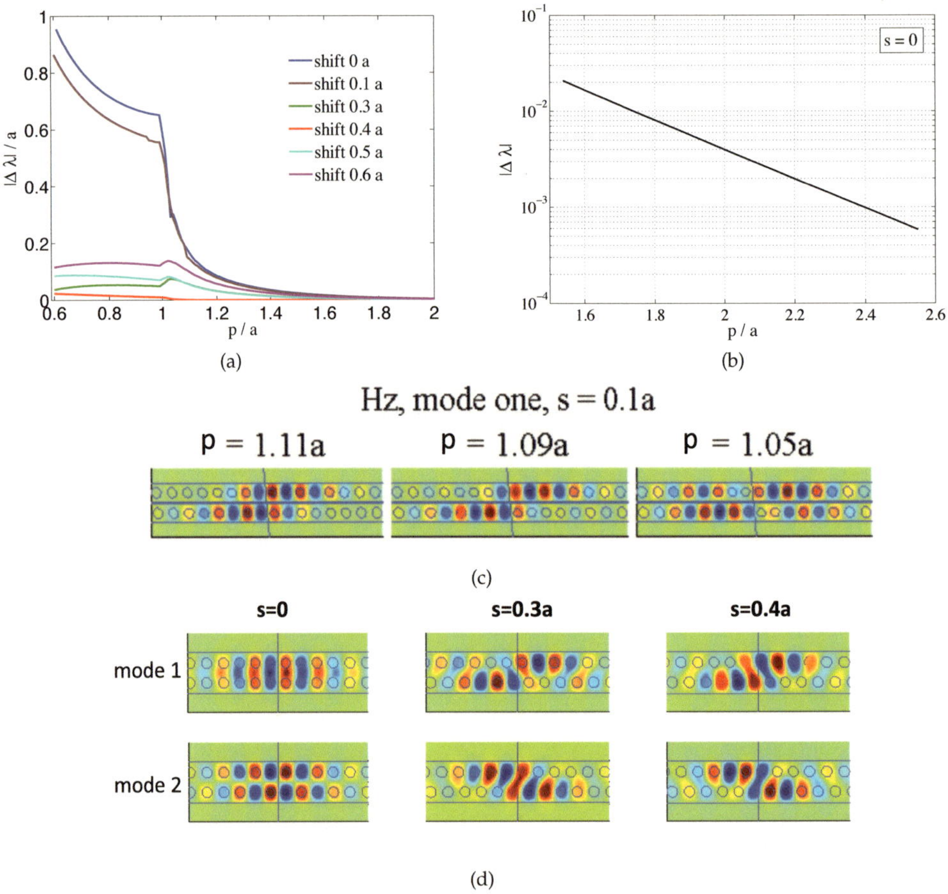

Figure 9. (a) Increase of eigenwavelength difference with shortening of transversal separation p between the nanobeams, different shifts are printed in the legend. (b) $|\Delta\lambda|$ in logarithmic scale for non-shifted nanobeams. (c) Evolution of magnetic field Hz in two slightly shifted nanobeams (s=0.1a) when closing the air gap between them, i.e. separation p is reduced from 1.1a to 1.05a. (d) H_z field distribution in two connected nanobeams with $p = 0.9a$ at $s = 0$, $s = 0.3a$ and $s = 0.4a$ shifts. 2D simulation is made with $\Delta x = \Delta y \simeq 0.004a$. $1a$-wide y-buffer (*arctanh* squeezed) is covered by PMLs on half, 3 grid cells come for x-PMLs.

eigenwavelength difference is plotted showing the highest values for non-shifted resonators and significantly smaller $|\Delta\lambda|$ for non-zero shifts. Due to the symmetric positioning of 0.3a and 0.5a shift values around the degeneracy point, the wavelength differences given by these shifts are equal to each other for the nanobeams separated by $p > 1.1a$. When the nanobeams are far enough transversally and the coupling strength is small, $|\Delta\lambda|$ depends on s by the order law seen in Fig. 9b. In Fig. 9b for the illustration purpose the spectral splitting for non-shifted nanobeams is plotted in a logarithmic y-scale. Other nonzero shifts, except for the degeneracy point, also give similar straight lines in the $|\Delta\lambda|$-log scale if separation p is big enough to correspond to the weak coupling regime.

When the resonators are moved closer so that the interaction between the nanobeams intensifies, all fields, including magnetic one, spread through the air gap. The picture of the mode profiles alters compared to $p = 1.5a$, see Fig. 9c,d. Eigenmode fields extend over the whole cavity even when the hole positions in the upper and lower cavities are effectively shifted in the out-of-phase configuration, i.e. $s = 0.4a$. In Fig. 9c the evolution of mode 1 at small shift 0.1a is shown when closing the air gap. Drastic changes occur with the mode 1 profile when the nanobeams are approaching each other: yet very intensive field at $p = 1.11a$ is pushed out from the cavity center at $p = 1.05a$. Thus for nonlinear applications special care should be paid to the extremely thin slots between the nanobeams. Fig. 9d collects field patterns for both modes at different shifts when air-dielectric interfaces are absent and we actually have one cavity consisting of two chirped arrays. Note that although field maps in stitched nanobeams are really complex they are all 180° rotation symmetric relative to the center point. In the degeneracy point $s = 0.4a$ field is sitting in two nanobeams here. So from the numerical experiment we see that mode picture can look two different ways at the degeneracy: with field concentrated in one of the nanobeams or spreading through both of cavities.

4.1.2. *Q and λ dependence on the longitudinal shift in 3D*

In 3D we consider the same nanobeam design from Fig. 7a and discuss more Q and λ curves with the rearrangement of the nanobeams. The 3D Q-factor of a single nanobeam cavity is around $4 \cdot 10^6$ and the operating wavelength $\lambda = 3.21a$ for the TE mode, Fig. 6. To achieve fine sampling in 3D problems the symmetry domain reduction is applied were it is possible to satisfy memory requirements. The system of two nanobeam resonators with a longitudinal shift loses plane-reflection symmetry, and the whole domain should be considered so benefits of stretched meshes are fully used here. Discretization in the cavity center is set to $0.07a - 0.1a$ with a sparser mesh in the rest of the structure (up to 3 times stretching along x-coordinate). In 3D again $1a$-wide buffer layers along y and z-direction are covered by PMLs on $1/3$, the squeezing function is $x/(1-x)$. No air buffer is used along x-direction, just PMLs comprising 3 grid cells. 3D Q-factor computations are done on a 48 Gb station with the maximum execution time approaching 2 hours per single run. Correct averaging of refractive index at boundaries is an important issue in 3D simulations, where the fine resolution as in the 2D case cannot be achieved. It is important to note that as $p \to 1$ a unit Yee cell between two nanobeams might contain two boundaries and then the averaging should be done taking into account both frontiers simultaneously.

Moving nanobeams closer to each other leads to stronger coupling and pronounced increase of the eigenwavelength difference between the doublet of supermodes, see Fig. 10a. At $p = a$ the system is changed abruptly as the gap between the two nanobeams disappears, so the structure consists now of a single high-dielectric bar with two parallel rows of holes in it. That explains a characteristic peak in the wavelength dependence of the mode eigenfrequency plotted in Fig. 10a. Mode 1 (even) has higher wavelength than mode 2 (odd) for the whole range of separations p as can be easily seen from a simple perturbation theory [41]. When the gap between nanobeams is closed and the y-dimension is further reduced, the effective refractive index of the system and hence the eigenwavelengths are also decreased [42]. Remarkably, when varying the separation no significant variation in the Q-factor is seen. The Q-factor value is close to 10^6 for both even and odd modes, see Fig. 10a.

Now we analyze the effect of longitudinal shift s. In Fig. 10b the wavelengths and Q-factors of the fundamental eigenmodes vs. shift s for separations $p = 1.2a$ and $p = 1.5a$ are plotted. As the shift grows from zero, mode detuning is reduced and, independently on separation

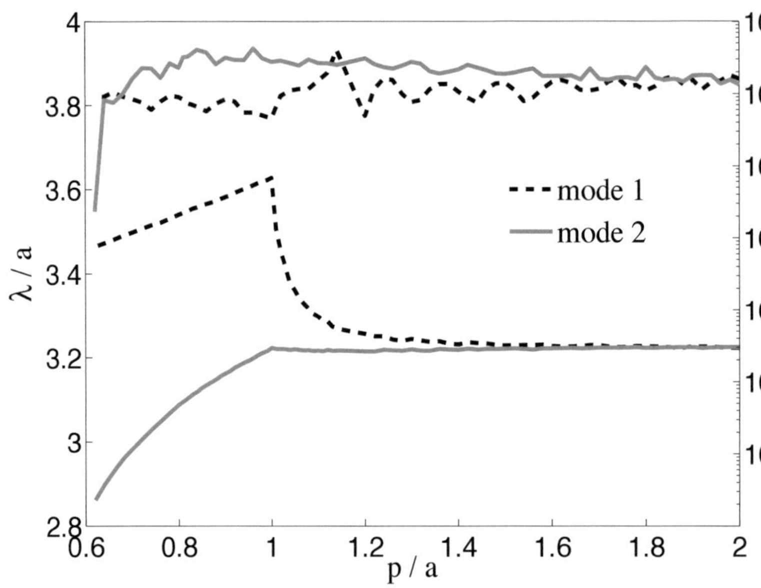

(a)

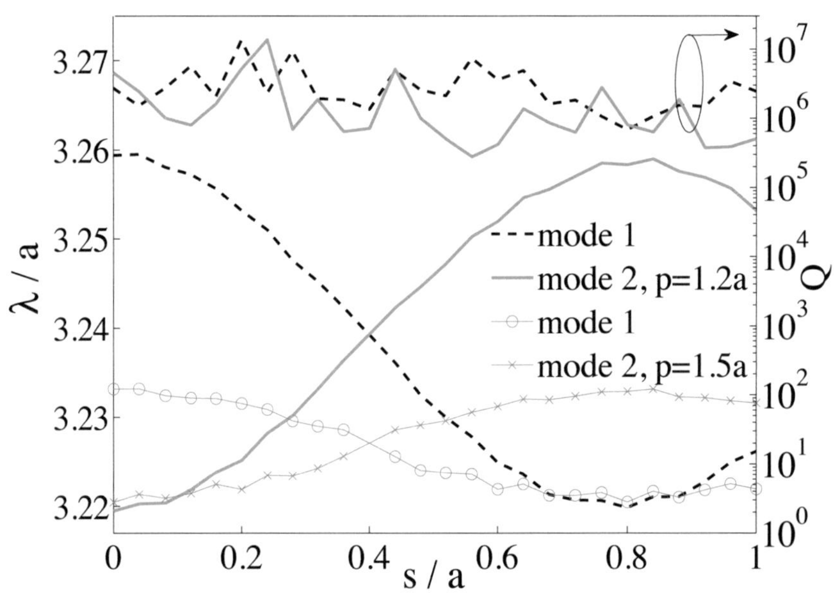

(b)

Figure 10. Tuning of resonance wavelengths of coupled modes by (a) changing separation between non-shifted ($s = 0$) nanobeams; (b) changing the longitudinal shift s for two separations ($p = 1.2a$ and $p = 1.5a$ as indicated by labels). Right axis shows the 3D Q-factor values. For 3D simulations $1a$-buffers ($x/(1-x)$ squeezed) along both y and z-direction are covered by PMLs on $1/3$, x-PMLs are 3 cells thick. $\Delta x = \Delta y = \Delta z$ take values between $0.07a - 0.1a$ in the center of the nanobeam, Δx is stretched 3 times in the mirror part.

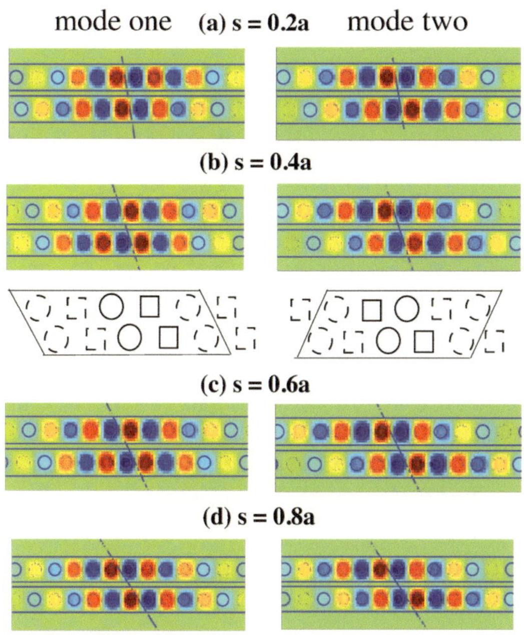

Figure 11. Magnetic field H_z of even and odd modes in coupled nanobeams for $p = 1.2a$ at different longitudinal shifts.

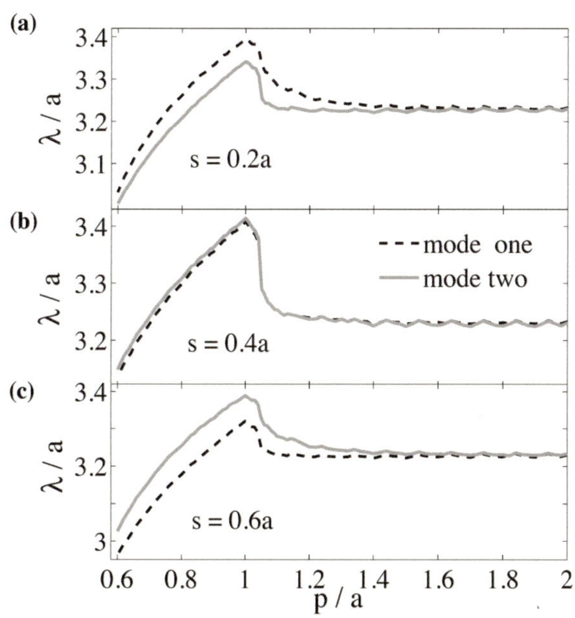

Figure 12. Splitting of resonance wavelengths of even and odd modes versus separation p for (a) $s = 0.2a$; (b) $s = 0.4a$; (c) $s = 0.6a$. For 3D simulations $1a$-buffers $(x/(1-x)$ squeezed) along the y and z-directions are covered by PMLs on $1/3$, x-PMLs are 3 cells thick. $\Delta x = \Delta y = \Delta z = 0.1a$ in the center of the nanobeams, Δx is stretched 3 times in the mirror part.

p, the modes become degenerate at around $s \simeq 0.4a$. It is different with resonators with unmodified lattice constant [43], where the degeneracy shift was exactly $0.5a$ independently on the rod radii variation. Thus, this is primarily variation of the lattice constant that is responsible for the specific value of the degeneracy shift. Note that $0.4a \simeq 0.5a_6$, i.e. the shift at the degeneracy point is approximately equal to a half of the average lattice constant in the cavity region.

In Fig. 11b for $s = 0.4a$ we plot another picture at the degeneracy point compared to Fig. 7b with the magnetic field nodes in one nanobeam opposing field lobes in another. These degenerate modes can be schematically sketched with diagrams in Fig. 11b. Standing wave profiles with slow spatial decay from the center of the cavity towards the periphery allow neglecting low-intensity outer regions and then central parts of the patterns are identical upon reflection, making the occurrence of the degeneracy point (geometrically, the central parts of the defect regions for two modes also satisfy reflection symmetry if chirped hole diameters approximated to be the same in the middle). Had mode profiles less gradual changes in the succession of field minima and maxima along the nanobeams (as shown by bold circles and squares in the diagrams), the formation of the degeneracy would be hardly possible.

For the shifts larger than $s = 0.4a$ the mode detuning is increased, reaching a maximum at around $s \simeq 0.8a$ where eigenwavelength difference approaches the same value as at $s = 0$. The revival of coupling at $s \simeq 0.8a$ is again due to the gradually chirped nanobeam design and field profiles extended along the nanobeams. Thus mode detuning depends on the shift almost periodically, and the cavity modes 1 and 2 are adiabatically transformed as the parameter s is varied from 0 to a, see Fig. 11d, where it is shown that modes 1 and 2 exchange their parity going from $s = 0$ to $s = 0.8$. Most important, the Q-factor values remain of the same order of magnitude as for a single cavity.

In Fig. 12 we compare the eigenmodes wavelength dependencies on separation p for three longitudinal shifts $s = 0.2a, 0.4a, 0.6a$. The upper panel shows the reduced spectral detuning of the modes for the intermediate shift $s = 0.2a$. The plot in Fig. 12b shows that for non-overlapping nanobeams the modes are almost exactly degenerate at $s = 0.4a$ for any transverse separation p. After the degeneracy point, at $s = 0.6a$ modes 1 and 2 swap their wavelengths.

By comparing Fig. 8 against Figs. 10 and 12 we see that 2D and 3D simulations give essentially similar dependencies for λ and Q on the longitudinal shift s and transversal separation p of the cavities, indicating the possibility to successfully design coupled nanobeam cavity systems in 2D. This is because the physics of side-coupling of dielectric nanobeam cavities is relatively simpler than, for example, the coupling of metallic split-ring resonators where essentially the three-dimensional interplay of magnetic and electric excitations is important.

4.2. Three coupled nanobeams

In multiple side-coupled nanobeam cavities modes can also be tuned by longitudinal shift. Degeneration of modes in structures containing many elements amounts to the absence of parasitic coupling between the neighboring units. Instead of increasing the distance between optical components usually employed to minimize the cross-talk, the longitudinal

http://dx.doi.org/10.5772/55096

shift can be proposed to create a dense photonic integrated circuit. Moreover, by taking large enough transversal separation the degeneracy wavelength of coupled nanobeams (up to three resonators in our tests) can be tuned to a single nanobeam resonance wavelength. This potentially allows adjacent waveguiding components, all together, and each separately, to operate at the same wavelength. For example, a compact single-wavelength switch matrix can be created on the basis of a nanobeam-switcher with nonlinearity [20]. Another field of application is building an array of nanobeam cavities to form a quantum optical network where many identical resonators should be placed closely one to another on a chip [44]. Additionally, by controlling the mode coupling it becomes possible to tailor the optical field across an array of multiple nanocavities for applications in particle trapping [27] and optomechanical interactions [18, 19].

4.2.1. *Weak coupling regime*

We did 2D simulations (letting the nanocavities be infinitely high) to catch basic features of the mode tunability. As usually, $\Delta x = \Delta y \simeq 0.004a$, $1a$-wide buffer layers along y-direction are squeezed with the $x/(1-x)$ function covered by PMLs on 1/2. No air buffer is used along x-direction, PMLs comprise 3 grid cells. For three side-coupled nanobeam cavities their relative alignment can be characterized by separations p_2, p_3 and longitudinal shifts s_2, s_3 of the second and third cavities. As an example we consider equally spaced ($p_2 = p_3 = p$) nanobeam cavities, only the middle one being shifted: $s_2 = s$, $s_3 = 0$. Dependence of the modes wavelength detuning (relative to a single cavity) on shift s is given in Fig. 13a for $p = 2.3a$. We observe behavior similar to the case of two nanobeams. Specifically, all three modes become degenerate at $s \simeq 0.4a$; by varying p we can control the wavelength of the degenerate modes, and it coincides with the wavelength of a single cavity, Fig. 13a. Mode profiles for the non-shifted system ($s = 0$) are shown in Fig. 13b. Note that mode 3 is localized at the outer cavities, so its wavelength is not sensitive to the middle cavity shift as observed in Fig. 13a. There is nice mechanical analogy with modes of three weakly coupled pendulums: in mode 1, all three pendulums are swinging in phase; in mode 2, two outward pendulums move forward while the middle one moves backward; in mode 3, the central pendulum is at rest and two others are moving oppositely.

At the degeneracy point ($s \simeq 0.4a$) the eigenmode profiles are primarily localized at individual cavities, see Fig. 13c. In Fig. 13c the mode profiles at $s = 0.41a$ reveal complete vanishing of field in neighboring nanobeams whereas for two coupled nanobeams $s = 0.4a$ shift was more likeable to be called the exact degeneracy shift value. In fact, it is quite difficult to detect the precise value of the degeneracy shift as it requires extremely fine steps in s and long simulation times; besides, accuracy of computation is also limited by the finite-difference description. However, small deviations from the exact degeneracy do not change field mapping significantly as solutions to Maxwell's equations are all smooth functions.

Wavelength detuning is much less pronounced at $p = 2.3a$ than at smaller separation. The reason to choose the separation $p = 2.3a$ is that coupling between the nanobeams is already weak and the energy splitting becomes symmetric relative to the initial energy level as follows from the standard perturbation approach. And the degeneracy wavelength of an array of nanobeams is the same as the isolated nanobeam eigenwavelength (note, that this is not true for the case of strong coupling at $p = 1.5a$).

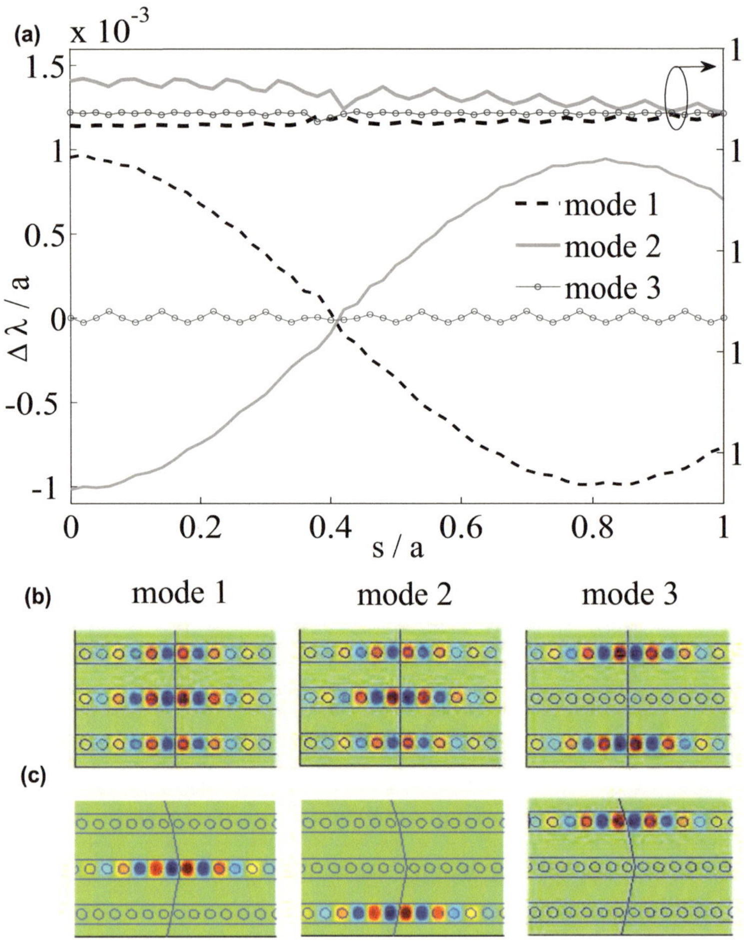

Figure 13. Modes in the three side-coupled nanobeam cavities with transversal displacement between the neighboring nanobeams being equal to $2.3a$. The middle nanowire is longitudinally shifted, the other two being kept stationary. (a) Spectral detuning of the three modes from the single nanobeam cavity wavelength $\lambda = 3.9965a$. Right y-axis—the Q-factor of the three coupled modes (for an isolated nanobeam cavity, $Q = 1.5 \cdot 10^8$). H_z profiles for the modes in the three (b) unshifted and shifted (c) by $0.41a$ nanobeams. For 2D calculation $\Delta x = \Delta y \simeq 0.004a$, $1a$-wide y-buffer is $x/(1-x)$ squeezed with half of the buffer covered by PMLs, x-PMLs are 3 grid cells thick.

Although we are in the weak coupling regime, the spectral splitting corresponding to $p = 2.3a$ is about 0.2%, which for telecom wavelength 1.5μm amounts to 3 nm spread in wavelengths. A comparable shift in the resonance wavelength is induced by inclusion of the nonlinear material in the nanobeam. This allows cavity operation as a switcher totally transmitting or suppressing the signal depending on turning on/off the nonlinearity [20]. Thus 0.2% energy difference for multiple nanobeams placed at $p = 2.3a$ on a photonic integrated chip introduces parasitic coupling hindering the single-wavelength operation.

http://dx.doi.org/10.5772/55096

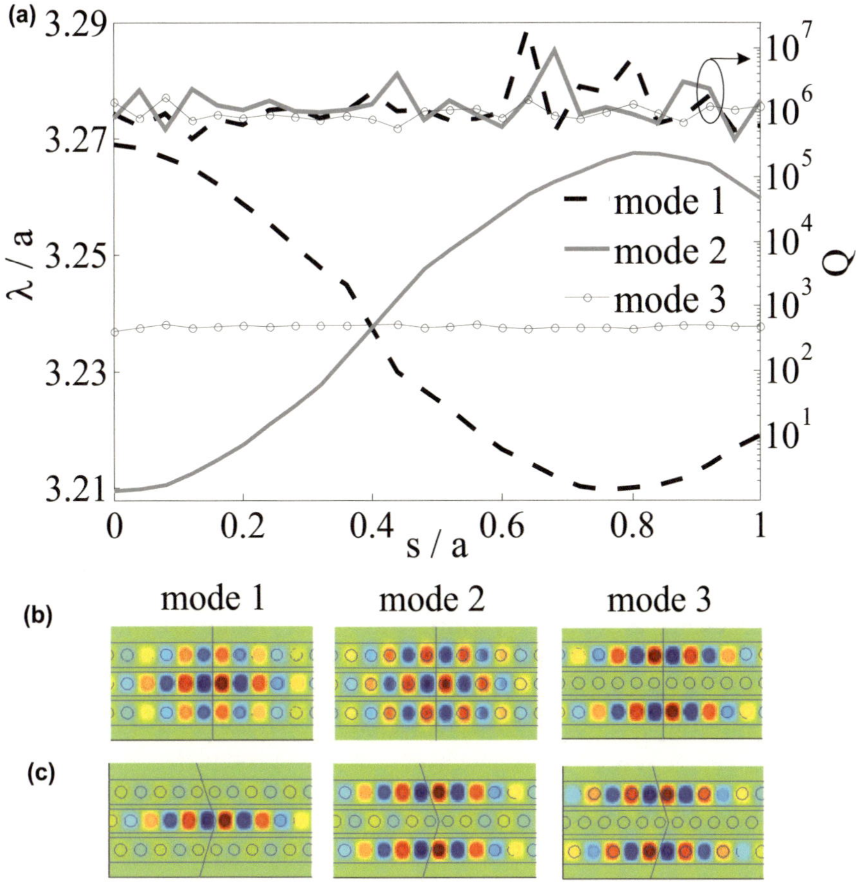

Figure 14. Modes in the three side-coupled nanobeam cavities. The middle nanobeam is longitudinally shifted, the other two being kept stationary. Transverse separation between the neighboring nanobeams is $p = 1.2a$. (a) λ and Q of the three modes. (b) H_z profiles for the modes in the three non-shifted nanobeams. (c) H_z profiles of the degenerate modes at $s = 0.4a$. For 3D simulation $1a$-buffer $(x/(1-x)$ squeezed) along y and z-direction is covered by PMLs on $1/3$, x-PMLs are 3 cells thick. $\Delta x = \Delta y = \Delta z \simeq 0.09a$ in the center of the nanobeam, Δx is stretched 3 times in the mirror part.

If we further suppose the nonlinear core of a nanobeam cavity, e.g. taking into account the refractive index changing due to nonlinearity, then the degenerate modes in the array of nonlinear cavities can be also tuned to a single resonator wavelength by p variation. A single-wavelength operating switching array can be build on the basis of such nanobeam cavities that will work equally well for applied single- or multiple-channel excitation. For instance, in the case of the single-cavity nonlinear operation based on 0.03% change in refractive index, the frequency shift due to presence of neighboring 'passive' nanobeams is estimated to be around a negligible $2 \cdot 10^{-4}$%.

We have verified that mode degeneracy also occurs in four side-to-side coupled nanobeam cavities. Thus we expect mode degeneracy happening in multiple side-coupled cavities when they have staggered longitudinal shifts, such that neighboring nanowires are shifted longitudinally by $s \simeq 0.4a$. We also foresee that the Q-factor in a multi-cavity structure should remain of the same order of magnitude as that of a single nanocavity.

4.2.2. Strong coupling regime

The strong coupling regime is investigated by doing 3D simulations with symmetry planes enforced through the domain reduction. For three nanobeams dependence of mode detuning on shift s of the middle nanobeam is given in Fig. 14a for $p = 1.2a$. Here three modes become degenerate at $s \simeq 0.4a$ with wavelength $\lambda \simeq 3.235a$ bigger than the single 3D nanobeam eigenwavelength $3.21a$. 3D Q-factors for three coupled cavities are found to remain of the same order as the longitudinal shift is varied, see Fig. 14a. Mode profiles for the non-shifted system ($s = 0$) are shown in Fig. 14b. When symmetry planes are enforced the eigenmode profiles at the degeneracy point ($s \simeq 0.4a$) look as in Fig. 14c.

In general, in the degeneracy point we presented the results of our numerical calculations based on the direct solution of Maxwell's equations and without taking linear combinations with data obtained. Our calculations for systems with two and three nanobeams gave two variants of field maps at the degeneracy point: 180° rotational (2 resonators) or reflection (3 resonators) symmetrical pictures with mode profiles exhibiting field localization in individual resonators. Simulations with enforced PEC/PMC perfect conducting planes for the three nanobeams give degenerate modes strictly reflection-symmetrical. However, we can guess that these pictures do not help in understanding physical reality better than do profiles with field localization in individual nanobeams implying complete vanishing of interaction between the cavities. Therefore, the characteristic degeneracy profile with field extinction in one of the cavities that is clearly predominant in all 2D and 3D simulations for all range of separations should be more expectable in the experiments than other linear combinations of the degenerate mode profiles.

5. Conclusions

We have suggested and showed numerically that a longitudinal shift in nanobeam cavities significantly alters coupling efficiency between multiple closely packed resonators. Whereas the concept of the longitudinal offset between cavities was previously developed for dielectric rod arrays at microwave frequencies [43] and micro-scale ring resonators at optical wavelengths [32], we have demonstrated here new possibilities for light control at nanoscale. Frequency detuning of coupled modes depends nontrivially on the longitudinal shift of the cavities, in particular, the modes become degenerate for a certain shift, a feature impossible in non-shifted resonators. At this shift of about a half the averaged lattice constant in the defect region, the magnetic field nodes in one nanobeam oppose the field lobes in the other. The degeneracy occurs for a broad range of separations between two or multiple side-coupled cavities. The quality factor of coupled nanobeam cavities stays close to that of a single cavity, indicating good practical prospects for such structures.

Author details

Aliaksandra M. Ivinskaya[1], Andrei V. Lavrinenko[2],
Dzmitry M. Shyroki[3] and Andrey A. Sukhorukov[4]

1 Department of Micro- and Nanotechnology, Technical University of Denmark, Lyngby, Denmark
2 Department of Photonics Engineering, Technical University of Denmark, Lyngby, Denmark
3 Institute of Optics, Information and Photonics, University Erlangen-Nürnberg, Erlangen, Germany
4 Nonlinear Physics Centre and Centre for Ultra-high bandwidth Devices for Optical Systems (CUDOS), Research School of Physics and Engineering, Australian National University, Canberra, Australia

References

[1] K. Hennessy, A. Badolato, M. Winger, D. Gerace, M. Atatüre, S. Gulde, S. Fält, E. L. Hu, and A. Imamoglu. Quantum nature of a strongly coupled single quantum dot-cavity system. *Nature*, 445(7130):896–899, February 2007.

[2] V. Rao and S. Hughes. Single quantum dot spontaneous emission in a finite-size photonic crystal waveguide: Proposal for an efficient chip single photon gun. *Physical Review Letters*, 99(19):193901, November 2007.

[3] T. Yoshie, A. Scherer, J. Hendrickson, G. Khitrova, H. M. Gibbs, G. Rupper, C. Ell, O. B. Shchekin, and D. G. Deppe. Vacuum Rabi splitting with a single quantum dot in a photonic crystal nanocavity. *Nature*, 432:9–12, November 2004.

[4] Y. Zhang, M. W. McCutcheon, I. B. Burgess, and M. Loncar. Ultra-high-Q TE/TM dual-polarized photonic crystal nanocavities. *Optics Letters*, 34(17):2694–2696, September 2009.

[5] B.-S. Song, S. Noda, T. Asano, and Y. Akahane. Ultra-high-Q photonic double-heterostructure nanocavity. *Nature Materials*, 4(3):207–210, February 2005.

[6] Z. Zhang and M. Qiu. Small-volume waveguide-section high Q microcavities in 2D photonic crystal slabs. *Optics Express*, 12(17):3988–3995, August 2004.

[7] O. Painter, J. Vučkovič, and A. Scherer. Defect modes of a two-dimensional photonic crystal in an optically thin dielectric slab. *Journal of the Optical Society of America B*, 16(2):275–285, February 1999.

[8] B.-S. Song, S.-W. Jeon, and S. Noda. Symmetrically glass-clad photonic crystal nanocavities with ultrahigh quality factors. *Optics Letters*, 36(1):91–93, January 2011.

[9] K. J. Vahala. Optical microcavities. *Nature*, 424(6950):839–846, August 2003.

[10] Y. Tanaka, T. Asano, and S. Noda. Design of photonic crystal nanocavity with Q equal to 10^9. *Journal of Lightwave Technology*, 26(11):1532–1539, June 2008.

[11] Y. Gong, B. Ellis, G. Shambat, T. Sarmiento, J. S. Harris, and J. Vuckovic. Nanobeam photonic crystal cavity quantum dot laser. *Optics Express*, 18(9):8781–8789, April 2010.

[12] P. B. Deotare, M. W. McCutcheon, I. W. Frank, M. Khan, and M. Loncar. High quality factor photonic crystal nanobeam cavities. *Applied Physics Letters*, 94:121106–3, 2009.

[13] Ph. Lalanne C. Sauvan, G. Lecamp and J. P. Hugonin. Modal-reflectivity enhancement by geometry tuning in photonic crystal microcavities. *Optics Express*, 13:245–255, 2005.

[14] P. B. Deotare, M. W. McCutcheon, I. W. Frank, M. Khan, and M. Loncar. Coupled photonic crystal nanobeam cavities. *Applied Physics Letters*, 95(3):031102, 2009.

[15] N.-V.-Q. Tran, S. Combrié, and A. De Rossi. Directive emission from high-Q photonic crystal cavities through band folding. *Physical Review B*, 79(4):041101, January 2009.

[16] S.-H. Kim, S.-K. Kim, and Y.-H. Lee. Vertical beaming of wavelength-scale photonic crystal resonators. *Physical Review B*, 73(23):235117, June 2006.

[17] Y. Gong and J. Vuckovic. Photonic crystal cavities in silicon dioxide. *Applied Physics Letters*, 96(3):031107, 2010.

[18] M. Eichenfield, R. Camacho, J. Chan, K. J. Vahala, and O. Painter. A picogram- and nanometre-scale photonic-crystal optomechanical cavity. *Nature*, 459(7246):550–555, May 2009.

[19] Q. Lin, J. Rosenberg, D. Chang, R. Camacho, M. Eichenfield, and K. J. Vahala. Nano-optomechanical structures. *Nature Photonics*, 4:236–242, April 2010.

[20] M. Belotti, M. Galli, D. Gerace, L. C. Andreani, G. Guizzetti, A. R. Md Zain, N. P. Johnson, M. Sorel, and R. M. De La Rue. All-optical switching in silicon-on-insulator photonic wire nano-cavities. *Optics Express*, 18(2):1450–1461, January 2010.

[21] A. Yariv, Y. Xu, R. K. Lee, and A. Scherer. Coupled-resonator optical waveguide: a proposal and analysis. *Optics Letters*, 24(11):711–3, June 1999.

[22] D. Gerace, H. E. Türeci, A. Imamoglu, V. Giovannetti, and R. Fazio. The quantum-optical Josephson interferometer. *Nature Physics*, 5(4):281–284, March 2009.

[23] D. G. Angelakis, L. Dai, and L. C. Kwek. Coherent control of long-distance steady-state entanglement in lossy resonator arrays. *Europhysics Letters*, 91(1):10003, July 2010.

[24] K. A. Atlasov, M. Felici, K. F. Karlsson, P. Gallo, A. Rudra, B. Dwir, and E. Kapon. 1D photonic band formation and photon localization in finite-size photonic-crystal waveguides. *Optics Express*, 18(1):117–22, January 2010.

[25] K. Foubert, L. Lalouat, B. Cluzel, E. Picard, D. Peyrade, F. de Fornel, and E. Hadji. An air-slotted nanoresonator relying on coupled high Q small V Fabry-Perot nanocavities. *Applied Physics Letters*, 94(25):251111, 2009.

[26] B. Cluzel, K. Foubert, L. Lalouat, J. Dellinger, D. Peyrade, E. Picard, E. Hadji, and F. de Fornel. Addressable subwavelength grids of confined light in a multislotted nanoresonator. *Applied Physics Letters*, 98(8):081101, 2011.

[27] S. Mandal, X. Serey, and D. Erickson. Nanomanipulation using silicon photonic crystal resonators. *Nano Letters*, 10(1):99–104, January 2010.

[28] F. Intonti, S. Vignolini, F. Riboli, M. Zani, D. S. Wiersma, L. Balet, L. H. Li, M. Francardi, A. Gerardino, A. Fiore, and M. Gurioli. Tuning of photonic crystal cavities by controlled removal of locally infiltrated water. *Applied Physics Letters*, 95(17):173112, 2009.

[29] H. Altug, D. Englund, and J. Vuckovic. Ultrafast photonic crystal nanocavity laser. *Nature Physics*, 2:484–488, 2006.

[30] A. E. Miroshnichenko, S. Flach, and Y. S. Kivshar. Fano resonances in nanoscale structures. *Reviews of Modern Physics*, 82:2257–2298, 2010.

[31] T. Baba. Slow light in photonic crystals. *Nature Photonics*, 2(8):465–473, August 2008.

[32] J. D. Domenech, P. Munoz, and J. Capmany. Transmission and group-delay characterization of coupled resonator optical waveguides apodized through the longitudinal offset technique. *Optics Letters*, 36(2):136–8, January 2011.

[33] S. Mookherjea and A. Yariv. Optical pulse propagation and holographic storage in a coupled-resonator optical waveguide. *Physical Review E*, 64(6):066602, November 2001.

[34] S. Mookherjea and A. Yariv. Second-harmonic generation with pulses in a coupled-resonator optical waveguide. *Physical Review E*, 65(2):026607, January 2002.

[35] S. Vignolini, F. Riboli, F. Intonti, D. S. Wiersma, L. Balet, L. H. Li, M. Francardi, A. Gerardino, A. Fiore, and M. Gurioli. Mode hybridization in photonic crystal molecules. *Applied Physics Letters*, 97(6):063101, 2010.

[36] M. Bayer, T. Gutbrod, J. Reithmaier, A. Forchel, T. Reinecke, P. Knipp, A. Dremin, and V. Kulakovskii. Optical modes in photonic molecules. *Physical Review Letters*, 81(12):2582–2585, September 1998.

[37] D. M. Shyroki, A. M. Ivinskaya and A. V. Lavrinenko. Modeling of nanophotonic resonators with the finite-difference frequency-domain method. *IEEE Transactions on Antennas and Propagation*, 59:4155–4161, 2011.

[38] M. Notomi and H. Taniyama. On-demand ultrahigh-Q cavity formation and photon pinning via dynamic waveguide tuning. *Optics Express*, 16(23):18657–66, November 2008.

[39] D. Chigrin, A. V. Lavrinenko, and C. M. Sotomayor Torres. Nanopillars photonic crystal waveguides. *Optics Express*, 12(4):617–622, 2004.

[40] M. J. Steel, T. P. White, C. M. de Sterke, R. C. McPhedran, and L. C. Botten. Symmetry and degeneracy in microstructured optical fibers. *Optics Letters*, 26:488–490, 2001.

[41] L. D. Landau and E. M. Lifshitz. *Quantum Mechanics.* MA Addison Weslay, USA, 1958.

[42] J. D. Joannopoulos, S. G. Johnson, J. N. Winn, and R. D. Meade. *Photonic Crystals: Molding the Flow of Light.* Princeton Univ. Press, 2001.

[43] S. Ha, A. A. Sukhorukov, A. V. Lavrinenko, and Y. S. Kivshar. Cavity mode control in side-coupled periodic waveguides: Theory and experiment. *Photonics and Nanostructures - Fundamentals and Applications*, 8(4):310–317, May 2010.

[44] H. J. Kimble. The quantum internet. *Nature*, 453(7198):1023–30, June 2008.

Chapter 3

Birefringence in Photonic Crystal Structures: Toward Ultracompact Wave Plates

Wenfu Zhang and Wei Zhao

Additional information is available at the end of the chapter

http://dx.doi.org/10.5772/55142

1. Introduction

Birefringence is an important optical effect of materials that having different refractive indices for different polarizations of light. Birefringence and related optical effects play an important role in quantum and nonlinear processes and also have been widely used in modern optical devices, such as optical sensors, light modulators, liquid crystal displays, crystal filters, medical diagnostics, and wave plates (WPs). Among these, WP is one of the most essential elements in many optical modules and equipment, and will certainly have many applications in future photonic integrated circuits (PICs).

According to the difference of generation mechanism, birefringence could be divided into two types: natural birefringence and artificial birefringence, which are microscopic and macroscopic anisotropy induced, respectively [1]. Generally, artificial birefringence is larger than natural birefringence and has designable characteristics. Many artificial structures with high birefringence have been proposed and studied recently. Table 1 gives the comparison of birefringence, dispersion and loss of different structures containing multilayer film (MF) [2], nanowires (NW) [3], metamaterials (MM) [4], plasmonic nanoslits array (PNA) [1], multi-slotted dielectric waveguides (MSDW) [5], bulk photonic crystal (PhC) [6-8], two dimensional (2D) PhC waveguide (PhCW) [9, 10], and periodic dielectric waveguide (PDW) which also named as one dimensional (1D) PhCW [11, 12]. Among these artificial materials, the PhC related structures are more important not only for the excellent birefringence properties but also the great importance of PhC in PICs.

PhCs are structures with periodic arrangement of dielectrics or metals, which provide the ability of molding the flow of light in it [13-15]. Due to the unique guiding properties of PhC structure, such as the photonic band gap guidance, it is foreseen as one of the key artificial materials for next generation PICs. This chapter gives a thorough review of the birefringence properties of PhC structures containing 1D PhCW (PDW), 2D PhCW and bulk PhC. The applications of the giant birefringence of PhC structures in both low-order WPs and high-order WPs are studied in details. This chapter is organized as follow. In Sec. 2, an overview of numerical algorithms used in this chapter is given. In Sec. 3, the birefringence properties are studied taking 2D PhCWs, 2D bulk PhCs, and 1D PhCWs as examples. High performance WPs are designed based on different PhC structures in Sec. 4. Discussions and conclusions are given in Sec. 5.

	MF	NW	MM	PNA	MSDW	PhCW	Bulk PhC	PDW
Δn	~0.3	~0.8	~3.2	2.7	1.0	0.111	0.938	1.5
Achromatic	-	-	No	No	-	Yes	Yes	Yes
Loss	Low	-	High	High	Low	Low	High	Low

* This table is from reference [12]

Table 1. Birefringence of different artificial structures

2. Numerical algorithms

Many numerical algorithms which solve the partial differential equations can be used in computational photonics. For PhCs, two categories of problems are most important [15]: one is frequency-domain eigenvalue problem which refers to find the band structure ω(k) and associated electromagnetic fields; the other one is time-domain calculation which is related to obtain the fields E(x, t) and H(x, t) propagating in time. For the frequency-domain eigenproblems, plane wave expansion method (PWE) is the most popular method to get the dispersion relation (band structure) of specific PhC geometries [15, 16]. Correspondingly, Finite-difference time-domain method (FDTD) [17, 18] is the most important time-domain simulation approach for PhCs to get the propagating fields and transmission/reflection spectrum.

2.1. Plane Wave Expansion method (PWE)

PWE is used to solve the Maxwell equations by formulating an eigenvalue problem. The master equation for PhCs can be deduced from Maxwell equations as follow:

$$\hat{\Theta}\,\mathbf{H}(\mathbf{r}) = \left(\frac{\omega}{c}\right)^2 \mathbf{H}(\mathbf{r}) \tag{1}$$

where ω is angular frequency, c is the vacuum speed of light, r is the position vector, and H(r) is the macroscopic magnetic field. $\hat{\Theta}$ is a linear Hermitian operator and $\hat{\Theta}H(r)$ is written as [15]

$$\hat{\Theta}\,\mathbf{H}(\mathbf{r}) = \nabla \times \left(\frac{1}{\varepsilon(\mathbf{r})} \nabla \times \mathbf{H}(\mathbf{r}) \right) \quad (2)$$

where ε(r) is the relative permittivity of dielectric materials. By using the plane wave basis, the Bloch eigenmodes and band structures of the perfect periodic structures could be easily obtained by solving the master equation. However, for the non-periodic or quasi-periodic structures, such as line defect PhCWs, and the structures without periodically in all dimensions, the supercell technique must be used by choosing a large computational cell as the periodic element which helps to get the isolated electromagnetic modes. By using the supercell method, both point and line defect modes of PhCs can be solved. Taking a PhCW as example, the structure is square lattice dielectric rods in air. The radius of the dielectric rods is r=0.2a, where a is the lattice constant, and the permittivity is ε=8.9. The line defect is introduced by removing one row of dielectric rods in the ΓX direction. Fig. 1 gives the dispersion curves calculated by supercell method. The square lattice PhCW has single transverse-magnetic (TM) guided mode in the normalized frequency range of 0.32-0.446 a/λ, which locates in the photonic band gap (PBG) range of TM polarization mode. This reveals that the PhCW has PBG guided TM mode but no guided mode for TE polarization.

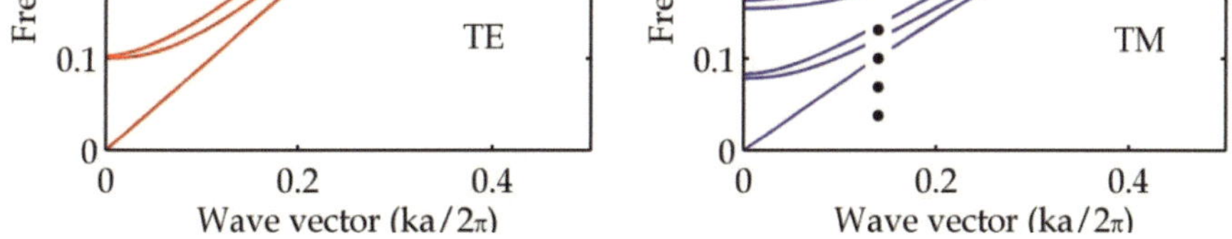

Figure 1. Dispersion curves of line defect square lattice PhCW. The inset is the 1×9a supercell used in PWE calculations. The parameters of the PhCW are chosen as follow: permittivity of high and low index materials are ε_H=8.9 and ε_L=1, respectively and the radius of dielectric rods is r=0.2a.

The PWE is suitable to calculate the band structure of PhC, but not conventional to get the transmission spectra of PhC. When the loss and transmitting properties of PhC are required, FDTD method is often used for the advantages of that broadband response can be accurately obtained in only one simulation run.

2.2. Finite-Difference Time-Domain method (FDTD)

The FDTD method is one of the grid-based differential time-domain numerical modeling methods. The time-derivative parts of Maxwell equations in partial differential form are written:

$$\begin{aligned} \frac{\partial \mathbf{B}}{\partial t} &= -\nabla \times \mathbf{E} - \mathbf{J}_B \\ \frac{\partial \mathbf{D}}{\partial t} &= -\nabla \times \mathbf{H} - \mathbf{J} \end{aligned} \tag{3}$$

where B is the electric displacement, D is the magnetic induction fields, J is the electric-charge current density, and J_B is an imaginary magnetic-charge current density for calculation convenience. By central-difference approximations, e.g. standard Yee grid [19], the electric and magnetic fields governed by Eq. (3) are discretized both in time and space. By properly selecting of initial excitation current J or J_B, the resulting finite-difference equations are solved in a leapfrog manner: the electric field vector components in a volume of space are solved at a given instant of time; then the magnetic field vector components in the same spatial volume are solved at the next instant of time; and the process is repeated over and over again until the desired transient or steady-state electromagnetic field behavior is fully evolved.

Although a lot of FDTD solver packages are available in literature, we use the free software MEEP [18] which is developed by MIT's researchers in this chapter. The transmission spectrum of PhC structures can be easily obtained by the FDTD solver. Still take the square lattice PhCW studied in Fig. 1 as example, the transmission spectrum for the TM polarization of the PhCW with length of 27*a* is shown in Fig. 2. From the figure, the PhCW is low loss in the frequency range of 0.32-0.446 a/λ which is consistent with the simulation results taken by PWE method.

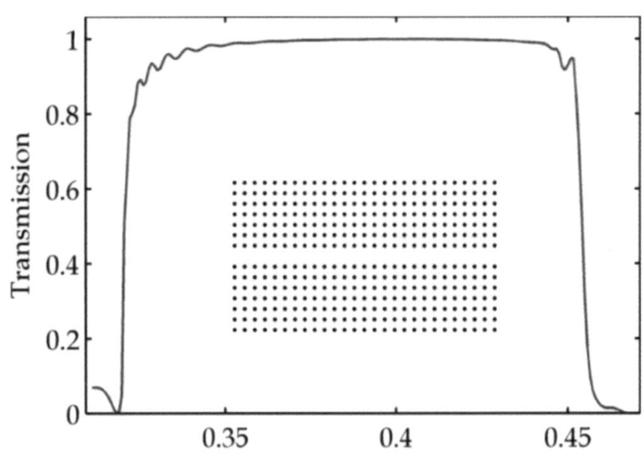

Figure 2. Transmission spectrum for the TM mode of the line defect square lattice PhCW. The inset is the PhCW structure used in FDTD simulation and the parameters of PhCW are as same as that in Fig. 1.

2.3. Spatial Fourier Transform method (SFT)

Except for the low loss of the PhC structures, large birefringence is essential for realizing the ultracompact WPs. The frequency-dependent effective mode indices for both TE and TM modes of PhC structure should be calculated to obtain the birefringence of the structures. When the PhC structure has single mode in the operating frequency range, such as PBG guided TM mode, in the PhCW shown in Fig. 1, the mode indices can be obtained from the dispersion curves calculated by PWE method. But, for the PhC structures with quasi-periodic and non-periodic structures or some special PhCW without PBG guided modes but total internal reflection (TIR) modes, the dispersion curves of the guided modes are difficultly obtained by the conventional PWE method. In this chapter, the SFT method [20] is used when the birefringence of this category of structure needs to be calculated.

The SFT method is based on the spatial Fourier transform spectrum of the electromagnetic field distributions of the waveguide mode along the propagating direction. Assuming $u_\omega(x, y_0)$ is the field distribution of waveguide mode at the given plane $y=y_0$ along the propagating direction x, where u is E or H, ω is the angular frequency of the mode, x is the propagation direction, and y is the direction perpendicular to x in the plane. The field can be expanded by the plane wave basis as $u_\omega(x, y_0) = \sum_n u_{n,\omega}\exp(j\beta_{n,\omega}x)$, where $u_{n,\omega}$ and $\beta_{n,\omega}$ are the field component and the propagation constant of the nth mode at frequency ω, respectively. The propagation constant $\beta_{n,\omega}$ can be extracted from the peaks of the SFT spectrum of $u_\omega(x, y_0)$. For the Bloch mode of the periodic structure, the wavevector β still can be obtained from the SFT analysis for that the peaks of SFT spectrum always located at $\beta+m(2\pi/a)$, where m is an integer [20].

To verify the SFT method, the field distribution of TM mode in the PhCW studied in Fig. 1 and Fig.2 are calculated by FDTD method and the snapshot of E_z fields at the frequency of 0.40 a/λ is shown in Fig. 3 (a). The peak of the SFT spectrum (as shown in Fig. 3 (b)) is located at 0.292 $a/2\pi$ which is as same as the Bloch wave vector obtained by PWE methods. The dispersion diagrams obtained by PWE and SFT method in frequency range of 0.34-0.42 a/λ are almost same as shown in Fig. 3 (c), which shows the validity of the SFT technique in the mode calculation for the PhCW structures.

3. Birefringence of PhC structures

Birefringence is related to the effective index difference of two orthogonal polarization modes and can be expressed as $\Delta n=n_p-n_s$, where p and s are TE (TM) and TM (TE), respectively, in this chapter. It is the basis on which the ultracompact WP is realized. Fortunately, most of the PhC structures have large birefringence for the large index difference of the composed materials of PhCs. Researching results show that the 2D PhCW having birefringence of ~0.07 and ~0.1 for square and triangular lattice air holes in high index materials, respectively [10]. For the 2D bulk PhCs, experimental results reveal that the birefringence is as high as 0.2 [8], and the theoretical analysis results are higher. The 1D PhCWs give giant birefringence which is higher than 1 [12]. In this section, the birefringence properties of different PhC structures, containing 2D PhCW, 2D bulk PhC and 1D PhCW, will be reviewed in details.

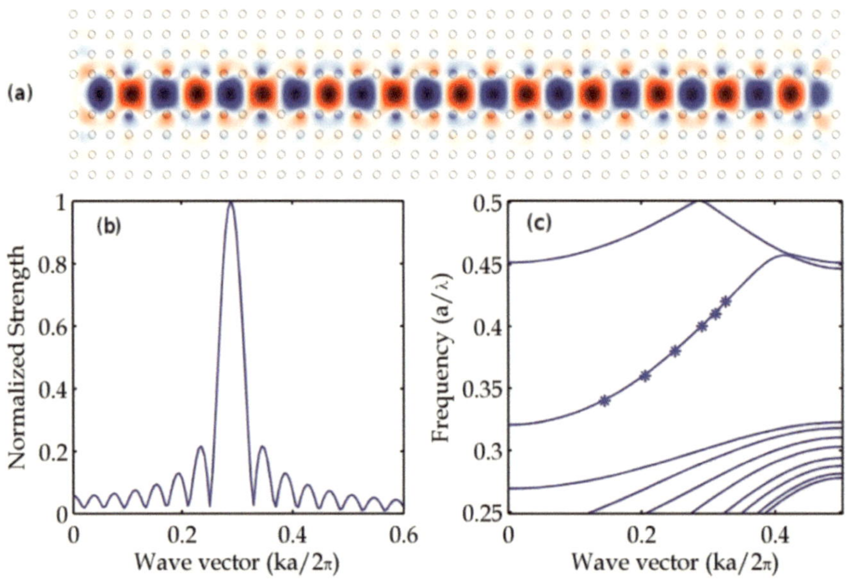

Figure 3. The snapshot of E_z field distribution for the TM guided mode of the square lattice PhCW at frequency of 0.40 a/λ; (b) SFT spectrum of the E_z field shown in (a); (c) Dispersion curves (also shown in Fig. 1) of the TM polarization, the solid lines are obtained by PWE method and the stars represent the results calculated by the SFT method.

3.1. Birefringence of 2D PhCW

For birefringence related applications, both TE and TM polarized light must propagate with low loss in the 2D PhCW which is formed by introducing of line defect in the perfect bulk PhC at a given direction. Different guided mechanisms could be used to confine light in the 2D PhCWs [21]. The widely studied mechanism in the literature is the PBG guiding as that shown in Fig. 1. It is difficult to realize 2D PhCW supporting both TE and TM PBG guided modes, only if carefully choosing the materials and geometry structures. Studied results show that the 2D PhCW can also guide the light as that in the conventional dielectric slab waveguide by total internal refection (TIR) guiding mechanism. Compound of PBG and TIR effects could also make low loss guiding for both TE and TM mode in the 2D PhCW [21-23].

3.1.1. Birefringence of 2D PhCW with PBG guided modes

The PhCW with hybrid triangular and honeycomb lattices can support both TE and TM modes [24, 25]. The structure is shown in Fig. 4 and has been optimized as follow: The permittivity of the background high index material is 11.56, and the radii of the large and small air holes are r_1=0.27$_a$ and r_2=0.15$_a$, respectively. The waveguide is formed by a line defect of elliptical air holes with major and minor axes of d_a=0.4$_a$ and d_b=0.2$_a$, respectively. Two lines of small air holes with radius of r_2 are added to get single TE and TM propagation.

Figure 4. The structure of PhCW with hybrid triangular and honeycomb lattices, which supports PBG guided TE and TM modes. [24, 25]

The band structure of PhCW shown in Fig. 4 is calculated by PWE method with $1{\times}4\sqrt{3}a$ supercell as shown in Fig.5 (a) [24]. The single TE and TM mode region is 0.558-0.569 a/λ, which located in the absolute band gap of the bulk PhC. The effective indices of the TE and TM modes are obtained from the dispersion curves and the birefringence Δn in the single mode region are calculated and shown in Fig. 5 (b). The birefringence can be larger than 0.5, which is much higher than the natural birefringence in the birefrigenct crystals. However, the structure has large dispersion for that the birefringence varies from 0.2 to 0.56 in the operating frequency region.

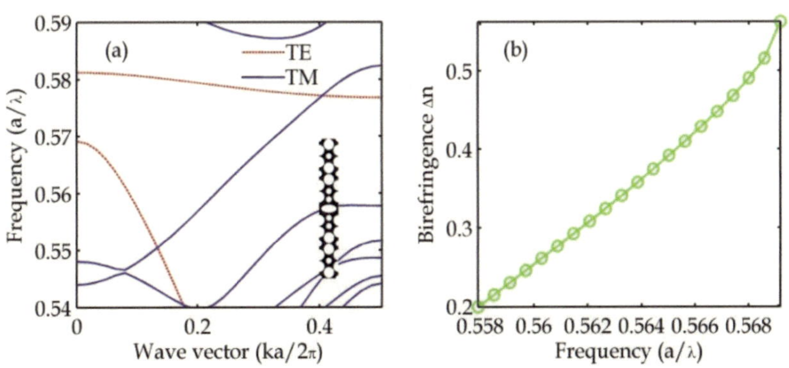

Figure 5. (a) Dispersion curves of PhCW shown in Fig. 4. The inset is the supercell used in PWE calculations. (b) Birefringence (Δn) of PhCW in the single mode region.

3.1.2. Birefringence of 2D PhCW with hybrid PBG and TIR guided modes

The PhCWs with both square and triangular lattice air holes in high index materials can support low loss transmitting of TE and TM polarizations with the help of TIR guided mechanism [21-23]. Taking a square lattice PhCW as example, the waveguide is formed by introducing of a line defect in ΓX direction in the perfect PhC which has square lattice air holes in high index material with permittivity of 12.96. Calculated dispersion curves of the PhCW are shown in Fig. 6 for both TE and TM polarizations. The PhCW has single TE guided mode in the normalized frequency range of 0.248-0.272a/λ, which locates in the PBG region of TE polarization mode. This reveals that the PhCW has PBG guided TE mode. However, there are no PBG guided modes in this frequency range for TM polarization as shown in Fig. 6 (b). To fully understand the guiding properties of the PhCW, the transmission spectra for both TE and TM polarization of the PhCW with length of 21a are calculated by FDTD method and shown in Fig. 7 (a). New information could be obtained from the figure: 1) Wider frequency range for low-loss guiding TE mode is achieved which reveals that the guiding mechanisms of the TE mode are not only PBG but also TIR; 2) Although there is no band-gap for TM mode in the frequency range of 0.24-0.30 a/λ, the TM polarization also can propagate with low-loss in this frequency range by the TIR mechanism. From the insets of Fig. 7 (a), EM fields are confined well in the line defect waveguide region for both TE and TM polarizations. For the effective indices of TM guided modes can't be obtained directly from the dispersion curves calculated by the PWE method, the SFT method is used to study the birefringence properties of the PhCW with hybrid PBG and TIR guided modes. By launching the continuous wave (CW) source with single frequency in the low loss frequency range of 0.255-0.268 a/λ, the EM field distributions for the guided mode of both TE and TM polarization are calculated by FDTD method. The propagation constants of guided modes, which are used to the calculating of effective indices, are found by seeking the peaks of the SFT spectra of the EM fields. Fig. 7 (b) shows the birefringence property of the 2D square lattice PhCW with hybrid PBG and TIR guided modes. The birefringence of the 2D square lattice PhCW is much lower than that shown in Fig. 5 (b), however, the PhCW has advantages of easily design and high tolerance of distortion in fabrication. Except for the square lattice structure, the PhCW with triangular lattice air holes in high index material is also can be used as birefringent waveguide for the WP applications. Previously studied results shows that the birefringence in the triangular lattice PhCW is a little higher than that in the square lattice structures [10].

3.2. Birefringence of 2D bulk PhCs

Actually, the typical bulk PhC itself is strongly anisotropic artificial material which provides large birefringence. Different from the PhCW, the PhC as birefringent media must work outside the PBG for that the PhC are highly reflection material for the EM wave located in the PBG frequency range. The birefringence of bulk PhC is measured for the hexagonal lattice structure in microwave band and the experimental measurement birefringence is below 0.20 [8]. By increasing the index difference of the materials, the birefringence of the bulk PhC could be much stronger. Taking a bulk 2D structure as example, the PhC is composed by parallel cylinder dielectric rods in air, in which the dielectric rods have radius of r=0.37a and permittivity of ε=8.9 [26]. Choosing frequency range of 0.1-0.2 a/λ as the operating band which is below the first forbidden band (0.2495-0.2778 a/λ) of the TM polarization, the EM fields

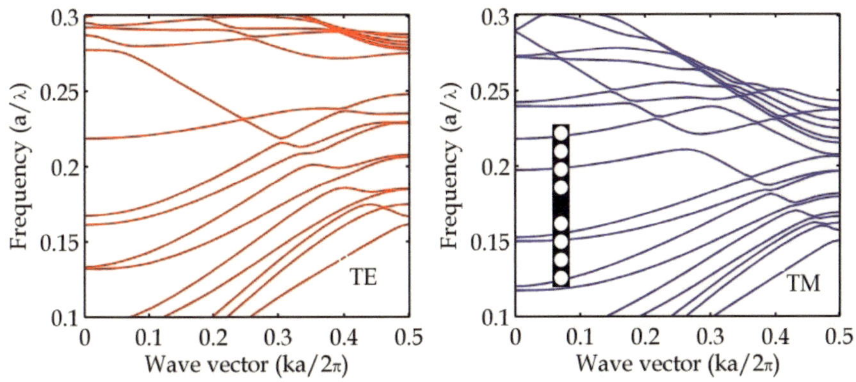

Figure 6. Dispersion curves of line defect square lattice PhCW. The inset is the 1×9*a* supercell used in PWE calculations. The parameters of the PhCW are chosen as follow: permittivity of high and low index materials are 12.96 and 1, respectively and the radius of air holes is r=0.40a.

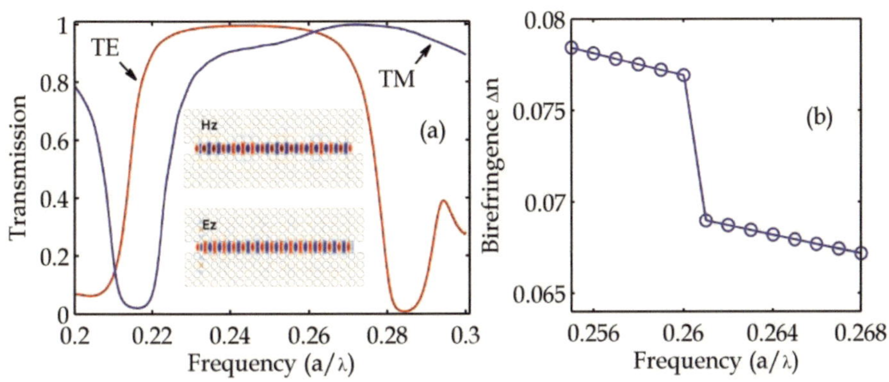

Figure 7. (a) Transmission spectra of the line defect square lattice PhCW. The insets are the EM field distributions for TE (H_z) and TM (E_z) polarization, respectively, at the frequency of 0.263 *a*/λ. (b) Birefringence (Δ*n*) of the PhCW calculated by SFT method. Structure used in FDTD and SFT simulation and the parameters of PhCW are as same as that in Fig. 6

propagate in the bulk PhC are shown in Fig. 8 (a). By using the SFT method, the birefringence of the studied bulk PhC are calculated and shown in Fig. 8 (b). The largest birefringence of the bulk PhC is about 0.8, which is much larger than that of the PhCW for the transmission path is almost homogeneous in the line defect of the PhCW but periodic in the bulk PhC.

For the birefringence related applications, the bulk PhC is a good candidate for the large birefringence in it, however, there are still some disadvantages [27]: 1) Lacking of effective light confining in the propagating plane makes beam divergence, as shown in Fig. 8 (a), which will spread the EM fields into the adjacent devices and cause crosstalk if there are many components packaged compactly, such as in PIC, to fulfill complicated functions. 2) The attenuation of light in the bulk PhC is high for the high scattering loss in it. These two problems should be solved in the practical applications such as WPs.

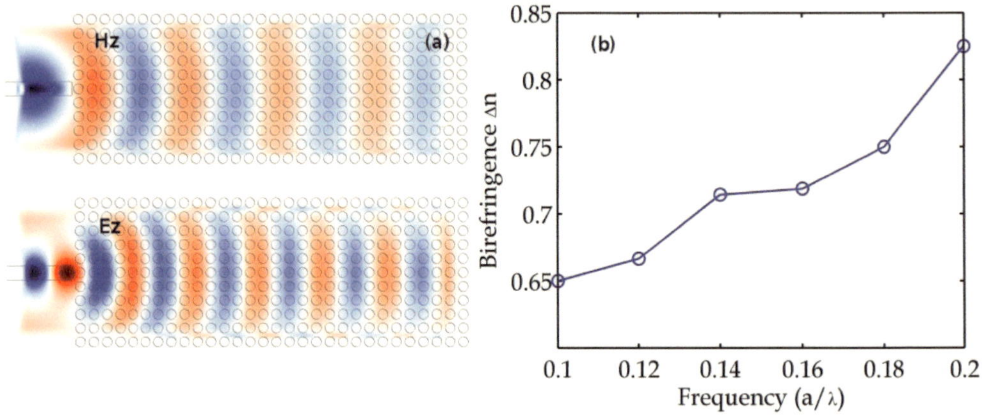

Figure 8. (a) EM fields propagate in the bulk PhC for TE (H_z) and TM (E_z) polarization, respectively, at the frequency of 0.10 a/λ. The CW source is excited in the conventional planar waveguide adjacent to the bulk PhC. (b) Birefringence (Δn) of the PhC calculated by SFT method. The calculated structure is a 2D bulk PhC with square lattice dielectric rods in air and the permittivity and the radius of dielectric rods are ε=8.9 and r=0.37a, respectively.

3.3. Birefringence of 1D PhCW

As shown in Fig. 9 (a), the 1D PhCW here refers in particular to the PDW [28-30], also known as nanopillar periodic waveguide [31-35] or coupled periodic waveguide [36], which has periodicity only in the light propagating direction. The 1D PhCW has attracted a lot of research interests for the simpler structure comparing with the 2D PhCW, and can be used in slow light [37], laser [35], low-loss waveguide [28-30, 38, 39], micro-resonator cavities [40], splitters for polarization and frequency [41-44], etc. The dispersion curves of the 1D PhCW could be examined by PWE method as what shown in Fig. 9 (b). There are guided modes under the light line and only single TE and TM modes are supported by the structure when the frequency is under 0.2065 a/λ. In the single mode frequency region, the TE and TM guided modes at the same frequency have different propagation constants which reveal the 1D PhCW is a birefringent media [12].

The birefringence properties of the 1D PhCW with cylinder dielectric rods in air are shown in Fig. 10, and all birefringence values are calculated in the single mode frequency band. The 1D PhCW has giant birefringence, which is larger than 1.5, when the permittivity of dielectric rods is 12.96. The higher the permittivity of dielectric rods is, the larger the birefringence is. The birefringence varies rapidly in the frequency band nearby the upper edge (slow light region) of the first TM guided mode when the 1D PhCW having relatively small dielectric rods (e.g. r=0.45a and r=0.50a). The largest birefringence appears at the edge of the slow light band. For the frequency outside the slow light band, the largest values of birefringence are almost equal for the 1D PhCW with different size of dielectric rods if the values of permittivity are same. There are *flat* sections in which the birefringence varies slowly ($d\Delta n/d\omega$~0) for the radius of 1D PhCW is be equal or greater than 0.50a, especially for r=0.50a as shown in the zoom-in curves in Fig. 10 (d). This reveals that the 1D PhCW has broadband achromatic birefringence. For example, the birefringence is between 1.1505 and 1.1530 in the frequency band of 0.167-0.188a/

http://dx.doi.org/10.5772/55142

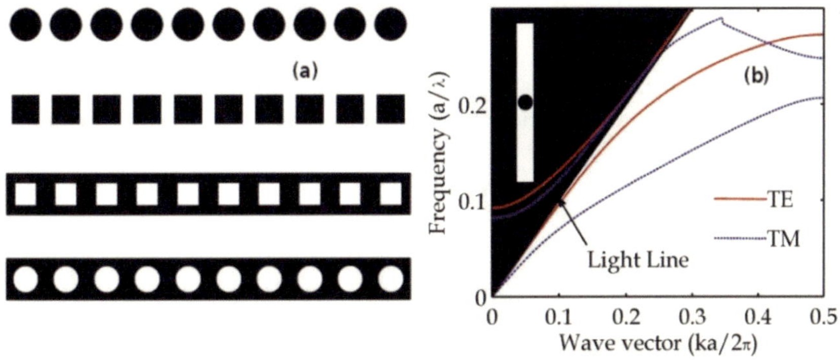

Figure 9. (a) Top view of the structures of different type of 1D PhCWs: cylinder and square dielectric rods in air, and square and cylinder air holes in dielectric waveguide, respectively, from top to bottom. (b) Dispersion curves of 1D PhCW with dielectric rods in air for both TE and TM polarizations. The inset is the 1×9*a* supercell used in PWE calculation, and the permittivity and the radius of dielectric rods are ε=8.9 and *r*=0.45*a*, respectively

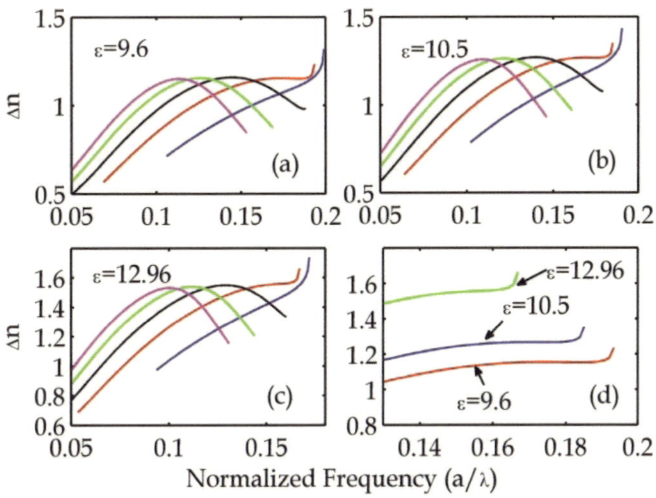

Figure 10. Birefringence of 1D PhCW with different parameters of (a) ε=9.6, (b) ε=10.5, and (c) ε=12.96, respectively. The blue, red, black, green and magenta lines represent the radius of dielectric rod equals 0.45*a*, 0.50*a*, 0.55*a*, 0.60*a*, and 0.65*a*, respectively. (d) Zoon-in picture of birefringence of 1D PhCW with radius of *r*=0.50*a*. [12]

λ when the dielectric rods have permittivity of ε=9.6, and for ε=10.5, the achromatic band is 0.165-0.178 *a*/λ in which the birefringence is 1.264-1.265 [12].

Other types of 1D PhCW have large birefringence too. Taking the 1D PhCW with square air holes in dielectric waveguide as example, the birefringence is around 1 when the length of side of the square is *w*=0.5*a*, and the permittivity and width of the dielectric waveguide are ε=12.96 and *a*, respectively [11]. Generally speaking, the 1D PhCW is better for birefringence related applications than the 2D PhCWs and bulk PhC for it has simple structure, low-loss and most important higher birefringence.

4. Ultracompact WPs based on PhC structures

One of the most important devices for birefringence related applications is WP which is worked as phase retarder. The phase difference ($\Delta\varphi$) between TE and TM polarization after propagating a distance of L can be expressed as $\Delta\varphi$=$2\pi\Delta nL/\lambda$, where Δn and λ are birefringence and wavelength, respectively. When the phase difference has relationship of $\Delta\varphi$=$m\pi/2$ and $\Delta\varphi$=$m\pi$, where m is an integer, the phase retarder is names as quarter-wave plate (QWP), and half-wave plate (HWP), respectively. Low-order (m is a small value) QWPs and HWPs have been investigated both theoretically and experimentally by different PhC structures. Although there are works to achieve $\pi/2$ and π phase retarding by reflecting EM wave at the stop band of perfect bulk PhC [45], most of the WPs are realized by the birefringence effect of the PhC structures. Compact transmissive QWPs and HWPs have been firstly analyzed theoretically by prof. Li [26] using bulk 2D PhC and soon after experimentally realized by Dr. Soli in microwave frequency [8, 46-49]. The WPs based on the birefringence effect of slab and bulk 2D PhCW are studied by numerical calculating in 2007 [9] and 2009 [10]. The ultracompact WPs with ultra-broadband achromatic phase difference are investigated by air-holes and dielectric rods 1D PhCW, recently [11, 12].

WPs with broadband achromatic phase difference are widely used in practice for the most of polarization-phase controlling devices require frequency independent phase retarding. Beyond that, compact in size is essential for that the original intention is realizing phase retarding in PICs for PhC structures based WPs. From the expression of the phase difference, the value of Δn is determinant of the size of WPs and the higher Δn is, the more compact the WP is, this gives the reason of the requirement of giant birefringence. It is a little complicated to achieve broadband achromatic phase difference for that $\Delta\varphi$ depends not only on Δn, but L, and λ as well. Slow-varying or constant maintaining of $\Delta\varphi$ relies on the envelope of $\Delta n/\lambda$, and the value of L. So, it is quite necessary for achromatic WPs that Δn should grows in a slow and linear fashion with wavelength in broadband. Meanwhile, choose L as small as possible, under the premise of that the requisite phase retarding could be realized, to avoid the enhancement of the non-uniformity of $\Delta n/\lambda$ at different wavelength. This is the cause of the low-order WPs are preferred in practical broadband applications. However, there are still some cases requiring $\Delta\varphi$ changing rapidly with frequency, i.e. high dispersion, in some particular applications such as polarization scrambling and depolarization. High-order WPs with large value of L is preferred to implement these functions. Except for operating band, dispersion, and size discussed above, the loss and compatibility should also be take into consideration in practically to evaluate the performance of WPs.

For the large and designable of birefringence in PhC structures, high performance WPs can be realized, such as low-order broad-band achromatic and high-order compact QWPs and HWPs. This section will focus on the low-order WPs based on 2D PhCW, 2D bulk PhC and 1D PhCW, respectively, and the compact high-order WPs based on the so called formed birefringence are discussed, too.

4.1. Low-order WPs based on 2D PhCW

For the simplicity of the 2D PhCW with hybrid PBG and TIR modes, the square lattice air holes type of PhCW studied in Fig. 6 and Fig. 7 is used to design the low-order WPs. The structure

http://dx.doi.org/10.5772/55142

used in the FDTD simulation is shown in Fig. 11 (a). A waveguide broadband Gaussian pulse source with width of a is excited at x=-6.5a, and point detectors located at different positions record the electric fields by which the phase shift (Φ) of TE and TM polarizations are obtained. Fig. 11 (b) and (c) give the phase shifts at x=13.1a and x=27.1a, respectively, on the initial phase which obtained at x=-0.5a where the phase differences ($\Delta\varphi$) between TE and TM modes are zero at the operating frequency band. The unwrapping $\Delta\varphi$ are shown in Fig. 11 (d) and (e), correspondingly. The values of $\Delta\varphi$ are about $\pi/2$ (QWP) and π (HWP) at x=13.1a and x=27.1a, respectively, with high phase accuracy of ±0.005π in the frequency range of 0.2632-0.2642 a/λ for both QWP and HWP. The relative achromatic bandwidth is about 0.38%. The length of 2D PhCW that introduce $\pi/2$ phase difference between TE and TM polarization is about $L_{\pi/2}$=27.1a-13.1a =14a, which is in good agreement with that obtained by SFT method, e.g. the length is about 13.9a when ω=0.264 a/λ for the birefringence at this wavelength is about Δn= 0.0682 as shown in Fig. 7 (b). So, the length of the zero-order QWP and HWP is about 3.7λ and 7.4λ, respectively. Although not shown here, the phase characteristics can also be verified in separate frequency by launching CW source in the waveguide and recording the EM field variation in time. The size of WPs can be reduced, although not too much, by the triangular

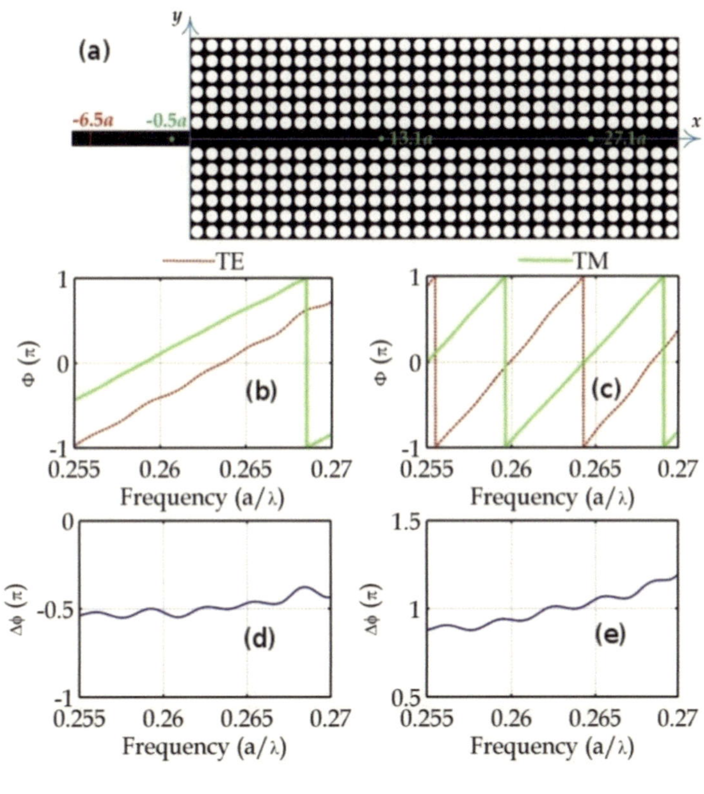

Figure 11. (a) Structure used in simulations. The parameters of the 2D PhCW are as same as that in Fig. 6 and Fig. 7. [(b), (c)] The phase (Φ) at (b) x= 13.1a and (c) x=27.1a, respectively. [(d), (e)] Unwrapping phase difference ($\Delta\varphi$) at (d) x= 13.1a and (e) x=27.1a, respectively. [10]

lattice 2D PhCW for the birefringence in it is larger than the square lattice one. However, it must be admitted that the PhCW has no advantages comparing with the other PhC structure in size and achromatic bandwidth, the only merit may be the guiding and confining of light is perfect in 2D PhCW.

4.2. Low-order WPs based on bulk 2D PhCs

The bulk 2D PhC has larger birefringence than the 2D PhCW, therefore, the WPs realized by bulk PhC has smaller size. However, the beam divergence is severe in the perfect PhC structure as discussed before and the divergent beam in nonwaveguiding structure will interfere with other components for large number of devices integrated in ultrasmall space in practical PICs [27]. Interference and scattering loss caused by beam divergence can be solved by the so called self-collimating (SC) effect under which light beam can propagate with no diffraction in perfect PhCs [50-52].

Polarization independent SC propagation is need in PhC for the WP applications. As shown in the inset of Fig.12 (a), the designed PhC is square lattice air holes in dielectric materials, and the permittivity of host material and radius of air holes are ε=11.0224 and r=0.315a, respectively [27]. The band structure is calculated by PWE method and shown in Fig. 12 (a). The equal frequency contours near the frequency band marked as green shallow region in Fig. 12 (a) are plotted in Fig. 12 (b) and (c) for TM and TE polarization, respectively. From the figures, both TE and TM have ultra-flat equifrequency surface in the frequency band of 0.273-0.281 a/λ. As

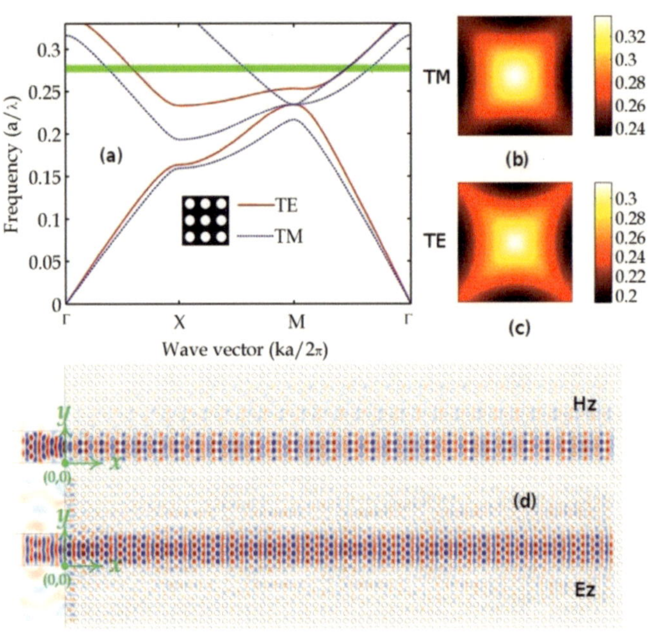

Figure 12. (a) Band structure of the square lattice PhC with polarization independent SC effect. [(b), (c)] Equal frequency contours for TM (b) and TE (c) polarizations. (d) Snapshots of the EM fields distributions at the frequency of 0.274 a/λ for TE (H_z) and TM (E_z) polarizations.

the direction of light propagation is always normal to the equifrequency surface, the light will propagate in the PhC without divergence along the <0 1> direction for both TE and TM polarizations, which can be seen clearly in Fig. 12 (d).

By launching broadband Gaussian pulse source in the coupling waveguide (refer Fig. 12 (d)), the transmission behaviors of the SC beam in the PhC are quantified and the spectra are shown in Fig. 13 (a) and (b) for TE and TM polarization, respectively. In the polarization independent SC band, the transmission efficiencies are above 75% for both TE and TM polarizations when the length is shorter than 70*a*. This is a significant improvement over the conventional PhC without SC guiding effect. Same as what have done in the section 4.1, the phase shift for TE and TM polarizations are calculated by using the recorded EM fields. Fig. 13 (c) and (d) show the phase shifts (Φ) and unwrapping phase difference ($\Delta\varphi$) at positions of 10*a* and 12*a*, respectively. During the simulation, the initial phases are obtained at $-4a$. In a wide band region of 0.273-0.281 a/λ, the values of $\Delta\varphi$ are almost constants of $\pi/2$ (QWP) and π (HWP) when the lengths of the PhC are 10*a* and 12*a*, respectively. For both QWP and HWP, the phase accuracy is about $\pm 0.01\pi$, and the transmission efficiencies are above 96%. The relative spectral bandwidth of the designed WP is about 3%, which is about the half of that of dense wavelength division multiplexing (DWDM) optical communication systems. Although the bandwidth of the WPs is not wide enough to cover all the frequency range of DWDM systems, it is as wide as 45 nm and is wide enough for many applications. The fact that should be indicated is that the 10*a* and 12*a* length WPs are not the zero-order ones. The lower-order WPs may have broader achromatic frequency band.

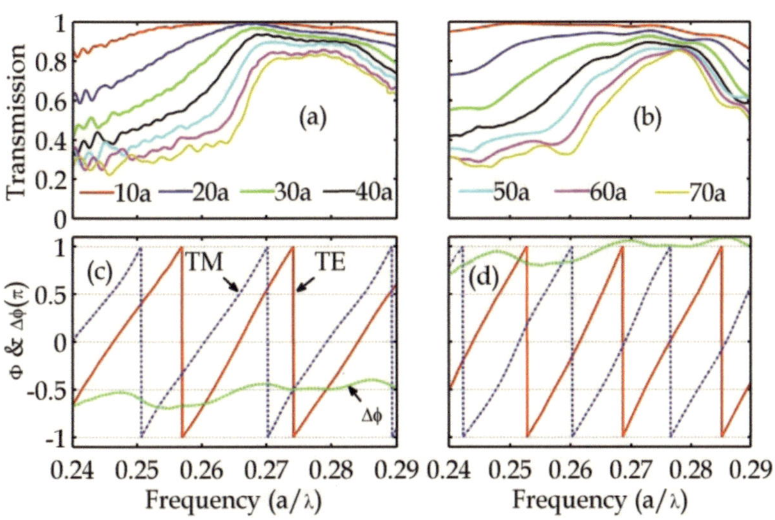

Figure 13. [(a), (b)] Transmission spectra for TE (a) and TM (b) polarizations. [c, d] Phase (Φ) and phase difference ($\Delta\varphi$) at x=10*a* and x=12*a*, respectively. [27]

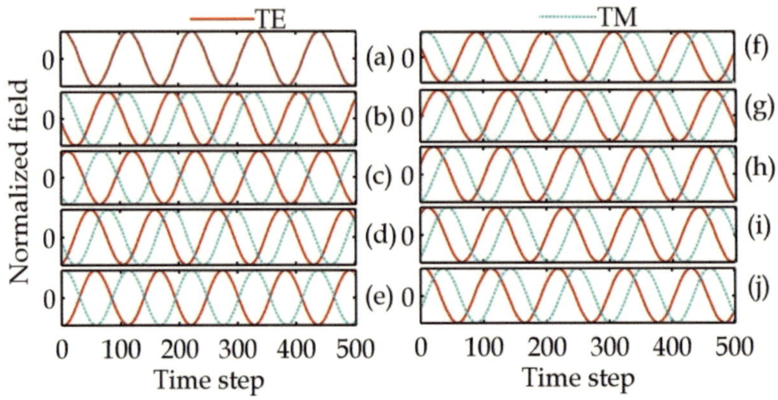

Figure 14. Normalized EM fields variations with time for TE and TM polarizations at different space locations and EM wave frequencies. [(a)-(e)] With the same frequency of 0.275 a/λ and at different positions of x=-4a (a), 10a (b), 12a (c), 50.4a (d) and 50.7a (e), respectively. [(f)-(j)]. At the same location of x=10a, but with different frequency of ω=0.273 a/λ (f), 0.274 a/λ (g), 0.276 a/λ (h), 0.278 a/λ (i), and 0.280 a/λ (j), respectively. [27]

To verify the phase characteristics of the WPs designed above, a CW source is launched in the dielectric waveguide, and a set of point monitors are inserted along the x direction in the center of the light beam in y direction to detect the electric fields (E_y for TE and E_z for TM) at different times. The phases of TE and TM at the reference position x=–4a are identical at the frequency 0.275 a/λ [see Fig. 14 (a)], which reveals zero phase difference between TE and TM polarizations. After propagating to 10a, as shown in Fig. 14 (b), the phase difference between TE and TM is about $\pi/2$. Furthermore, at 12a, TE and TM reverse in phase [see Fig. 14 (c)]. These results certify that the WPs work effectively at the frequency 0.275 a/λ. Higher-order QWP and HWP also can be realized (as shown in Fig. 14 (d) and (e)) for the light can propagate a relatively long way in PhC with the help of SC effect. To verify the broadband characteristic of the WPs, one monitor is set at 10a, and the frequency of source is changed. Fig. 14 (f)–(j) show that the 10a length QWPs have an almost fixed phase difference of $\pi/2$ (accuracy of ±0.01π) in a wide frequency range. Although not shown in the figures, similar broadband characteristics have been obtained for 12a length HWPs with the same simulation method.

4.3. Low-order WPs based on 1D PhCW

Comparing with the 2D PhCW, 1D PhCW has larger birefringence, so that it is very suitable for the ultracompact low-order WP applications. Both 1D PhCW with dielectric rods in air and air holes in dielectric waveguides have been used to design high performance WPs.

4.3.1. Low-order WPs based on 1D PhCW with dielectric rods in air

The birefringence properties of 1D PhCW with dielectric rods in air have been studied thoroughly in section 3.3. Although there is broadband achromatic birefringence, it is not enough to achieve wide band achromatic phase difference for $\Delta\varphi$ is also determined by L and λ. Actually, the structure supporting achromatic $\Delta\varphi$ is $r \geq 0.55a$ which is different from the critical value of 0.50a for achromatic birefringence.

For the ultrahigh birefringence, 2π phase difference ($\Delta\varphi=2\pi$) can be introduced by 1D PhCW with short length even smaller than the operating wavelength λ if Δn is large enough. Taking the 1D PhCW with $\varepsilon=12.96$ and $r=0.55a$ as an example, if the operating frequency is $\omega=0.145$ a/λ, 2π phase difference is introduced by the structure with length of $L=4.63a$, which is smaller than the wavelength of $\lambda=6.9a$, for the birefringence at the frequency is $\Delta n=1.488$ from Fig. 10. By using the FDTD simulation scheme, the characteristics of this 1D PhCW based WPs are verified. The normalized EM fields for TE (E_y) and TM (E_z) polarizations are shown in Fig. 15 (a)-(j). The phases of TE and TM are identical at location of $x=3.4a$ as shown in Fig. 15 (a). After propagating to 4.5a, as shown in Fig. 15(b), the phase difference between TE and TM is $\pi/2$. Furthermore, the phase differences are π [see Fig. 15(c)], $3\pi/2$ [see Fig. 5(d)], and 2π [see Fig. 5(e)] at the positions of 5.55a, 6.55a, and 8.0a, respectively. These results certify that the 1D PhCW with length of 4.5a and 5.55a (the real lengths are about 0.6525λ and 0.80475λ, respectively) can be served as the first-order QWP and HWP at frequency of 0.145 a/λ, respectively. The phase difference of 2π is introduced by the length of $L_{2\pi}$=8.0a-3.4a=4.6a, and this is in good agreement with the theoretical value of 4.63a calculated above. To verify the broadband achromatic properties of the phase difference, CW sources with different frequencies are launched into the PDW, and the detector is located at $x=4.5a$ unchanged. Fig. 15(f)–(j) show that the phase differences are fixed at $\pi/2$ in frequency range of 0.140–0.150a/λ. The relative bandwidth is about 28% which is much wider than that of the 2D PhCW and bulk PhC. If taking the central frequency 0.145 a/λ as λ=1550 nm (a=224.8 nm), the real frequency bands are from 1498nm to 1605 nm, which almost covers the whole telecommunication band containing C, L and S. [12].

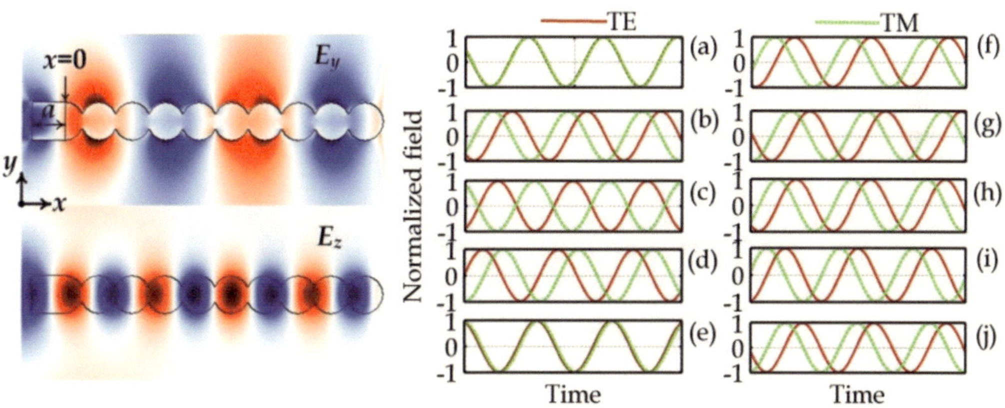

Figure 15. [Left] Snapshots of the EM fields for TE (E_y) and TM (E_z) polarizations. The simulation structure is shown in the figure as background. [Right] Normalized EM fields variations with time for TE and TM polarizations at different space locations and EM wave frequencies. The left column is ω=0.145 a/λ at different locations of 3.4a (a), 4.5a (b), 5.55a (c), 6.55a (d), and 8.0a (e), respectively. The right column is x=4.5a with different frequency of 0.140 a/λ (f), 0.142 a/λ (g), 0.145 a/λ (h), 0.148 a/λ (i), and 0.150 a/λ (j), respectively. [12]

Except for the waveguide dispersion, the material dispersion also can affect the effective indices of the waveguide modes, and then affect the birefringence. To study the phase difference of 1D PhCW with dispersive material, Silicon (Si) is chosen as the material of dielectric rods for the permittivity of Si can be obtained by Sellmeier formula:

$$\varepsilon = 11.6858 + \frac{0.939816}{\lambda^2} + \frac{0.000993358}{\lambda^2 - 1.22567} \tag{4}$$

where the wavelength λ is in μm. The radius of the Si rods is 0.55a. The phase difference ($\Delta\varphi$) of Si 1D PhCW with length of L=4a and L=6a are shown in Fig. 16. Although the material dispersion affects the birefringence of 1D PhCW, there is broadband achromatic phase difference for the Si 1D PhCW. The achromatic bandwidth is larger than 100nm with excellent phase accuracy of ±1°. The achromatic band can be tuned by changing the lattice constant a, and it is red shift by increasing the value of a. The bandwidth is affected by the length of 1D PhCW. Taking a=240nm as example, with the same accuracy of ±1°, the bandwidth is about 125nm (1495-1620nm, 301±1°) when L=4a, and about 100nm (1506-1606nm, 452±1°) when L=6a. Just as discussed before, the achromatic band is narrower when the length of 1D PhCW is longer. For the Si 1D PhCW, the material dispersion does affect the phase difference of TE and TM modes propagate in it, but does not restrict the realization of broadband achromatic WPs.

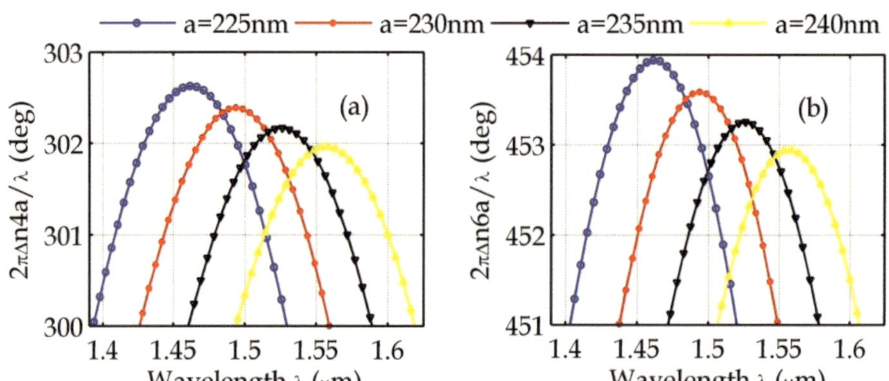

Figure 16. Phase difference of Si 1D PhCW with L=4a (a) and L=6a (b), respectively. The radius of Si rods is r=0.55a, and the indices of Si are calculated by Eq. (4). [12]

4.3.2. Low-order WPs based on 1D PhCW with air holes in dielectric waveguide

Another type of 1D PhCW is the periodic air holes in dielectric waveguide as shown in Fig. 9 (c) and (d). This type of structure is easier integrated with other components for the most of devices in PIC are constructed and interconnected by waveguide. Here, the 1D PhCW with square air holes is used to design high performance low-order WPs. The width of the waveguide is as same as lattice constant a, the length of square's side is w=0.5a, and the permittivity of the high index material is 13. Although not shown, the birefringence of this 1D PhCW is larger than 1 in the single mode frequency range of 0.1-0.15 a/λ.

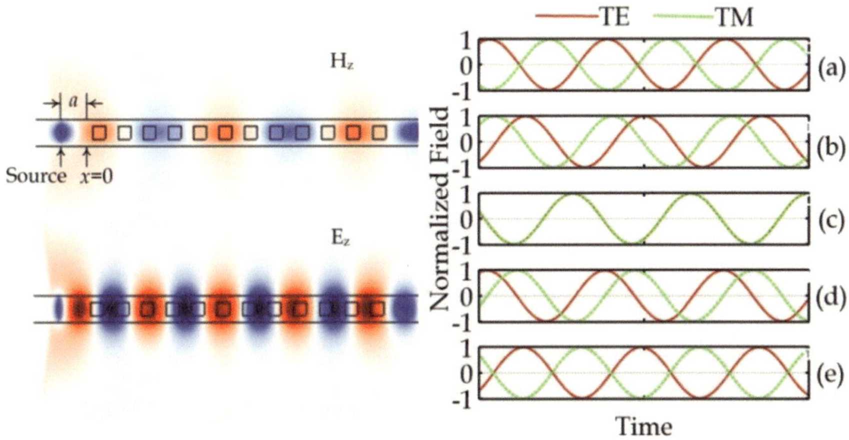

Figure 17. [Left] Snapshots of EM fields at frequency of 0.14 a/λ for TE (H_z) and TM (E_z) polarizations.[Right] Normalized EM field variations with time for TE and TM modes with same frequency of 0.14 a/λ but at different space locations of 2.9*a* (a), 4.45*a* (b), 6.4*a* (c), 7.9*a* (d), and 89.6*a* (e), respectively. [11]

The phase differences between TE and TM polarizations are directly studied by FDTD simulation method. A CW source with frequency of ω=0.14 a/λ is excited in the waveguide at position of x=-a, and several point monitors are placed at different positions along x direction at the center of the propagation beam in y direction. The snapshots of the EM fields in the 1D PhCW are shown in Fig. 17 (left) and the recorded electric fields at different positions are shown in Fig. 17 (a)-(e). From the figures, the phase differences ($\Delta\varphi$) at x=2.9a, x=4.45a, x=6.4a, x=7.9a, and x=9.6a are π, $3\pi/2$, 2π, $5\pi/2$, and 3π, respectively. As a result, the structures with length, which contains length of adjacent coupling waveguide and 1D PhCW, of 3.9a, 5.45a, 8.9a, and 10.6a can serve as HWP, QWP, QWP, and HWP, correspond real length of 0.546λ, 0.763λ, 1.246λ, and 1.484λ, respectively. The phase difference of 2π is introduced by the distance of 6.7a and this is in good agreement with the value of 6.5a calculated by PWE method.

The shortage of the structure studied above, referring to nontaper structure, is that the transmission loss is relatively high as shown in Fig. 18 (a) and (b).To reduce the loss, a taper structure is used to design high performance WP and the structure is shown at the top of Fig. 18.The transmission efficiencies are effectively improved by the taper 1D PhCW with w_1=0.35a, and w_2=0.5a, especially for TM polarization. The transmission efficiency is more than 90% in the frequency range of 0.139-0.148 a/λ for both TE and TM polarizations. The efficiency can be improved further by optimizing the taper structure. The phase differences are around $\pi/2$ for different frequencies from 0.139 a/λ to 0.148 a/λ as shown in Fig. 18 (c)-(f). The taper with total length of 5a can serve as broadband QWP, although the phase accuracy is not very high ($\pm 0.02\pi$). The phase accuracy also can be improved by carefully tuning the parameters of the taper.

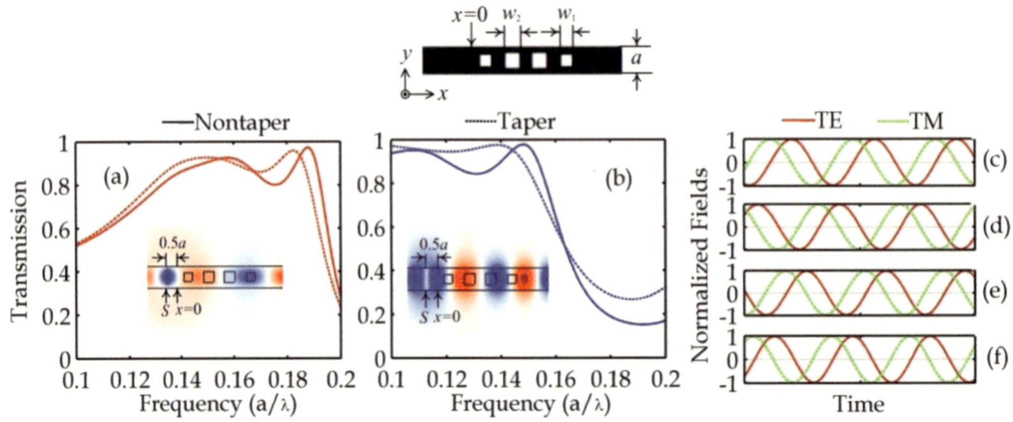

Figure 18. [a-b] Transmission spectra for square air hole 1D PhCW with conventional structure and taper structure for both TE (a) and TM (b) polarizations. The insets in (a) and (b) are snapshots of the EM fields propagating in the taper structure at frequency of 0.14 *a*/λ for TE and TM polarizations, respectively. The structure of the taper is shown at the top and the wave source is excited at *x*=-0.5*a*.[c-f] Normalized electric fields of E_y (TE) and E_z (TM) in the taper structure at positions of *x*=4*a* with frequency of 0.139 *a*/λ (c), 0.142 *a*/λ (d), 0.145 *a*/λ (e), and 0.148 *a*/λ (f), respectively. [11]

4.4. High-order WPs based on formed birefringence effect

Except for the low-order achromatic WPs, high-order WPs are also useful in some special applications. By increasing the length of the PhC structures, high-order WPs can be realized, but it is not inadvisable when the birefringence is not large enough and the loss increases rapidly with the increasing of length. Another method to realize high-order WPs proposed before is so-called formed birefringence which takes full advantage of birefringence and optical path difference between TE and TM polarizations [27, 10, 12].The schematic diagram is shown in Fig. 19 (a). The key point of the formed birefringence is bringing path difference of two orthogonal polarizations into use to enhance the phase difference between them. As that shown in Fig. 19 (a), the lengths of path for polarization 1 (TE or TM) and polarization 2 (TM or TE) are denoted as L_1 and L_2, respectively. The effective indices are n_1 and n_2, correspondingly, and supposing that $n_2>n_1$. The total phase difference between two polarizations after propagating through two paths can be expressed as:

$$\Delta\varphi_{\text{tol}} = \frac{2\pi}{\lambda}\left(n_2 L_2 - n_1 L_1\right) + \varphi_0 \tag{5}$$

where φ_0 is the total phase difference caused by other factor such as reflection in the interfaces. Eq. (5) can be further written as

$$\Delta\varphi_{\text{tol}} = \frac{2\pi}{\lambda}\Delta n L_1 + \frac{2\pi}{\lambda} n_2 \Delta L + \varphi_0 \tag{6}$$

where $\Delta n = n_2 - n_1$ is the birefringence and $\Delta L = L_2 - L_1$ is the length difference of paths. The three items of the phase difference in Eq. (6) are caused by birefringence, path difference and other factors, respectively.In most case, the effective index is larger than the birefringence ($n_2 > \Delta n$), so, the total phase difference will be enormously enhance if using the path difference effectively.

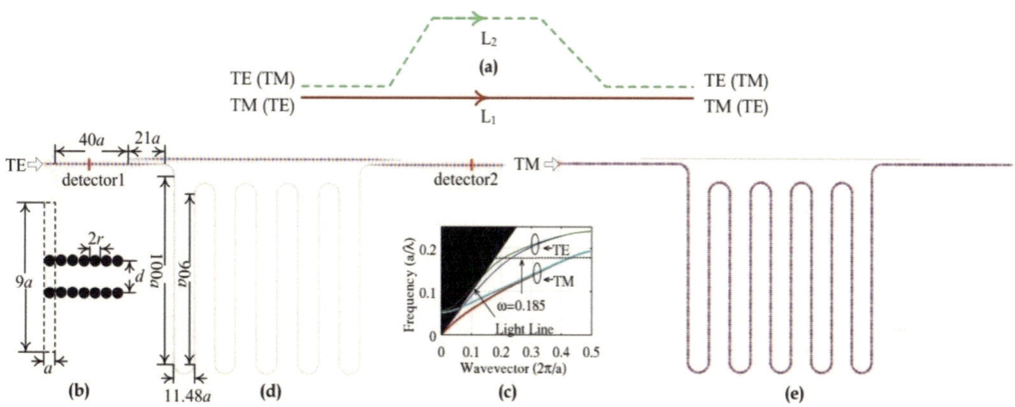

Figure 19. (a) The schematic diagram of formed birefringence (b) Structure of the direct coupler used as PBS and PBC and the supercell used in PWE calculations. (c) Band structure of the direct coupler. [(d),(e)] Snapshots of the EM fields at frequency of 0.185 a/λ for (d) TE (H_z) and (e) TM (E_z) polarizations. [12]

The high-order WP by formed birefringence effect is designed based on dielectric rod type of 1D PhCW with ε=9.6 and r=0.5a. The structure is composed by two coupling waveguides, which serve as input and output ports, and two direct couplers, which are used for polarization beam splitting and combining. As shown in Fig. 19 (b), the direct coupler is constituted by two parallel 1D PhCW placed closely with distance of d=2.8a. The dispersion curves of the two parallel rows of dilectric rods are calculated by PWE method with 1×9a supercell (shown in Fig. 19 (b)) and are shown in Fig. 19 (c). From the figure, the first and second TM modes are superposed but two lowest TE modes are separate with each other at the frequency of 0.185 a/λ. According to the direction coupling length equation, $L_c = \pi/\Delta k$, the coupling length is an infinitely large number for TM polarization for the wave vector difference (Δk) of first and second modes are zero. For the TE modes, the first and second modes can coupled to each other completely after propagating a length of L_c which is determined by Δk. The length of the direct coupler is optimized to L_c=21a to couple the mode from one waveguide to another in broadband. From the dispersion curve of the 1D PhCW, the effective index of TM mode (n_{TM}) is larger than that of the TE mode (n_{TE}), so, the propagation route of the TM mode is chosen as a zigzag path to make the TM mode having larger propagating length, which is shown in the background of Fig. 19 (d) and (e). To reduce the loss of the bend, the bend diameter is chosen as a relatively large value of 11.48a. The other parameters of the whole structure of designed high-order WP can also be found in Fig. 19. The whole size of the simulation is about 240×120a.The snapshots of the EM fields at the frequency of 0.185 a/λ for TE (H_z) and TM (E_z) polarization are shown in Fig. 19 (d) and (e), respectively. According to Eq. (6) and by using the parameters of Δn=1.1505, L_1=240a, ΔL=980a, n_2=n_{TM} =2.4427, and ω=0.185 a/λ, the phase difference caused by birefringence and formed birefringence are about 102π and 886π, respectively. Neglecting φ_0 (generally, it is a small value), the total phase difference is about $\Delta\varphi_{tol}$=1000π, and it can serve as a 500th-order WP. It can depolarize a laser with full width at half maximum that is wider than 3.1 nm [53] and can be used in the depolarization of a Raman pump laser diode and super-luminescence LED used in the fiber gyro. Taking λ=1550nm as an example, the lattice constant a is about 287nm. Therefore, the whole size of the 500th-order WP based on formed birefringence is about 69×34.5 μm.

Except for the 1D PhCW, the bulk PhC with polarization independent SC effect and the 2D PhCW are also can be used to design high-order WPs based on formed birefringence effect. The polarization beam splitter and combiner can be realized by direct coupler in 2D PhCW structure, and by reflecting and transmitting mirror in bulk PhC structure. Actually, the 80th order WP has been designed by the 2D PhCW with triangular lattice air holes in high index dielectric material [10].

5. Conclusion

In this chapter, the birefringence properties of three types of PhC structures, containing 2D PhCW, 2D bulk PhC, and 1D PhCW, have been studied thoroughly. High performance WPs based on the birefringence of these three types of PhC structures have been proposed. The comprehensive remarks about the three PhC structures are shown in Table 2.

For the 2D PhCW, the birefringence is not very high, so that the size of the WPs based on it is in wavelength magnitude. Although the achromatic bandwidth is not very large, 2D the PhCW provide perfect guiding for the light in it.

For the 2D bulk PhC, the birefringence in it is much larger than that in 2D PhCW. The disadvantage of the bulk PhC is the beam divergence which will cause scattering loss and signal crosstalk. This problem has been improved by the SC effect in this chapter. The WPs based on the 2D PhC with polarization independent SC effect have compact size and broad achromatic bandwidth (about 45nm).

For the 1D PhCW, giant birefringence, which is even larger than 1.5, can be realized in special structures. The 1D PhCW is very suitable for WP applications for its giant birefringence, low loss and compact in size. The achromatic bandwidth of the WPs based on 1D PhCW can be larger than 100nm with excellent phase accuracy of ±1°. Meanwhile, the size of the WP is in sub-wavelength magnitude.

Besides, the high-order WPs based on so called formed birefringence is proposed too. Additional phase differences are introduced by the path difference of different polarizations. The 500th order WP is designed based on the formed birefringence by using the 1D PhCW.

http://dx.doi.org/10.5772/55142

	Birefringence	Achromatic phase bandwidth	High orders	Length ($\Delta\varphi=2\pi$)	Loss
2D PhCW	~0.1	~6nm (±0.005π)	80	3.7λ	Low
Bulk PhC	~0.8	~45nm (±0.01π)	-	-	Relatively high
1D PhCW	~1.5	"/>100nm (±1°)	500	0.67λ	Low

Table 2. Comparing of the properties of three types of PhC structures

The WP is one of the basic elements in optical devices. The proposed PhC structure based WPs have a lot potential applications in future PICs for sensing, optical communications and measurements.

Acknowledgements

The authors would like to gratefully acknowledge W. –P. Huang at McMaster University, J. – W. Mu at MIT and J. Liu at Xi'an University of Posts and Telecommunications for their fruitful suggestion. The study was supported by CAS/SAFEA International Partnership Program for Creative Research Teams, West Light Foundation of The Chinese Academy of Sciences, and National Natural Science Foundation of China (Grant No. 61275062).

Author details

Wenfu Zhang and Wei Zhao

State Key Laboratory of Transient Optics and Photonics, Xi'an Institute of Optics and Precision Mechanics, Chinese Academy of Sciences, Xi'an, China

References

[1] Hsu, S., Lee, K., Lin, E., Lee, M., & Wei, P. Giant birefringence induced by plasmonic-nanoslit arrays, *Appl. Phys. Lett.*, 95: 013105 (2009).

[2] Weber, M. F., Stover,C. A., Gilbert, L. R., Nevitt, T. J., & Ouderkirk,A. J. Giant birefringent optics in multilayer polymermirrors, *Science*, 287: 2451 (2000).

[3] Muskens,O. L.,Borgström, M. T., Bakkers, E. P. A. M., & Gómez Rivas, J. Giant optical birefringence in ensembles of semiconductor nanowires, *Appl. Phys. Lett.*, 89: 233117 (2006).

[4] Weis, P., Paul, O., Imhof, C., Beigang, R., & Rahm, M. Strongly birefringent metamaterials as negative index terahertz wave plates, *Appl. Phys. Lett.*, 95: 171104 (2009).

[5] Yang, S., Cooper, M. L., Bandaru, P. R., & Mookherjea, S. Giant birefringence in multi-slotted silicon nanophotonic waveguides, *Opt. Express*, 16: 8306 (2008).

[6] Genereux, F., Leonard, S. W., & Van Driel,H. M. Large birefringence in two-dimensional silicon photonic crystals, *Phys. Rev. B, Condens. Matter,* 63: 161101 (2000).

[7] Xiao, X., Hou, B., Wen, W., & Sheng, P. Tuning birefringence by using two-dimensional photonic band structure, *J. Appl. Phys.*, 106: 086103 (2009).

[8] Solli, D. R., McCormich, C. F., Chiao, R. Y., & Hickmann, J. M. Birefringence in two-dimensional bulk photonic crystals applied to the construction of quarter waveplates, *Opt. Express* 11: 125 (2003).

[9] Bayat, K., Chaudhuri, S. K., & Safavi-Naeini, S. Design and simulation of an ultracompact integrated wave plate using photonic crystal slab waveguide, *Proc. SPIE* 6468: 64680N (2007).

[10] Zhang, W., Liu, J., Huang, W.-P., & Zhao, W. Birefringence and formed birefringence in photonic crystal line waveguides, *Proc. SPIE* 7516: 751603 (2009).

[11] Zhang, W., Liu, J., Huang, W. –P., & Zhao, W. Ultracompact wave plates by air holes periodic dielectric waveguides, *Proc. SPIE* 7609: 76091H (2010).

[12] Zhang, W., Liu, J., Huang, W. –P., & Zhao, W. Giant birefringence of periodic dielectric waveguides, *IEEE Photon. J.*, 3: 512 (2011).

[13] Yablonovitch, E. Inhibited spontaneous emission in solid-state physics and electronics, *Phys. Rev. Lett.*, 58: 2059 (1987).

[14] John, S. Strong localization of photons in certain disordered dielectric superlattices, *Phys. Rev. Lett.*, 58:2486 (1987).

[15] Joannnopoulos, J. D., Johnson, S. G., Winn, J. N. & Meade, R. D. *Photonic Crystals: Molding the Flow of Light,Second Edition,* Princeton University Press, 2008.

[16] Johnson, S. G. & Joannopoulos, J. D. Block-iterative frequency-domain methods for Maxwell's equations in aplanewave basis, *Opt. Express,* 8: 173 (2001).

[17] Taflove, A. & Hagness, S.C. *Computational Electrodynamics: The Finite-Difference Time-Domain Method,* Artech House Publishers, 2005.

[18] Oskooi, A. F., Roundy, D., Ibanescu, M., Bermel, P., Joannopoulos, J. D., & Johnson, S. G. MEEP: A flexible free-software package for electromagnetic simulations by the FDTD method, *Comput. Phys. Commun.,* 181: 687 (2010).

[19] Yee, K. Numerical solution of initial boundary value problems involving Maxwell's equations in isotropic media, *IEEE Transactions on Antennas and Propagation,* 14: 302. (1966).

[20] Jafarpour, A., Reinke, C. M., Adibi, A., Xu, Y., & Lee, R. K. A new method for the calculation of the dispersion of nonperiodic photonic crystal waveguides, *IEEE J. Quantum Electron.,* 40: 1060 (2004).

[21] Adibi, A., Xu, Y., Lee, R. K., Yariv, A.& Scherer, A. Guiding mechanisms in dielectric-core photonic-crystal waveguides, *Phys. Rev. B*, 63: 033308 (2001).

[22] Borel, P. I., Frandsen, L. H., Thorhauge, M., Harpoth, A., Zhuang,Y. X., Kristensen, M., & Chong, H. M. H. Efficient propagation of TM polarized light in photonic crystal components exhibiting band gaps forTE polarized light, *Opt. Express,*11: 1757 (2003).

[23] Zhang, W., Liu, J., & Zhao, W. Design of a compact photonic-crystal-based polarization channel drop filter, *IEEE Photon. Technol. Lett.*, 21: 739 (2009).

[24] Tsuji, Y., Morita, Y., & Hirayama, K. Photonic crystal waveguidebased on 2-D photonic crystal with absolute photonic band gap, *IEEEPhoton. Technol. Lett.*, 18: 2410 (2006).

[25] Morita, Y., Tsuji, Y., & Hirayama, K. Proposal for a compactresonant-coupling-type polarization splitter based on photonic crystalwaveguide with absolute photonic bandgap, *IEEE Photon. Technol.Lett.*, 20: 93 (2008).

[26] Li, L.-M. Two-dimensional photonic crystals: candidatefor wave plates, *Appl. Phys. Lett.*, 78: 3400 (2001).

[27] Zhang, W., Liu, J., Huang, W. –P., & Zhao, W. Self-collimating photonic-crystal wave plates, *Opt. Lett.*, 34: 2676 (2009).

[28] Fan, S., Winn, J. N., Devenyi, A., Chen, J. C., Meade, R. D., & Joannopoulos, J. D. Guided and defect modes inperiodic dielectric waveguides, *J. Opt. Soc. Amer. B, Opt. Phys.*, 12: 1267 (1995).

[29] Luan, P., & Chang,K. Transmission characteristics of finite periodic dielectric wave-guides, *Opt. Express*, 14: 3263 (2006).

[30] Lee, K., Chen, C., & Lin, Y. Transmission characteristics of various bent periodic dielectric waveguides, *Opt.Quantum Electron.*, 40: 633 (2008).

[31] Chigrin, D. N., Lavrinenko, A. V., & Torres, C. M. S. Nanopillars photonic crystal waveguides, *Opt. Express*, 12: 617 (2004).

[32] Chigrin, D. N., Lavrinenko, A. V., & Torres, C. M. S. Numerical characterization of nanopillar photonic crystalwaveguides and directional couplers, *Opt. Quantum Electron.*, 37: 331 (2005).

[33] Chigrin, D. N., Zhukovsky, S. V., Lavrinenko, A. V., & Kroha, J. Coupled nanopillar waveguides optical properties andapplications, *Phys. Stat. Sol. (A)*, 204: 3647 (2007).

[34] Zhukovsky, S. V., Chigrin, D. N., & Kroha, J. Low-loss resonant modes in deterministically aperiodic nanopillar waveguides, *J. Opt. Soc. Amer. B, Opt. Phys.*, 23: 2265 (2006).

[35] Zhukovsky, S. V., Chigrin, D. N., Lavrinenko, A. V. & Kroha, J. Selective lasing in multimode periodic and non-periodic nanopillar waveguides, *Phys. Stat. Sol. (B)*, 244: 1211 (2007).

[36] Boucher, Y. G., Lavrinenko, A. V., & Chigrin, D. N. Out-of-phase coupled periodic waveguides: A"couplonic"approach, *Opt. Quantum Electron.*, 39: 837 (2007).

[37] Sukhorukov, A. A., Lavrinenko, A. V., Chigrin, D. N., Pelinovsky, D. E., & Kivshar,Y. S. Slow-light dispersion in coupled periodic waveguides, *J. Opt. Soc. Amer. B, Opt. Phys.*, 25: C65 (2008).

[38] Bock, P. J., Cheben, P., Schmid, J. H., Lapointe, J., Delâge, A., Janz, S., Aers, G. C., Xu, D.-X., Densmore, A., & Hall,T. J. Subwavelength grating periodic structures in silicon-on-insulator: A new type of microphotonic waveguide, *Opt.Express*, 18: 20251 (2010).

[39] Bock, P. J., Cheben, P., Schmid, J. H., Lapointe, J., Delâge, A., Xu, D.-X., Janz, S., Densmore,A., & Hall,T. J. Subwavelength grating crossings for silicon wire waveguides, *Opt. Express*, 18: 16146 (2010).

[40] Zhang, Y., Huang, W., & Li, B. Fabry–Pérotmicrocavities with controllable resonant wavelengths in periodic dielectric waveguides, *Appl. Phys. Lett.*, 93: 031110 (2008).

[41] Luan P., & Chang, K. Periodic dielectric waveguide beam splitter based on co-directional coupling, *Opt. Express*, 15: 4536 (2007).

[42] Zheng, G., Li, H., Jiang, L., Jia, W., Wang, H. & Li, X. Design of an arbitrarily bent beam splitter for optical interconnections based on co-directional coupling mechanism, *J. Opt. A: Pure Appl. Opt.*, 10: 125303 (2008).

[43] Huang, W., Zhang, Y. & Li, B. Ultracompact wavelength and polarization splitters in periodic dielectric waveguides, *Opt. Express*, 16: 1600 (2008).

[44] Zeng, S., Zhang, Y., & Li, B. Self-imaging in periodic dielectric waveguides, *Opt. Express*, 17: 365 (2009).

[45] Dai, Q. F., Li, Y. W., & Wang, H. Z. Broadband two-dimensional photonic crystal wave plate, *Appl. Phys. Lett.*, 89: 061121 (2006).

[46] Solli, D. R., McCormich, C. F., Chiao, R. Y., & Hickmann, J. M. Experimental demonstration of photonic crystal waveplates, *Appl. Phys. Lett.* 82: 1036 (2003).

[47] Solli, D. R., & Hickmann, J. M. Photonic crystal based polarization control devices, *J. Phys. D: Appl. Phys.*, 37: R263 (2004).

[48] Solli, D. R., McCormich, C. F., & Hickmann, J. M. Polarization-dependent reflective dispersion relations of photonic crystals for waveplate mirror construction, *J. Lightwave Technol.*, 24: 3864 (2006).

[49] Solli, D. R., & Hickmann, J. M. Engineering an achromatic photonic crystal waveplate, *New J. Phys.*, 8: 132 (2006).

[50] Kosaka, H., Kavashima, T., Tomita, A., Notomi, M., Tamamura, O., Sato, T., & Kavakami, S. Self-collimating phenomena in photonic crystals, *Appl. Phys.Lett.* 74: 1212 (1999).

[51] Yu, X. & Fan, S. Bends and splitters for self-collimated beams in photonic crystals, *Appl. Phys. Lett.*, 83: 3251 (2003).

http://dx.doi.org/10.5772/55142

[52] Zabelin, V., Dunbar, L. A., Le Thomas, N., & Houdr, R. Self-collimating photonic crystal polarization beam splitter, *Opt. Lett.*, 32: 530 (2007).

[53] Azami, N., Villeneuve, E., Villeneuve, A., & Gonthier, F. All-SOP all-fibre depolariser, *Electron.Lett.*, 39: 1573 (2003).

Chapter 4

Propagation of Electromagnetic Waves in Anisotropic Photonic Structures

V.I. Fesenko, I.A. Sukhoivanov, S.N. Shul'ga and
J.A. Andrade Lucio

Additional information is available at the end of the chapter

http://dx.doi.org/10.5772/54847

1. Introduction

In this chapter we discuss plane-wave propagation in a layered arbitrarily anisotropic media.One-dimensional (1D) arbitrary layered structure is usually formed by stacking together layers of several different materials with some specific thickness d_j and refractive index n_j as depicted on Fig.1.

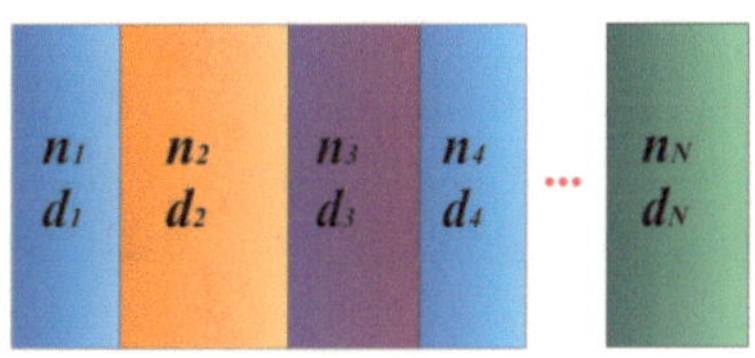

Figure 1. Arbitrary layered structure

Nowadays layered photonic structures (LPSs) are key of optoelectronic and microwave devices such as, phased-array antennas, microcavities and mirrors [1,2], filters of xWDM systems [3], waveguide structures, photodetectors, sensors and others. In case of active devices the layered structures are usually used in form of superlattices [4-6], multiple quantum wells [7,8] and asymmetric multiple quantum wells [9].

Different kinds of materials are used today for design LPSs, such as linear and nonlinear [3,10] dielectric materials, anisotropic or bi-anisotropic materials [11], chiral media [12], metamaterials [13], etc. If one use these materials one can effectively control emission, propagation and detection of the electromagnetic waves, and develop new designs of the photonic devices and those one for other parts of the electromagnetic spectrum. For instance, very active research worldwide is concentrated currently on the THz range, attempting to overcome the so called problem of the terahertz gap [14].

The analysis of propagation of the electromagnetic waves in periodic [3,11,13,15], quasiperiodic [12,16,17] and random [18] layered media is a problem which extends over the all fields in the modern physics. Optics is the area where it is crucial to calculate spectral characteristics, absorbance coefficients, polarization properties and other features of the multilayer structures in a wide spectral range and at various thicknesses or material properties of constituents. Even in the fiber optics, where usually propagation characteristics of the optical pulses [19, 20] are considered, the spectral characteristics become of the principal interest when optical channel incorporates such inhomogeneities as fiber Bragg gratings and fiber knots, and microresonators based on them.

In addition the use of the optical control techniques for phased-array antennas [21-22] promises to alleviate many of the problems associated with traditional electronic steering systems. The unique properties of layered anisotropic photonic structures (for example see [11]) are suitable for this application.

Here we discuss the optical properties of one-dimensional arbitrarily anisotropic photonic multilayers. The main objective of the chapter is the obtaining of a solution to the numerical problem of the electromagnetic plane wave interaction with arbitrarily anisotropic and arbitrarily inhomogeneous one-dimensional photonic structures. It is well known that many of novel technological designs have resulted from analysis of the properties of materials and creation of new structural configurations for them. In order to develop a new structural configuration with unique properties, one needs to thoroughly understand the characteristics of the structure. This can be accomplished by applying an advanced computational engine.

Today several numerical techniques are commonly used to compute the spectral characteristics of the layered photonic structures and electromagnetic field distribution in the interiors [3,12-15,23-31]. The most known and, probably, most usable are the finite element method (FEM) [3,24], the transfer matrix method (TMM) [25,26], the finite difference time-domain method (FDTD) [15,27] and the beam propagation method (BPM) [28]. Unfortunately, some methods are entirely disregarding the anisotropy and the inhomogeneity of the constituent materials. The most of methods which do account for material anisotropy require that the permittivity tensor be diagonal. Others allow for nondiagonal tensors, but require that the off-diagonal elements be small in comparison to the diagonal terms [28]. Although these techniques are adequate for many layered structures, they cannot be easily applied to multilayers in which the anisotropy is arbitrary oriented along an oblique axis on the random layers. In contrast to these numerical methods the method discussed here makes it possible the analyzing of inhomogeneousone-dimensional anisotropic multilayers with an arbitrary permittivity tensor and the optical axis arbitrarily oriented on any layer of LPS.

A general theory of electromagnetic propagation in periodic anisotropic layered media has been treated by a number of authors [26, 29-31]. The present chapter describes the efficient physico-mathematical model pertinent to one-dimensional optical-range microstructures based upon anisotropic materials. The electromagnetic field scalarization procedure [29-30] is used after the initial vector electromagnetic diffraction problem is reduced to the boundary problem for two scalar potentials. As a result, a set of linear algebraic equations are obtained.By solving them we find the unknown transmission and reflection spectra for the structure under study. The major advantage of the proposed method is that homogeneous, piecewise homogeneous and continuously inhomogeneous flat-layered anisotropic media can be analyzed on the same footing.

The chapter is organized as follows: in Section 2 we present the method of calculation employed in the chapter, which is based on electromagnetic field scalarization procedure in conjunction with the finite-difference method; Section 3 is devoted to the presentation of the numerical results, together with the discussion of their main features; we summarize our study and conclude the chapter in Section 4.

2. Theoretical model

In this section we present the mathematical background for calculation spectral characteristics of anisotropic layered media. The presented theory is applicable to any anisotropic layers with arbitrary orientation of the optical axes on each layer and for arbitrary angle of incidence.

The structure under consideration is schematically depicted in Fig. 2. Let's introduce the Cartesian coordinate system x, y, z such that the z axis is directed vertically upward. In this coordinate system, an inhomogeneous anisotropic layered medium is represented by the single layer that occupies the domain $-b<z<0$, $-\infty<x$, $y<+\infty$. The upper free half-space $z>0$ and substrate $z<-b$ are homogeneous and isotropic and have permittivities ε_0, μ_0 and ε_c, μ_c, respectively. In general case, all layers in this geometry are lossy media.

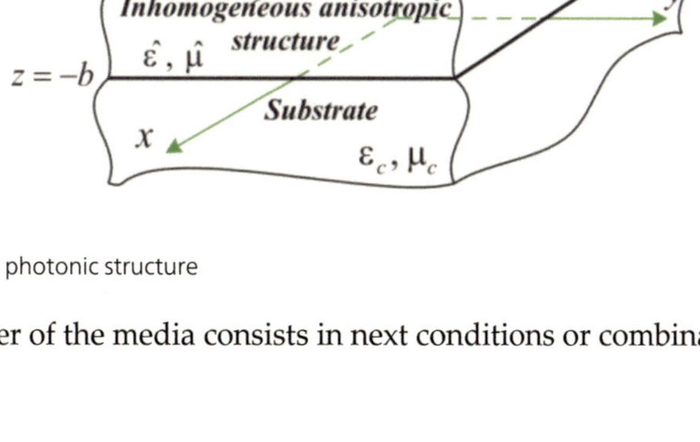

Figure 2. The benchmark photonic structure

The layered character of the media consists in next conditions or combination of them:

- permeability $\hat{\mu}$ and permittivity $\hat{\varepsilon}$ are continuous functions of the variable z. These complex-valued tensors $\hat{\eta}=\hat{\varepsilon}$, $\hat{\mu}$ can be expressed in Cartesian coordinates as:

$$\hat{\eta} = \begin{bmatrix} \eta_{xx} & \eta_{xy} & \eta_{xy} \\ \eta_{yx} & \eta_{yy} & \eta_{yz} \\ \eta_{zx} & \eta_{zy} & \eta_{zz} \end{bmatrix}; \tag{1}$$

- the media are piecewise continuous, i.e. there are boundary surfaces $z=z_j=const$, $(const>0)$, where properties of anisotropic media are varying stepwise;

- in the points $z=0$ and $z=-b$, $(b>0)$ the medium is bounded by homogeneous conducting planes or the planes permeable for the electromagnetic field. In our case the layered medium is confined by the impedance planes. These planes are characterized by impedance dyads $\hat{L}^{(a)}$ and $\hat{L}^{(u)}$:

$$\hat{L}^{(a,u)} = \begin{bmatrix} \hat{L}_{xx}^{(a,u)} & \hat{L}_{xy}^{(a,u)} \\ \hat{L}_{yx}^{(a,u)} & \hat{L}_{yy}^{(a,u)} \end{bmatrix}. \tag{2}$$

Here indices *a* (*above*) and *u* (*under*) correspond to upper $(z=0)$ and lower $(z=-b)$ boundaries, respectively.

We will start with Maxwell's equations, for complex field vectors $\vec{E}(\vec{R})$ and $\vec{H}(\vec{R})$, in the form:

$$\begin{aligned} \nabla\times\vec{E}(\vec{R}) - ik_0\hat{\mu}(z)\vec{H}(\vec{R}) &= -(4\pi/c)\vec{M}(\vec{R}), \\ \nabla\times\vec{H}(\vec{R}) + ik_0\hat{\varepsilon}(z)\vec{E}(\vec{R}) &= (4\pi/c)\vec{J}(\vec{R}), \end{aligned} \tag{3}$$

where k_0 is the wave number in free space; c is the velocity of light; $\vec{J}(\vec{R})$ and $\vec{M}(\vec{R})$ are electric and magnetic volume current densities, respectively; $\vec{R}=(x, y, z)$ is the radius vector. Here we assume the harmonic time dependence $\exp(-i\omega t)$ of the fields. Equations (3) are satisfied everywhere within the medium except the interfaces.

The electric and magnetic fields must satisfy the suitable boundary condition at the interfaces of the layered media:

- at the first, the tangential components of the electromagnetic field must be continuous at the all boundaries of the layered media:

$$\{z_0\times\vec{E}_\perp\}=0,\ \{z_0\times\vec{H}_\perp\}=0,\ (z=z_j, j=1,2,\ldots,N). \tag{4}$$

where $\vec{z}_0$ is unit vector along z axis; $\vec{E}_\perp \equiv \vec{E}_\perp(\vec{R})$, $\vec{H}_\perp \equiv \vec{H}_\perp(\vec{R})$ are electric and magnetic field components that are orthogonal to $\vec{z}_0$. Throughout the chapter we use braces {} for next operator designation $\{f(z)\} \equiv f(z+0) - f(z-0)$;

- at the second, we introduce impedance boundary conditions that are desired on the above and the bottom boundaries of the inhomogeneous anisotropic structure:

$$\begin{aligned} &\vec{E}_\perp + \hat{L}^{(a)}\vec{z}_0 \times \vec{H}_\perp = 0, (z=0); \\ &\vec{E}_\perp - \hat{L}^{(u)}\vec{z}_0 \times \vec{H}_\perp = 0, \ (z=-b). \end{aligned} \tag{5}$$

The next one, in what follows we assumed that external sources and electromagnetic field components are represented by spatial harmonics with wave vector $\vec{\kappa} = (\kappa_x, \ \kappa_y, \ 0)$

$$\begin{aligned} &\vec{J}(\vec{R}) \equiv \vec{J}(\vec{\kappa}, z)\exp(i\vec{\kappa}\cdot r), \\ &\vec{M}(\vec{R}) \equiv \vec{M}(\vec{\kappa}, z)\exp(i\vec{\kappa}\cdot r); \end{aligned} \tag{6}$$

$$\begin{aligned} &\vec{E}(\vec{R}) \equiv \vec{E}(\vec{\kappa}, z)\exp(i\vec{\kappa}\cdot r), \\ &\vec{H}(\vec{R}) \equiv \vec{H}(\vec{\kappa}, z)\exp(i\vec{\kappa}\cdot r). \end{aligned} \tag{7}$$

In expressions (6) – (7) $\kappa_{x,y}$ are the arbitrary complex constants, $\vec{J}(\vec{\kappa}, z)$, $\vec{M}(\vec{\kappa}, z)$ and $\vec{E}(\vec{\kappa}, z)$, $\vec{H}(\vec{\kappa}, z)$ are the vector amplitudes of the sources and the fields, respectively.

Now let's try to obtain general expressions for the transmittance and reflectance of a layered medium.

2.1. Reduction of the electromagnetic field diffraction problem to a boundary value problem for scalar potentials

Now let's consider in details the solving of the initial electromagnetic field diffraction problem. At first we should to introduce right-hand basis of vectors $\vec{a}_z$, $\vec{a}_l$, $\vec{a}_t$:

$$\begin{aligned} &\vec{a}_z = \vec{z}_0, \\ &\vec{a}_l = \vec{n}, \\ &\vec{a}_t = \vec{z}_0 \times \vec{n}. \end{aligned} \tag{8}$$

In writing eq (8) we have assumed that: $\vec{n} = \vec{\kappa} / \kappa$ is the unit vector, that is situated in the plane $z=0$; $\kappa = \sqrt{\kappa_x^2 + \kappa_y^2}$ is the branch of the square root, which is chosen such that condition $0 \le \arg\sqrt{(\cdot)} \le \pi$ shall be satisfied. These unit vectors obey the orthogonality relations:

$$\vec{a}_\sigma \cdot \vec{a}_\tau = 0, \quad (\sigma \neq \tau). \tag{9}$$

Then after scalar multiplication of the Maxwell's equations (3) by unit vectors $\vec{a}_z$, $\vec{a}_l$ and using simple conversion of vector algebra we obtain $E_{z,l} = \vec{a}_{z,l} \cdot \vec{E}(\vec{k}, z)$, $H_{z,l} = \vec{a}_{z,l} \cdot \vec{H}(\vec{k}, z)$ in terms of scalar potentials:

$$\begin{aligned} e(\vec{\kappa}, z) &= \vec{a}_t \cdot \vec{E}(\vec{\kappa}, z), \\ h(\vec{\kappa}, z) &= \vec{a}_t \cdot \vec{H}(\vec{\kappa}, z). \end{aligned} \tag{10}$$

In the basis of orthogonal vectors (8), vector amplitudes $\vec{E}(\vec{\kappa}, z)$, $\vec{H}(\vec{\kappa}, z)$ can be written down as:

$$\begin{aligned} \vec{E}(\vec{\kappa}, z) &= \vec{V}_\varepsilon(\vec{n}, z) e(\vec{\kappa}, z) - \vec{W}_\varepsilon(\vec{\kappa}) h(\vec{\kappa}, z) + (4\pi i / k_0 c) \hat{\alpha}_\varepsilon(\vec{n}, z) \cdot \vec{J}(\vec{\kappa}, z), \\ \vec{H}(\vec{\kappa}, z) &= \vec{V}_\mu(\vec{n}, z) h(\vec{\kappa}, z) - \vec{W}_\mu(\vec{\kappa}) e(\vec{\kappa}, z) + (4\pi i / k_0 c) \hat{\alpha}_\mu(\vec{n}, z) \cdot \vec{M}(\vec{\kappa}, z). \end{aligned} \tag{11}$$

The following symbols are used in (11): $\vec{V}_\eta(\vec{n}, z)$ and $\hat{\alpha}_\eta(\vec{n}, z)$, $(\eta = \varepsilon, \mu)$ are the vector functions and the dyad functions, respectively; $\vec{W}_\eta(\vec{\kappa})$– vector differential operator:

$$\vec{V}_\eta(\vec{n}, z) = \vec{z}_0 \times \vec{n} + 1/a_\eta(\vec{n}, z) \left[b_\eta(\vec{n}, z) \vec{n} + c_\eta(\vec{n}, z) \vec{z}_0 \right], \tag{12}$$

$$a_\varepsilon(\vec{n}, z) \hat{a}_\varepsilon(\vec{n}, z) \equiv a_t \times \varepsilon(z) \times a_t = (\varepsilon_{lz} \vec{n} - \varepsilon_{ll} \vec{z}_0) \vec{z}_0 + (\varepsilon_{zl} \vec{z}_0 - \varepsilon_{zz} \vec{n}) \vec{n}, \tag{13}$$

$$\vec{W}_\varepsilon(\vec{\kappa}) = (1/a_\eta(\vec{n}, z) k_0) \times \left[(\varepsilon_{zz} \vec{n} - \varepsilon_{zl} \vec{z}_0) i \partial_z + (\varepsilon_{zl} \vec{z}_0 - \varepsilon_{zz} \vec{n}) k_0 \right], \tag{14}$$

$$a_\varepsilon \to a_\mu, \quad \hat{a}_\varepsilon \to \hat{a}_\mu, \quad \hat{W}_\varepsilon \to \hat{W}_\mu, \quad (\varepsilon \to \mu). \tag{15}$$

In expressions (12) – (15), scalars $\varepsilon_{\sigma\tau} \equiv \varepsilon_{\sigma\tau}(\vec{n}, z)$, $\mu_{\sigma\tau} \equiv \mu_{\sigma\tau}(\vec{n}, z)$, $(\sigma, \tau = z, l, t)$ are components of the relevant dyads $\hat{\varepsilon}$, $\hat{\mu}$ (1):

$$\begin{aligned} \varepsilon_{\sigma\tau}(\vec{n}, z) &= \vec{a}_\sigma \cdot \varepsilon(z) \cdot \vec{a}_\tau, \\ \mu_{\sigma\tau}(\vec{n}, z) &= \vec{a}_\sigma \cdot \mu(z) \cdot \vec{a}_\tau. \end{aligned} \tag{16}$$

The scalar functions $a_\eta(\vec{n}, z)$, $b_\eta(\vec{n}, z)$, $c_\eta(\vec{n}, z)$, $(\eta = \varepsilon, \mu)$ that are used in (12) – (14) have the form:

$$\begin{aligned} a_\varepsilon &= \varepsilon_{zz}\varepsilon_{ll} - \varepsilon_{zl}\varepsilon_{lz}, \\ b_\varepsilon &= \varepsilon_{zt}\varepsilon_{lz} - \varepsilon_{zz}\varepsilon_{lt}, \\ c_\varepsilon &= \varepsilon_{lt}\varepsilon_{zl} - \varepsilon_{ll}\varepsilon_{zt}, \end{aligned} \tag{17}$$

$a_\varepsilon \to a_\mu, \quad b_\varepsilon \to b_\mu, \quad c_\varepsilon \to c_\mu, \quad (\varepsilon \to \mu).$

On the next step after scalar multiplication of the Maxwell's equations (3) by the unit vector $\vec{a}_t$, and using the expressions (11) we obtain:

$$\begin{aligned} D_{ss}(\vec{\kappa})h(\vec{\kappa},z) + D_{sp}(\vec{\kappa})e(\vec{\kappa},z) &= (4\pi/c)q_s(\vec{\kappa},z), \\ D_{pp}(\vec{\kappa})e(\vec{\kappa},z) + D_{ps}(\vec{\kappa})h(\vec{\kappa},z) &= (4\pi/c)q_p(\vec{\kappa},z). \end{aligned} \tag{18}$$

Expressions (18) represent the system of coupled ordinary differential equations for two scalar potentials $e(\vec{\kappa}, z)$ and $h(\vec{\kappa}, z)$ within the interval $-b<z<0$; the external sources are entered into the quantities $q_v(\vec{\kappa}, z)$; $D_{v\xi}(\vec{\kappa})$ are the scalar operators that depend on $\vec{\kappa}$ (v, $\xi = s, p$). These operators in explicit form are written as follows:

$$\begin{aligned} q_s(\vec{\kappa},z) &= ik_0\left[-\vec{z}_0 \times \vec{n} + \frac{1}{a_\mu(\vec{n},z)}\left(e_\mu \vec{z}_0 + d_\mu \vec{n}\right)\right]\cdot \vec{M}(\vec{\kappa},z) + \\ &+ \left[\left(\partial_z \varepsilon_{lz} + i\kappa\varepsilon_{ll}\right)\vec{z}_0 - \left(\partial_z \varepsilon_{zz} + i\kappa\varepsilon_{zl}\right)\vec{n}\right]\cdot \frac{\vec{J}(\vec{\kappa},z)}{a_\varepsilon(\vec{n},z)}, \\ q_s &\to q_p \quad \left(\varepsilon \leftrightarrow \mu, \quad \vec{J} \to \vec{M}, \quad \vec{M} \to -\vec{J}\right); \end{aligned} \tag{19}$$

$$D_{ss}(\vec{\kappa}) = \frac{\partial}{\partial z}\frac{\varepsilon_{zz}}{a_\varepsilon(\vec{n},z)}\frac{\partial}{\partial z} + i\kappa\left[\frac{\partial}{\partial z}\frac{\varepsilon_{lz}}{a_\varepsilon(\vec{n},z)} + \frac{\varepsilon_{zl}}{a_\varepsilon(\vec{n},z)}\frac{\partial}{\partial z}\right] + k_0^2\frac{|\hat{\mu}(z)|}{a_\mu(\vec{n},z)} - \kappa^2\frac{\varepsilon_{ll}}{a_\varepsilon(\vec{n},z)}; \tag{20}$$

$$ik_0^{-1}D_{sp}(\vec{\kappa}) = i\kappa\left[\frac{c_\varepsilon(\vec{n},z)}{a_\varepsilon(\vec{n},z)} + \frac{e_\mu(\vec{n},z)}{a_\mu(\vec{n},z)}\right] - \frac{\partial}{\partial z}\frac{b_\varepsilon(\vec{n},z)}{a_\varepsilon(\vec{n},z)} - \frac{\partial}{\partial z}\frac{e_\mu(\vec{n},z)}{a_\mu(\vec{n},z)}; \tag{21}$$

$$D_{ss} \to D_{pp}, \; D_{sp} \to D_{ps}, \; \left(\varepsilon \leftrightarrow \mu\right) \tag{22}$$

where:

$$\begin{aligned} d_\varepsilon &= \varepsilon_{zz}\varepsilon_{tl} - \varepsilon_{zl}\varepsilon_{tz}, \\ e_\varepsilon &= \varepsilon_{ll}\varepsilon_{tz} - \varepsilon_{lz}\varepsilon_{tl}, \\ d_\varepsilon &\to d_\mu, \; e_\varepsilon \to e_\mu, \; \left(\varepsilon \to \mu\right) \end{aligned} \tag{23}$$

After the substituting expressions (11) into condition (4) we obtain boundary conditions for scalar potentials $e(\vec{\kappa}, z)$, $h(\vec{\kappa}, z)$ that satisfied on all boundaries $z=z_j$:

$$\begin{gathered}\left\{e(\vec{\kappa},z)\right\}=0,\quad \left\{h(\vec{\kappa},z)\right\}=0,\\ \left\{1/a_\mu(\vec{n},z)\times\left[\left(\mu_{zz}\partial_z-i\vec{\kappa}\mu_{lz}\right)e(\vec{\kappa},z)-ik_0b_\mu(\vec{n},z)h(\vec{\kappa},z)\right]\right\}=0,\\ \left\{1/a_\varepsilon(\vec{n},z)\times\left[\left(\varepsilon_{zz}\partial_z-i\vec{\kappa}\varepsilon_{lz}\right)h(\vec{\kappa},z)-ik_0b_\varepsilon(\vec{n},z)e(\vec{\kappa},z)\right]\right\}=0.\end{gathered} \tag{24}$$

Substituting expressions (11), for vector amplitudes $\vec{E}(\vec{\kappa}, z)$, $\vec{H}(\vec{\kappa}, z)$, into impedance boundary conditions (5) we obtain two pair of equations for scalar potentials and their derivates. The first pair of equations for lower boundary $z=-b+0$ has the form:

$$\begin{gathered}\left[ik_0a_{pp}^{(u)}(\vec{\kappa})+b_{pp}^{(u)}(\vec{\kappa})\partial_z\right]e(\vec{\kappa},z)+ik_0a_{ps}^{(u)}(\vec{\kappa})h(\vec{\kappa},z)=0,\\ \left[ik_0a_{sp}^{(u)}(\vec{\kappa})+b_{sp}^{(u)}(\vec{\kappa})\partial_z\right]e(\vec{\kappa},z)+\left[ik_0a_{ss}^{(u)}(\vec{\kappa})+b_{ss}^{(u)}(\vec{\kappa})\partial_z\right]h(\vec{\kappa},z)=0,\end{gathered} \tag{25}$$

where:

$$\begin{aligned}
a_{pp}^{(u)}(\vec{\kappa})&=1+\frac{\kappa\mu_{lz}(\vec{\kappa},z)}{k_0a_\mu(\vec{n},z)}L_{tt}^{(u)}(\vec{\kappa}),\\
b_{pp}^{(u)}(\vec{\kappa})&=\frac{\mu_{zz}(z)}{a_\mu(\vec{n},z)}L_{tt}^{(u)}(\vec{\kappa}),\\
a_{ps}^{(u)}(\vec{\kappa})&=L_{tl}^{(u)}(\vec{\kappa})+\frac{b_\mu(\vec{n},z)}{a_\mu(\vec{n},z)}L_{tt}^{(u)}(\vec{\kappa}),\\
b_{ss}^{(u)}(\vec{\kappa})&=\frac{\varepsilon_{zz}(z)}{a_\varepsilon(\vec{n},z)},\\
b_{sp}^{(u)}(\vec{\kappa})&=\frac{\mu_{zz}(z)}{a_\mu(\vec{n},z)}L_{lt}^{(u)}(\vec{\kappa}),\\
a_{sp}^{(u)}(\vec{\kappa})&=\frac{b_\varepsilon(\vec{n},z)}{a_\varepsilon(\vec{n},z)}+\frac{\kappa\mu_{lz}(\vec{\kappa},z)}{k_0a_\mu(\vec{n},z)}L_{lt}^{(u)}(\vec{\kappa}),\\
a_{ss}^{(u)}(\vec{\kappa})&=\frac{\kappa\varepsilon_{lz}(\vec{\kappa},z)}{k_0a_\varepsilon(\vec{n},z)}-\frac{b_\mu(\vec{n},z)}{a_\mu(\vec{n},z)}L_{lt}^{(u)}(\vec{\kappa})+L_{ll}^{(u)}(\vec{\kappa}).
\end{aligned} \tag{26}$$

In expressions (26) the equality is used $L_{\sigma\tau}^{(u)}(\vec{\kappa})=\vec{a}_\sigma\cdot\hat{L}^{(u)}\cdot\vec{a}_\tau$, $(\sigma, \tau=z, l, t)$.

http://dx.doi.org/10.5772/54847

The boundary conditions for top plane ($z = -0$) are similar to (25). They are resulting from the following replacements in (25) – (26): $b_{pp}^{(u)}(\vec{\kappa}) \to -b_{pp}^{(a)}(\vec{\kappa})$, $a_{ps}^{(u)}(\vec{\kappa}) \to -a_{ps}^{(a)}(\vec{\kappa})$, $b_{sp}^{(u)}(\vec{\kappa}) \to -b_{sp}^{(a)}(\vec{\kappa})$, $b_{ss}^{(u)}(\vec{\kappa}) \to -b_{ss}^{(a)}(\vec{\kappa})$; in formulas for $a_{pp}^{(u)}(\vec{\kappa})$, $a_{sp}^{(u)}(\vec{\kappa})$, $a_{ss}^{(u)}(\vec{\kappa})$ upper index $u \to a$ and $\kappa \to -\kappa$.

2.2. Numerical solutions by finite-difference method

Now let's build a numerical solution of the problem of monochromatic plane wave diffraction on the inhomogeneous anisotropic layered structure. We will assume that the structure is piecewise homogeneous along the axis z and within each homogeneous layer the anisotropic material is gyrotropic one, or, particularly, an uniaxial material with arbitrary orientation of the optical axes.

The electromagnetic properties of the benchmark structure in a fixed point of the space are defined by permeability and permittivity dyads:

$$\begin{aligned} \hat{\varepsilon}(z) &= \varepsilon_{\perp}(z)\hat{I} + \left(\varepsilon_{||}(z) - \varepsilon_{\perp}(z)\right)\vec{a}\vec{a} - if(z)\vec{a}\times\hat{I}, \\ \hat{\mu}(z) &= \mu_{\perp}(z)\hat{I} + \left(\mu_{||}(z) - \mu_{\perp}(z)\right)\vec{b}\vec{b} - ig(z)\vec{b}\times\hat{I}. \end{aligned} \tag{27}$$

Here, $\varepsilon_{\perp}(z)$, $\varepsilon_{||}(z)$, $f(z)$ and $\mu_{\perp}(z)$, $\mu_{||}(z)$, $g(z)$ are twice differentiable functions of the variable z; $\hat{I}$ is the identity dyad; $\vec{a}$ and $\vec{b}$ are the unit vectors in the direction of the optical axes which have the following components in the Cartesian coordinate system:

$$\begin{aligned} \vec{a} \equiv \vec{a}(z) &= \left(\cos\theta_a \sin\varphi_a, \cos\theta_a \cos\varphi_a, \sin\theta_a\right), \\ \vec{b} \equiv \vec{b}(z) &= \left(\cos\theta_b \sin\varphi_b, \cos\theta_b \cos\varphi_b, \sin\theta_b\right). \end{aligned} \tag{28}$$

For the sake of clarity, the sloping angles θ_a, θ_b and the azimuthal angles φ_a, φ_b, which determine the optical axes direction, are shown in Fig. 3 and therewith we have:

$$\begin{aligned} -\pi/2 \le \theta_a, \theta_b \le \pi/2, \\ 0 \le \varphi_a, \varphi_b \le 2\pi. \end{aligned} \tag{29}$$

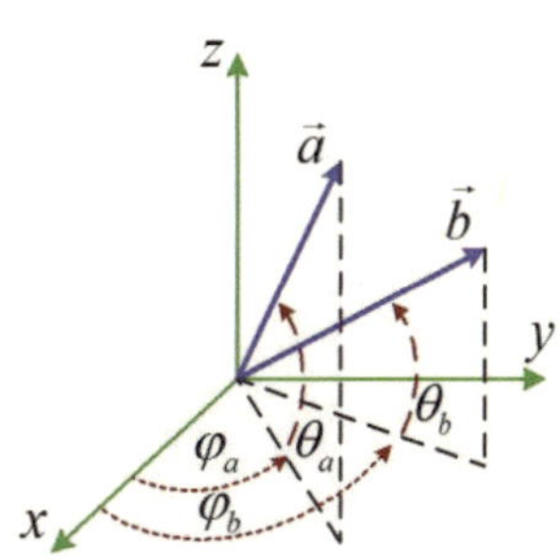

Figure 3. The optical axis orientation in the arbitrary layer

At the same time we consider that an incident (*s*- or *p*-polarized) plane wave arrives from the free half-space ($z>0$) in the direction of the unit vector $\vec{l}_{in}$ which is determined by the sloping angle θ and the azimuthal angle φ, as depicted in Fig.2. Its components in the Cartesian coordinate system are as follows:

$$\vec{l}_{in} = \left(\cos\theta\cos\varphi, \cos\theta\sin\varphi, -\sin\theta\right), \tag{30}$$

$$\begin{gathered} -\pi/2 \le \theta \le \pi/2, \\ 0 \le \varphi \le 2\pi. \end{gathered} \tag{31}$$

Let us assume the inhomogeneous anisotropic structures under consideration with thickness b are placed on isotropic (or anisotropic) homogeneous substrate with permittivity and permeability ε_c, μ_c. In general case, ε_c and μ_c are complex values. We assume also that anisotropic layer is bounded above ($z>0$) by free half-spice with ε_0 and μ_0. In the case presented here, the anisotropic layer is inhomogeneous in the thickness; it means, that structure's parameters are depended on the coordinate z, but they are invariable along the axes x and y. Generally these parameters are piecewise continuous functions of z.

Then scalar potentials $e(\vec{\kappa}, z)$, $h(\vec{\kappa}, z)$ beyond the anisotropic layer will be presented by the following expressions:

- in the free half-spice ($0<z<+\infty$):

$$\begin{aligned} e(\vec{\kappa},z) &= \left(e^{-i\gamma_0 z} + R_{pp}e^{i\gamma_0 z}\right)A_p + R_{ps}A_s e^{i\gamma_0 z}, \\ h(\vec{\kappa},z) &= \left(e^{-i\gamma_0 z} + R_{ss}e^{i\gamma_0 z}\right)A_s + R_{sp}A_p e^{i\gamma_0 z}; \end{aligned} \tag{32}$$

- in the substrate ($-\infty<z<-b$):

$$\begin{aligned} e(\vec{\kappa},z) &= \left[T_{ps}A_s + T_{pp}A_p\right]e^{-ik_0 n_c (z+b)}, \\ h(\vec{\kappa},z) &= \left[T_{ss}A_s + T_{sp}A_p\right]e^{-ik_0 n_c (z+b)}. \end{aligned} \tag{33}$$

Here $\gamma_0 = k_0\sin\theta$; $n_c = \sqrt{\varepsilon_c\mu_c - \cos\theta}$, ($0 \le \arg n_c \le \pi$); complex values A_p and A_s characterize electromagnetic (s or p –polarized) wave's components. The *s*- (*p*-) polarized light corresponds to an electric (magnetic) field be parallel to the layers.

In the expressions (32) – (33) we have introduced the complex reflection $R_{\upsilon\xi}$(υ, $\xi = p, s$) and transmission $T_{\upsilon\xi}$ coefficients, which depend on: wave number in the free space k_0; angles θ, φ and other geometrical and electrodynamical parameters of the problem. The coefficients with similar lower indices ($\upsilon = \xi$) describe conversion of the incident wave into the wave with the same polarization. By analogy, reflection and transmission coefficients with dissimilar lower indexes ($v \neq \xi$) describe conversion of the incident wave into the wave with the orthogonal polarization. In that notation, the left lower index ν corresponds to the polarization of the reflected/transmitted wave; the right lower index ξ corresponds to the polarization of the incident wave.

Notice, that the presence of the "crossed" reflection (R_{sp}, R_{ps}) and transmission coefficients (T_{sp}, T_{ps}), which are responsible for incident plane wave depolarization, is the specific properties of anisotropic media (see, for example [31]).

As it follows from the expressions (32) – (33):

- for the case of s – polarized incident plane wave $(A_s=1,\ A_p=0)$:

$$\begin{aligned} R_{ss} &= h(\vec{\kappa},0)-1, \\ R_{ps} &= e(\vec{\kappa},0), \\ T_{ss} &= h(\vec{\kappa},-b), \\ T_{ps} &= e(\vec{\kappa},-b); \end{aligned} \tag{34}$$

- for the case of p – polarized incident plane wave $(A_s=0,\ A_p=1)$:

$$\begin{aligned} R_{pp} &= e(\vec{\kappa},0)-1, \\ R_{sp} &= h(\vec{\kappa},0), \\ T_{pp} &= e(\vec{\kappa},-b), \\ T_{sp} &= h(\vec{\kappa},-b). \end{aligned} \tag{35}$$

The impedance boundary conditions (25) for the scalar potentials can be rewritten in the following form:

$$\begin{aligned} \partial h/\partial z &= ik_0\left(\lambda_{sp}^{(a)}e+\lambda_{ss}^{(a)}h\right)+f_s, \\ & \qquad\qquad (z=-0); \\ \partial e/\partial z &= ik_0\left(\lambda_{pp}^{(a)}e+\lambda_{ps}^{(a)}h\right)+f_p, \end{aligned} \tag{36}$$

$$\begin{aligned} \partial h/\partial z &= ik_0\left(\lambda_{sp}^{(u)}e+\lambda_{ss}^{(u)}h\right), \\ & \qquad\qquad (z=-b+0). \\ \partial e/\partial z &= ik_0\left(\lambda_{pp}^{(u)}e+\lambda_{ps}^{(u)}h\right), \end{aligned} \tag{37}$$

The values $\lambda_{\upsilon\xi}^{(a,u)}$, f_{υ} are depending on the angles θ and φ and have the form:

$$\begin{aligned}
\lambda_{pp}^{(a)} &= \frac{1}{\mu_{zz}(0)}\left[a_{\mu}(0)\sin\theta - \mu_{lz}(0)\cos\theta\right],\\
\lambda_{ps}^{(a)} &= \frac{b_{\mu}(0)}{\mu_{zz}(0)},\\
\lambda_{sp}^{(a)} &= -\frac{b_{\varepsilon}(0)}{\varepsilon_{zz}(0)},\\
\lambda_{ss}^{(a)} &= \frac{1}{\varepsilon_{zz}(0)}\left[a_{\varepsilon}(0)\sin\theta - \varepsilon_{lz}(0)\cos\theta\right];
\end{aligned} \tag{38}$$

$$\begin{aligned}
\lambda_{pp}^{(u)} &= -\frac{1}{\mu_{zz}(-b)}\left[a_{\mu}(-b)\frac{n_c}{\mu_c} + \mu_{lz}(-b)\cos\theta\right],\\
\lambda_{ps}^{(u)} &= \frac{b_{\mu}(-b)}{\mu_{zz}(-b)},\\
\lambda_{sp}^{(u)} &= -\frac{b_{\varepsilon}(-b)}{\varepsilon_{zz}(-b)},\\
\lambda_{ss}^{(u)} &= \frac{1}{\varepsilon_{zz}(-b)}\left[a_{\varepsilon}(-b)\frac{n_c}{\mu_c} + \varepsilon_{lz}(-b)\cos\theta\right];
\end{aligned} \tag{39}$$

$$\begin{aligned}
f_s &= -2ik_0 a_{\varepsilon}(0)\sin\theta A_s / \varepsilon_{zz}(0),\\
f_p &= -2ik_0 a_{\mu}(0)\sin\theta A_p / \mu_{zz}(0).
\end{aligned} \tag{40}$$

Let us build the finite-difference procedure that approximately describes the system of coupled differential equations (18) and the boundary conditions (36) – (37). Taking into account the fact, that the external sources are absent inside the anisotropic layer, the substitutions $q_p = q_s \equiv 0$ are required in formulas (18).

At first, we divide the segment $-b<z<0$ on the N equal parts. After that the grid step is immediately obtained as $\Delta b = b/N$, and then the grid's knot set $z_0, z_1, ..., z_N$ is defined by formula $z_j = j\Delta b - b$, $(j=0, 1, ..., N)$. Notice, that this knot set includes boundary points $z_0 = -b$ and $z_N = 0$ as well. The partial derivatives in the differential equations (18) are approximated by the central difference; in the boundary conditions (36) – (37), they are approximated by the left-hand difference and the right-hand difference, correspondingly. Then, we obtain the system of linear algebraic equations with dimension $2N+2$ for the unknown complex variables $x_k \equiv h(\vec{\kappa}, z_k)$, $y_k \equiv e(\vec{\kappa}, z_k)$, $(k=0, 1, ..., N)$. For the case of s-polarized wave this system has the form:

http://dx.doi.org/10.5772/54847

$$\begin{cases} x_N\left(1-ik_0b\lambda_{ss}^{(a)}\right)-x_{N-1}-ik_0b\lambda_{sp}^{(a)}y_N=f_{ss}, \\ -ik_0b\lambda_{ps}^{(a)}x_N+y_N\left(1-ik_0b\lambda_{pp}^{(a)}\right)-y_{N-1}=0; \end{cases} \tag{41}$$

$$\begin{cases} A_jx_{j+1}+B_jx_j+C_jx_{j-1}+D_jy_{j+1}+F_jy_j+G_jy_{j-1}=0, \\ P_jx_{j+1}+Q_jx_j+R_jx_{j-1}-K_jy_{j+1}-L_jy_j-M_jy_{j-1}=0, \end{cases} \\ (j=1,2,...,N-1); \tag{42}$$

$$\begin{cases} x_1-x_0\left(1+ik_0b\lambda_{ss}^{(u)}\right)-ik_0b\lambda_{sp}^{(u)}y_0=f_{ss}, \\ y_1-y_0\left(1+ik_0b\lambda_{pp}^{(u)}\right)-ik_0b\lambda_{ps}^{(u)}y_0=0. \end{cases} \tag{43}$$

Here, the first equation set (41) is corresponding to the finite-difference approximation of the boundary conditions (36) imposed at $z=0$, $A_s=1$, $A_p=0$. The next equation set (42) is the finite-difference approximation of the differential equations (18) and the next one equation set (43) is the finite-difference approximation of the boundary conditions (37) imposed $z=-b;\ f_{ss}=\Delta b f_s$.

The coefficients A_j, ..., M_j that enter the equations (42) are given by the expressions:

$$A_j=\frac{\varepsilon_{zz}(z_j)}{a_\varepsilon(z_j)}+\frac{1}{4}\left[\frac{\varepsilon_{zz}(z_{j+1})}{a_\varepsilon(z_{j+1})}-\frac{\varepsilon_{zz}(z_{j-1})}{a_\varepsilon(z_{j-1})}\right]+\frac{ik_0b\cos\theta}{2}\left[\frac{\varepsilon_{zl}(z_j)}{a_\varepsilon(z_j)}+\frac{\varepsilon_{lz}(z_{j+1})}{a_\varepsilon(z_{j+1})}\right],$$
$$B_j=-2\frac{\varepsilon_{zz}(z_j)}{a_\varepsilon(z_j)}+(k_0b)^2\left[\frac{|\hat{\mu}(z_j)|}{a_\mu(z_j)}-\cos^2\theta\frac{\varepsilon_{ll}(z_j)}{a_\mu(z_j)}\right],$$
$$D_j=\frac{ik_0b}{2}\left[\frac{d_\mu(z_j)}{a_\mu(z_j)}+\frac{b_\varepsilon(z_{j+1})}{a_\varepsilon(z_{j+1})}\right],$$
$$C_j=\frac{\varepsilon_{zz}(z_j)}{a_\varepsilon(z_j)}-\frac{1}{4}\left[\frac{\varepsilon_{zz}(z_{j+1})}{a_\varepsilon(z_{j+1})}-\frac{\varepsilon_{zz}(z_{j-1})}{a_\varepsilon(z_{j-1})}\right]-\frac{ik_0b\cos\theta}{2}\left[\frac{\varepsilon_{zl}(z_j)}{a_\varepsilon(z_j)}-\frac{\varepsilon_{lz}(z_{j-1})}{a_\varepsilon(z_{j-1})}\right], \tag{44}$$
$$F_j=-(k_0b)^2\cos\theta\left[\frac{c_\varepsilon(z_j)}{a_\mu(z_j)}-\frac{e_\mu(z_j)}{a_\mu(z_j)}\right],$$
$$G_j=-\frac{ik_0b}{2}\left[\frac{d_\varepsilon(z_j)}{a_\mu(z_j)}-\frac{b_\varepsilon(z_{j-1})}{a_\varepsilon(z_{j-1})}\right],$$
$$A_j\to K_j,\quad B_j\to L_j,\quad C_j\to M_j,\quad D_j\to P_j,\quad F_j\to Q_j,\quad G_j\to R_j,\quad (\varepsilon\leftrightarrow\mu).$$

The system of equations for the p-polarized incident plane wave, can be received from (41) – (43) by the substitutions $f_{ss}\to 0$ and $0\to f_{pp}=\Delta b f_p$ in the systems (41), (43). For the case, when anisotropic layer is uniaxial media: $f(z)=g(z)\equiv 0$.

As a result, a set of linear algebraic equations (41) – (43) with dimension $2N+2$ is derived. Obtained linear system of equations can be solved by standard techniques such as Gauss method. By solving it we find the unknown transmission and reflection factors for the structure under study.

3. Results and discussion

In this section we present the results of numerical simulations that illustrate the influence of anisotropy and inhomogeneities of the materials composed of layers in the structure being studied upon the mechanisms of diffraction of the incident plane electromagnetic wave.

All structures presented here are based on the porous silicon (PSi). Today, PSi plays an important role in a number of applications. These include microcavities [11,32], photonic crystals [33], waveguide structures [34], photodetectors [35], sensors [36], etc. Besides, PSi has the potential to be an optically active material in the case when an acceptable electro- or thermo-optic media is infiltrated into the pores [32]. Therefore, porous silicon is an excellent candidate for tunable optical interconnects and optical switches. For all these applications a strict control over the reflectance and transmission properties of PSi layers is required.

Today porous silicon attracts a great deal of attention because it's a material with great technological promise. The main advantages of PSi may be summarized as follows:

- PSi is a simple and low cost dielectric material that can be easy prepared;
- PSi is a promising material for photonic applications due to its excellent thermal and mechanical properties, obvious compatibility with standard Si-based technologies;
- PSi is a suitable material for the formation of arbitrary multilayers. PSi multilayers are almost arbitrary combination of layers with different thickness and porosity (refractive index), because these two parameters can be relatively easily controlled during the formation process of porous silicon.

It is well known that the PSi films can be produced by anode electrochemical etching of the monocrystalline silicon plates [37]. The nanometer-size pores tend to grow in the direction of electrochemical etching and, accordingly, nanocrystal formation sets in. The porosity P and the effective refractive index $n_{eff}=\sqrt{\varepsilon_{eff}}$ of PSi are controlled by the current density under electrochemical etching, because the effective refractive index of PSi is determined by the porosity and refractive index of the medium inside the pores. Thus, by periodically varying the magnitude of current density we are able to obtain the structure with alternating layers of different porosity and, consequently, with different refractive indices.

The bulk silicon crystal is not birefringent due to its cubic crystal symmetry. However, porous modification of silicon can exhibit strong in-plane anisotropy of the refractive index [37,38]. The observed birefringence depends on the porosity, the size of Si nanocrystals, the spacing between them and the dielectric properties of surrounding medium. As was demonstrated in

[39], PSi layers, with dimension of the pore about 10-30 nm, have properties of the negative uniaxial crystal with diagonal permittivity tensor

$$\varepsilon = \begin{bmatrix} n_0^2 & 0 & 0 \\ 0 & n_0^2 & 0 \\ 0 & 0 & n_e^2 \end{bmatrix}, \tag{45}$$

whose birefringence magnitude $\Delta n = n_o - n_e$ is up to 0.24. In expression (45) n_0 is the index of refraction for the waves polarized perpendicularly to the optical axis, which are called "ordinary" or "o – waves"; n_e is the index of refraction for the waves polarized parallel to the optical axis, which are called "extraordinary" or "e – waves". It is important to note that if $n_e > n_0$ the crystal is said to be positively uniaxial, in opposite case if $n_e < n_0$ the crystal is said to be negatively uniaxial. When a linearly polarized wave of arbitrary polarization direction enters an anisotropic medium, it will be split into two components polarized along the two allowed polarization directions which are determined by the direction of the wave vector relative to the axes of the indicatrix. As a result, each *s*- or *p*- polarized plane wave incident on such an anisotropic photonic structure will generate two reflected and two transmitted plane waves containing both *s*- and *p*- polarized planewaves. For the special cases when the principal axes of the layers are parallel or perpendicular to the fixed axes, the *s*-and *p*-polarized waves remain uncoupled [26].

Two types of PBG structures are investigated, namely, a distributed Bragg reflector (DBR) in other words photonic crystal (PhC) and a microcavity.

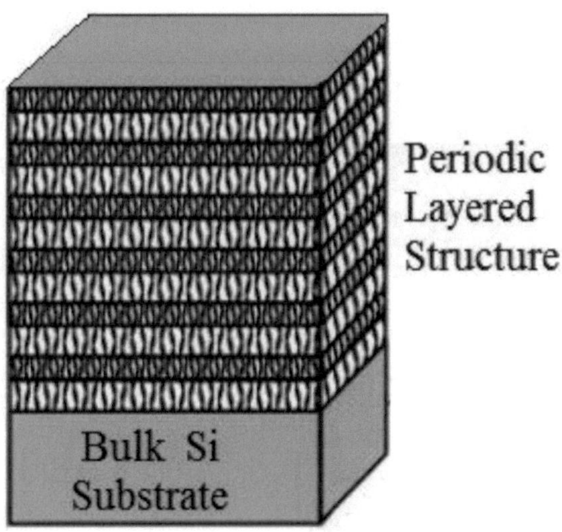

Figure 4. Schematic presentation of PSi-based PhC: the dark layers have high porosity (low refractive index) and the bright layers are of low porosity (high refractive index)

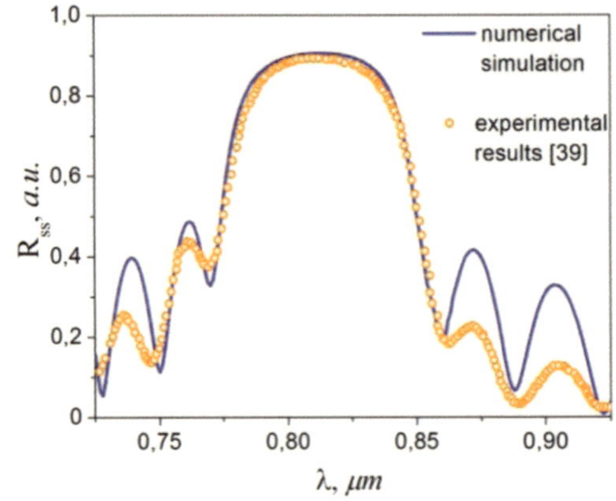

Figure 5. Reflectance spectra R_{SS} of the PhC: solid line – numerical simulation; scatter – experimental results [39]. The calculations involved the value of sloping and azimutal angles: $\theta = 70^0$, $\theta_a = 45^0$ and $\varphi = \varphi_a = 0^0$

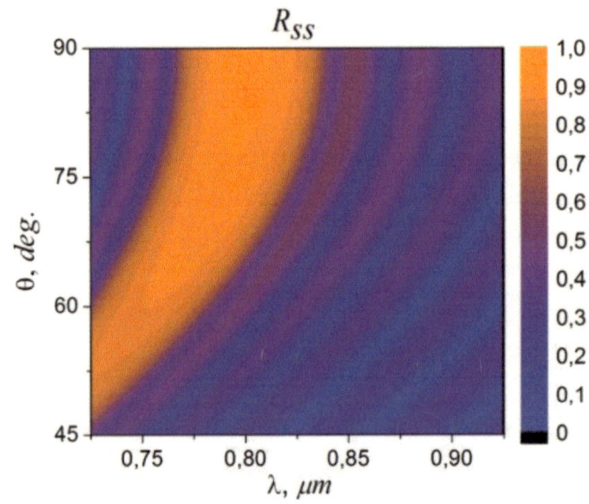

Figure 6. Reflectance spectra R_{SS} as function of incident sloping angle θ. In this case $\theta_a = 45^0$, $\lambda_c = 800\,nm$

The simplest multilayered PBG structure is one-dimensional photonic crystal as depicted on Fig. 4. It's well known that PhCs are class of optical media represented by the natural or artificial structures with periodic modulation of the refractive index. Such optical media have some peculiar properties which gives an oportunity for a number of applications to be implemented on their basis. The most important property which determines practical significance of the PhC is the presence of the omnidirectional photonic band gap. The PBG refers to the energy or frequency range where the light propagation is prohibited inside the PhC. As an example of such a PhC one can give a Bragg grating which is widely used as a distributed

http://dx.doi.org/10.5772/54847

reflector in vertical cavity surface emitting lasers. Besides, such structures are widely used as antireflecting coatings which allow dramatically decrease the reflectance from the surface and are used to improve the quality of lenses, prisms and other optical components.

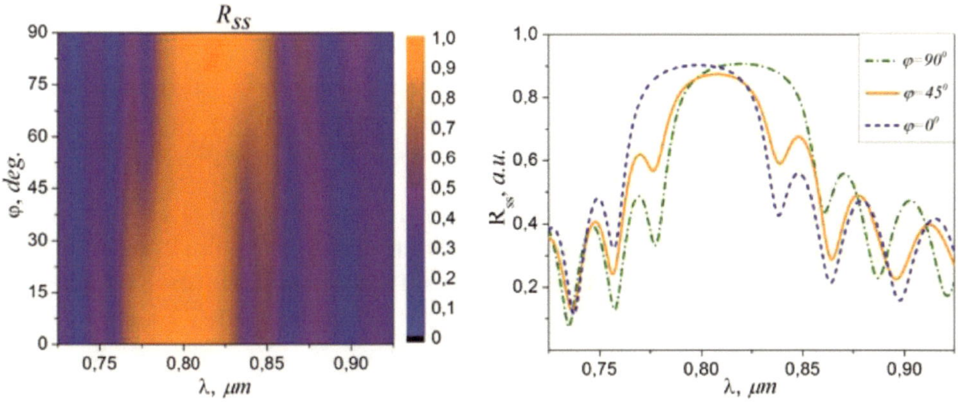

a b

Figure 7. Reflectance spectra R_{ss} as function of the incident azimuthal angle φ

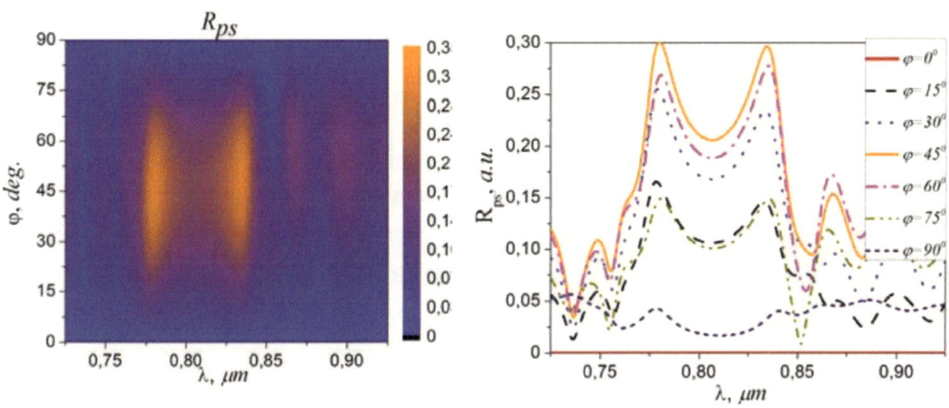

(a) (b)

Figure 8. Reflectance spectra R_{ps} as function of the incident azimuthal angle φ

First, we study light propagation in a one-dimensional photonic crystal, which was identical to PhC that was experimental investigated by Aktsipetrov *et al.* [39]. It consists of 25 pairs (building blocks) of lossy anisotropic layers with refractive indices: $n_{o1}=1.39+0.008i$, $n_{e1}=1.32+0.008i$, $n_{o2}=1.58+0.008i$, $n_{e2}=1.5+0.008i$. The materials are assumed to be nonmagnetic so that $\mu=1$ throughout the whole layered medium. The physical thickness of the layers was chosen such that the optical thickness of layers was equal to $\lambda_c/4$, where $\lambda_c=800\,nm$ is the Bragg wavelength corresponding to the photonic band gap (PBG) centre at the normal incidence ($\theta=90$). The numerical modeling were made for the case where the optical axes of all structure layers were oriented in one and the same direction.

The comparison of the results of the proposed numerical scheme with results of the cited experimental work [39] is presented on Fig. 5. The reflectance spectra demonstrate existence of a PBG with a reflectance of about 0.9 in the wavelength region of 770–860 nm. As we see theoretical result is in very good agreement with the experimental one. So proposed method can be successfully used for computation of the spectral characteristics of 1D anisotropic layered structures.

For all the structures discussed here, the reflection/transmission spectrum is strongly dependent on the incident sloping θ and azimuthal φ angles. The dependences of the reflection spectra of PSi-based distributed Bragg reflectors from variation of the sloping angle θ and the azimuthal angle φ of the incident *s*-polarized plane wave are shown in Fig. 6. – Fig. 8. In all cases we take $\lambda_c=800\,nm$.

It may be seen from the Fig. 6 that in agreement with the theory the width of the high reflectance region (width of the photonic band gap) is decreased with decreasing the sloping angle θ, and the central wavelength of PBG is shifted to the short wavelength region. In this case we assume that $\varphi=\varphi_a=0^0$ in result the conversion incident wave to wave with orthogonal polarization is absent.

The influence of the azimuthal angle φ on the conversion of the incident plane wave is show in Fig.7-8. The curves presented are calculated at $\theta=80^0$, $\theta_a=45^0$ and $\varphi_a=0^0$. Note that in a uniaxial crystal the maximum angular separation of the "o" and "e" waves, in other words the maximum conversion of the incident linearly *s*- or *p*- polarized plane wave into the cross-polarized wave, occurs when the wave vector has the angle $\varphi_{\max}\approx45^0$ with the optic axis [31]. Also we should note that the value of $\varphi_{\max}$ is proportional to $|n_o-n_e|$. Analyzing the figures 7-8, we can see that when the incident azimuthal angle φ varies:

- the reflectance spectra demonstrate conversion of the incident *s*-polarized plane wave into the wave with same (Fig. 7) and orthogonal (Fig. 8) polarization. It occurs within the wide range of the incident azimuthal angles;
- the spectra also show a shift in the PBG spectral position when the azimuthal angle is changed. The largest shift, about 20 nm, is observed under variation of the incident azimuthal angle φ from 0^0 to 90^0. This results are in agreement with [39];
- the maximum conversion of the incident *s*-polarized wave into the wave with orthogonal polarization occurs at the incident azimuthal angle $\varphi\approx45^0$ (see Fig. 8(b)), that is in conformity with [31]. In this case two reflection peaks with magnitude about 0.3 are clearly observed in the reflectance spectra R_{ps}.The first reflection peak with the central wavelength 780 *nm*

is corresponding to the short-wavelength PBG edge for case $\varphi=90^0$. The second one with the central wavelength 835 nm – to the long-wavelength PBG edge for case $\varphi=0^0$

The influence of the sloping and the azimuthal angles θ_a, φ_a (which determine direction of the optical axes) on the conversion incident s-polarized plane wave into wave with orthogonal polarization at the Bragg wavelength ($\lambda_c=800\ nm$) is shown in Fig. 9(a,b). Reflection coefficients R_{ps} is calculated at incident angles $\theta=80^0$ and $\varphi=0^0$. It is clear from this graph that:

- the maximum value of R_{ps} is around 0.4 (see Fig. 9(b)) and corresponds to the sloping angle $\theta_a=0^0$ and the azimuthal angles $\varphi_a=m\cdot\pi/4$, $(m=1,\ 3,\ 5,\ 7)$ that is in accordance with [31];
- in the case $\varphi_a=m\cdot\pi/2$, $(m=0,\ 1,\ 2,\ 3,\ 4)$the conversion of the incident wave into the wave with orthogonal polarizations is virtually absent $R_{ps}\approx 0$.

Finally, we can conclude that reflection coefficients of the investigated structures are very sensitive to the azimuthal angle φ_a and the incidence angles θ and φ. When the angles of incidence $\theta=90^0$ $\varphi=0^0$ and the azimuthal angle φ_a possesses values 0° or 90°, the s- and p-modes are almost uncoupled, and the value of the R_{ps} remains very small. The situation is similar to that one in the case of isotropic materials. With the exception of these particular values of φ_a, there is a great mixing between s-and p-modes that gives rise to the emergence of a large reflection coefficient R_{ps}.

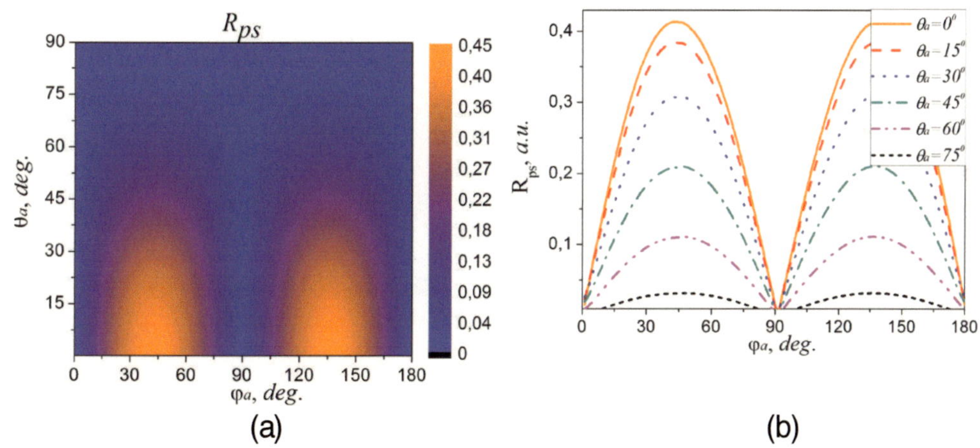

Figure 9. Reflection coefficient R_{ps} as function of the sloping and the azimuthal angles

The PSi-based mirrors studied above are periodic structures, but one important property of PhC is the presence of narrow resonance (localized defect modes) in the PBG region when a disorder is introduced in their periodic structure. Usually "defect" is a layer with half-wavelength optical thickness that inserted in the middle of the dielectric stack. Fig. 10 shows a scheme of a typical PSi-based microcavity structure consisting in an active layer sandwiched between two distributed Bragg reflectors (PSi mirrors). At that, the reflectance spectrum of the whole structure is changed. As a rule, when the parameters of defect layer are properly selected, sharp transmittance peak appears within the main reflectance band. The wavelength of the transmittance peak corresponds to the resonant wavelength of the defect. For instance, if the thickness of the defect layer is twice larger than it was in defectless structure, the transmittance peak appears at the Bragg wavelength of the corresponding defectless structure. Thus, if the radiation with wavelength equal to Bragg wavelength of defectless reflector falls at this structure it will pass the structure almost without the reflectance. If the defect thickness will be slightly different, the wavelength of the transmittance peak will be different as well.

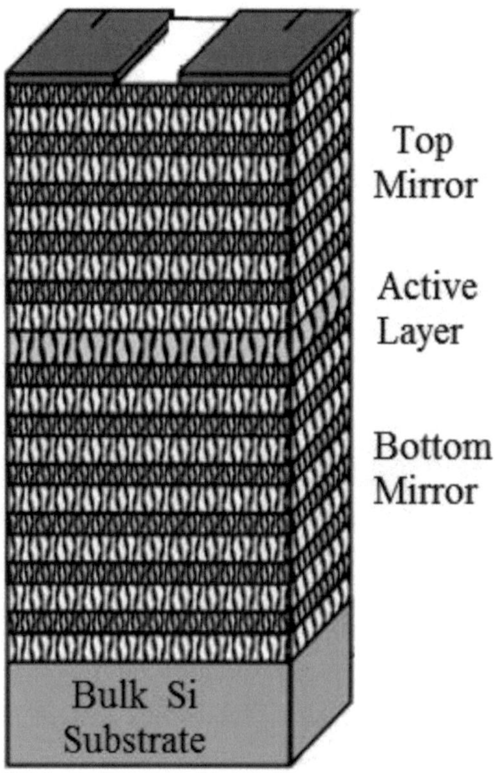

Figure 10. Schematic presentation of PSi-based microcavity

The advantage of using PSi microcavities is that the position of the transmittance peak is completely tunable by changing the properties of the central layer (i.e. porosity and thickness) during the electrochemical etching condition and by the infiltration of organic molecules (for example liquid crystals [32]).

Now, let's discuss the influence of anisotropy of layers on the optical properties of the microcavitiy. In cited case the PSi-based microcavity consist of two mirrors (each of them consists of 12 building blocks) separated by active layer with optical thickness $\lambda_c/2$ and low porosity.

The Fig. 11 shows the reflectance spectra of the microcavity for both *s*- (solid curve) and *p*- (dashed curve) polarized incident plane waves. It should be noted that spectra reveal the presence of photonic band gap with a reflectance of about 0.85-0.9 and microcavity mode within the PBG for both polarizations. Analyzing both spectra together, we can observe a spectral shift of the microcavity mode when the incident polarization is changed. This particular effect is also the manifests an attribute of birefringence and can be used in dividing the incident radiation polarization. In our case, the central wavelengths of resonance peaks are λ_{sc}=796 *nm* and λ_{pc}=817 *nm*for *s*- and *p*- polarization, respectively. Taking into account this result, we can consider that the variation in the incident radiation polarization brings about the shift $\Delta\lambda = |\lambda_{pc} - \lambda_{sc}|$ of the microcavity mode, and this shift may be as high as 21 *nm* for presented case.

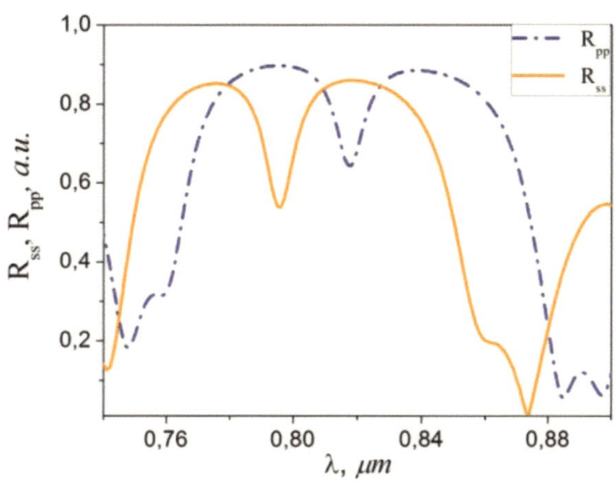

Figure 11. *s* - and *p*- polarized reflectance spectra of microcavity. In this case $\lambda_c = 800\,nm$, $\theta = 80^0$, $\varphi = \varphi_a = 0^0$, $\theta_a = 45^0$

Figure 12 plots the spectral shift of the microcavity mode for *p*-and *s*-polarized light, as a function of the of azimuthal angle φ_a. From this figure, some particular effects can be highlighted:

- the orientation of the optical axes of the layers with respect to incidence plane of *s*- or *p*-polarized waves influences on the quantity of resonant peaks within PBG area and its location on the wavelength scale. Particularly, with the increasing of azimuthal angle, the "additional" well recognized resonant peak is appeared in the stopbands for both polarizations of the incident plane waves. It correlates with results of the paper [40];

- as a general trend, the value of the "main" resonant peak is decreasing with increasing the azimuthal angle φ_a and it completely disappears when $\varphi_a = 90^0$. There against, the magnitude of the "additional" resonance peak is increasing with increasing φ_a and it has maximum value in the mentioned above case;

- both transmission peaks have identical magnitudes when azimuthal angle $\varphi_a \approx 48^0$ for both *s*- and *p*- polarization;

- the extreme cases, namely $\varphi_a = 0^0$ and $\varphi_a = 90^0$, correspond to presence of single-defect mode within PBG region, while in the other cases this mode splits into two defect modes. The comparison of the results presented in Fig. 12(a) and 12(b) shows that $\lambda_{s0} = \lambda_{p90} = 796\,nm$ and $\lambda_{s90} = \lambda_{p0} = 817\,nm$. Here we used next definition for central wavelengths – λ_{xy}. In this notation, the left subscript index $x = s,\ p$ is corresponding to polarization of incident wave; the right subscript index $y = 0,\ 90$ is corresponding to quantity of azimuthal angle φ_a;

- also, in the extreme cases, we observed the shift of the resonance peaks $\Delta\lambda = |\lambda_{p0} - \lambda_{p90}| = |\lambda_{s0} - \lambda_{s90}|$ about 21 nm to the short-wavelength and long-wavelength regions for *p*- and *s*-polarization, respectively. This result is in good agreement with the experimental results [39].

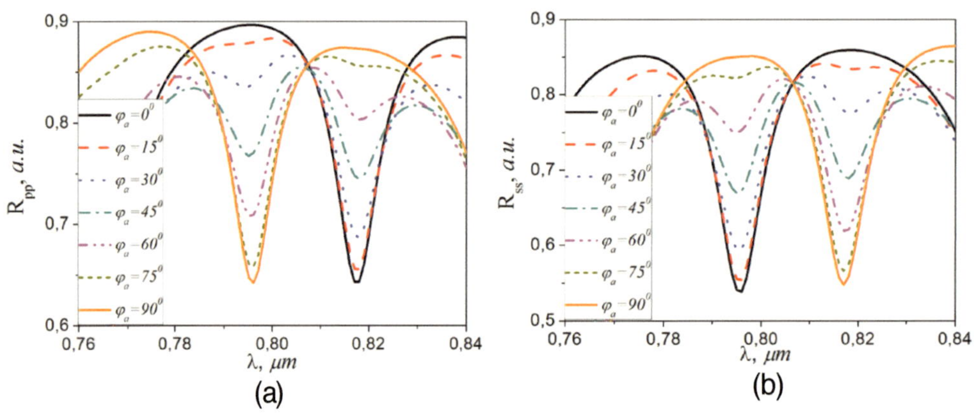

Figure 12. Microcavity resonance shift based on the azimuthal angle φ_a change. Reflectance spectra for *p*- and *s*- polarized incident plane wave are depicted on Fig. 12(a) and Fig. 12(b), respectively.

The Fig. 13 shows characteristics similar to those one presented on Fig.12, but for cross-polarized components R_{sp} and R_{ps}. In these figures we can see that within the PBGs areas the reflection coefficients R_{sp} and R_{ps} have local minimum that occur around the wavelength $\lambda \approx (\lambda_{p0} + \lambda_{p90})/2 \approx (\lambda_{s0} + \lambda_{s90})/2$.

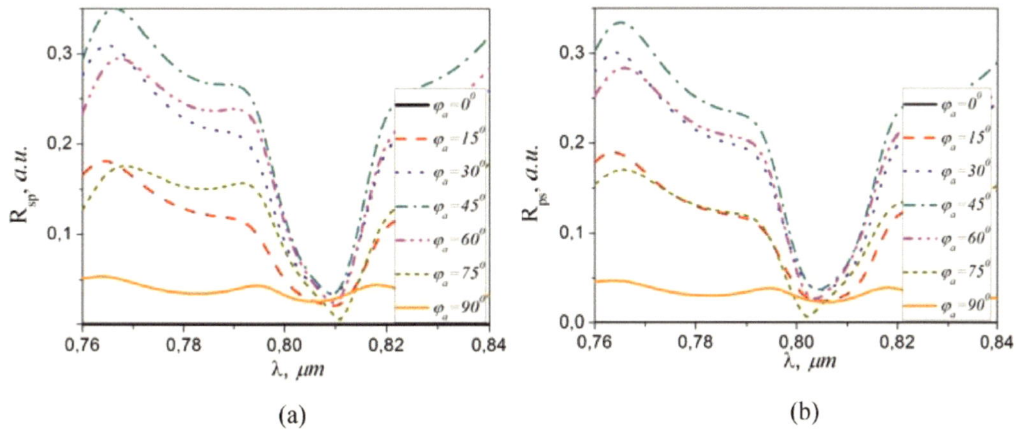

Figure 13. Reflectance spectra R_{sp} (a) and R_{ps} (b) as function of the azimuthal angle φ_a. In this case $\lambda_c = 800\,nm$, $\theta = 80^0$, $\varphi = 0^0$, $\theta_a = 45^0$.

4. Conclusions

In conclusion, the present chapter describes the mathematical background for calculation of spectral characteristics of the birefringent layered media. On the basis of the presented theoretical description and its numerical approximation the influence of material anisotropy of PSi-based layered photonic structures on their optical properties has been theoretically investigated. All these multilayer PBG structures have been designed for 0.8 μm applications.

The reflectance spectra of some photonic structures for both polarization of the incident plane wave are calculated. The agreement between the numerical calculations and the experiments [39] are obtained. It was shown numerically that anisotropy of layered media reduces to polarization transformation of the incident plane wave. Notably, the maximum conversion of incident plane wave into the wave with orthogonal polarization occurs when the optical axis of the structure have angles $\varphi_a = m \cdot \pi / 4$, $(m = 1,\ 3,\ 5,\ 7)$ with respect to the incidence plane of *s*- or *p*-polarization incident plane. In contrast to this, the conversion of the incident wave into the wave with orthogonal polarizations is absent when $\varphi_a = m \cdot \pi / 2$, $(m = 0,\ 1,\ 2,\ 3,\ 4)$.

Also from our numerical calculations, we conclude that the location of the PBG edges and location of microcavity modes within PBG region are different for *s*- and *p*- polarized waves and their spectral positions change under rotation of the optical axis of the structure with respect to the wave vector of the incident electromagnetic wave. The variation of polarization of the incident radiation brings about the shift of the microcavity mode, and this shift may be as high as 21 *nm* for discussed case. This particular effect can be used in the devices that divide the incident radiation according to polarization.

In addition the obtained results can be used in the designing of PSi-based photonic devices:

- spectral- and polarization-selective elements (filters);
- electrically tunable filters and optical switches. In this instance it is required that the pore be filled with liquid crystals;
- polarization converters;
- time-delay elements for optically controlled phased-array antenna system.

Author details

V.I. Fesenko[1], I.A. Sukhoivanov[2], S.N. Shul'ga[3] and J.A. Andrade Lucio[2]

1 Institute of Radio Astronomy of National Academy of Sciences of Ukraine and Kharkov National University of Radio Electronics, Ukraine

2 University of Guanajuato, Mexico

3 Kharkov National University, Ukraine

References

[1] Yakushev S.O., Shulika O.V., Petrov S.I., Sukhoivanov I.A. Chirp Compression with Single Chirped Mirrors and its Assembly. Microelectronics Journal 2008; 39(3-4) 690-695.

[2] Iakushev S.O., Shulika O.V., Lysak V.V., Sukhoivanov I.A. Air-Gap Silicon Nitride Chirped Mirror for Few-Cycle Pulse Compression. Optoelectronics and Advanced Materials. Rapid Communications 2008; 2(11) 686-688.

[3] Guryev I.V., Shulika O.V.,Sukhoivanov I.A., Mashoshina O.V. Improvement of Characterization Accuracy of the Nonlinear Photonic Crystals Using Finite Elements-Iterative Method. Applied Physics B – Lasers and Optics 2006; 84(1-2) 83-87.

[4] Klymenko M.V., Lysak V.V., Sukhoivanov I.A., Shulika A.V. Optical and Transport Properties of In0.49Ga0.51P/In0.49(Ga0.6Al0.4)0.51P Single Quantum Well Structure with Digital Alloy Barriers. Superlattices and Microstructures 2009; 46(4) 603-610.

[5] Klymenko M.V., Safonov I.M., Shulika O.V., Sukhoivanov I.A. Ballistic Transport in Semiconductor Superlattices with Non-Zero In-Plane Wave Vector. Phisica Status Solidi B 2008; 245(8) 1598-1603.

[6] Klymenko M.V., Safonov I.M., Shulika O.V., Sukhoivanov I.A., Michalzik R. Effective-Mass Superlattice as an Injector in Quantum Cascade Lasers. Optical and Quantum Electronics 2008; 40(2-4) 197-204.

[7] Shulika O.V., Safonov I.M., Sukhoivanov I.A., Lysak V.V. Quantum Capture Area in Layered Quantum Well Structures. Microelectronics Journal 2005; 36, 350-355.

[8] Safonov I.M., Sukhoivanov I.A., Shulika O.V., Lysak V.V. Piecewise-Constant Approximation of the Potential Profile of Multiple Quantum Well Intrinsic Heterostructures. Superlattices and Microstructures 2008; 43(2) 120-131.

[9] Lysak V.V., Sukhoivanov I.A., Shulika O.V., Safonov I.M., Lee Y.T. Carrier Tunneling in Complex Asymmetrical Multiple Quantum Well Semiconductor Optical Amplifiers. IEEE Photonics Technology Letters 2006; 18(12) 1362-1364.

[10] Tuz, V., Prosvirnin S., Zhukovsky S.Polarization Switching and Nonreciprocity in Symmetric and Asymmetric Magnetophotonic Multilayers with Nonlinear Defect.Physical Review A 2012; 85 043822(1)–043822(8).

[11] Fesenko V.I. and Sukhoivanov I.A. Polarization Conversion in Inhomogeneous Anisotropic Multilayer Structures. In AIOM 2012: Lasers, Sources, and Related Photonic Devices, OSA Technical Digest (CD) (Optical Society of America, 2012), paper JTh2A.7, AIOM 2012, 1-3 February 2012, San-Diego, USA.

[12] Tuz V. Optical Properties of a Quasiperiodic Generalized Fibonacci Structure of Chiral and Material Layers. Journal of the Optical Society of America B 2009: 26(4) 627-632.

[13] Nefedov I.S., Tretyakov S.A. Photonic Band Gap Structure Containing Metamaterial with Negative Permittivity and Permeability Physical Review E 2002; 66(3) 036611(1)-036611(5).

[14] Pereira M., Shulika O. (Eds) Terahertz and Mid Infrared Radiation. Generation, Detection and Applications. NATO Series B: Physics and Biophysics. Springer; 2011.

[15] Sukhoivanov I.A., Guryev I.V. Photonic Crystals Physics and Practical Modeling. Springer; 2009.

[16] Dal Negro L., Boriskina S.V. Deterministic Aperiodic Nanostructures for Photonics and Plasmonics Applications. Laser Photonics Review 2011; 1–41.

[17] Yakushev S.O., Sukhoivanov I.A., Shulika O.V., Lysak V.V., Petrov S.I.Modeling and Simulation of Interaction of the Ultrashort Laser Pulse with Chirped Mirror for Structure Design Improvement. Journal of Optoelectronics and Advanced materials 2007; 9(8) 2384-2390.

[18] John S. Localization of Light in Periodic and Disordered Dielectrics. In: Confined Electrons and Photons, Plenum Press; 1995.

[19] Iakushev S.O., Sukhoivanov I.A., Shulika O.V., Andrade-Lucio J.A., Perez A.G. Nonlinear Pulse Reshaping in Optical Fibers. In: Laser Systems for Applications, In-Tech; 2011.

[20] Iakushev S.O., Shulika O.V., Sukhoivanov I.A. Passive Nonlinear Reshaping Towards Parabolic Pulses in the Steady-State Regime in Optical Fibers. Optics Communication 2012; 285(21-22) 4493-4499.

[21] Corral J.L., Marti J., Fuster J.M. and Laming R.I. True Time-Delay Scheme for Feeding Optically Controlled Phased-Array Antennas Using Chirped-Fiber Gratings. IEEE Photonics Technology Letters 1997; 9(11) 1529-1531.

[22] Acoleyen K.V., Rogier H., Baets R. Two-Dimensional Optical Phased Array Antenna on Silicon-on-Insulator. Optics Express 2010; 18(13) 13655-13660.

[23] Komarevskiy N, Braginsky L, Shklover V, Hafner C, Lawson J. Fast Numerical Methods for the Design of Layered Photonic Structures with Rough Interfaces. Optic Express 2011; 19(6) 5489-5499.

[24] Mathew Ninan, Jiang Zhengyi, Wei Dongbin. Analysis of Multi-Layer Sandwich Structures by Finite Element Method. Advanced Science Letters 2011; 4(8-10) 3243-3248.

[25] Carretero L., Perez-Molina M., Acebal P., Blaya S., Fimia A. Matrix Method for the Study of Wave Propagation in One-Dimensional General Media. Optics Express 2006; 14(23) 11385-11391.

[26] Yeh P. Electromagnetic Propagation in Birefringent Layered Media. Journal of the Optical Society of America 1979; 69(5) 742-756.

[27] Taflove A., Hagness S.C. Computational Electrodynamics: The Finite-Difference Time-Domain Method, 2nd ed. Artech House; 2000.

[28] Thylen L. and Yevick D. Beam Propagation Method in Anisotropic media. Applied Optics 1982; 21(15) 2751–2754.

[29] Felsen L., Marcuvitz N. Radiation and Scattering of Waves. Wiley; 2003.

[30] Zhuck N.P. Electromagnetic theory of arbitrarily anisotropic layered media Part 1. Scalarization of field quantities. International Journal of Electronics 1993, 75(1) 141-148.

[31] Yariv A., Yeh P. Optical Waves in Crystals. Wiley; 1984.

[32] Tkachenko, G.V., Shulika, O.V. Thermal Tuning of a Thin-Film Optical Filter Based on Porous Silicon and Liquid Crystal. Ukrainian Journal of Physical Optics2010; 11(4) 260-268.

[33] Samuoliene N., Satkovskis E. Reflectivity Modeling of All-Porous-Silicon Distributed Bragg Reflectors and Fabry-Perot Microcavities. Nonlinear Analysis: Modeling and Control 2005; 10(1) 83 - 91.

[34] Charrier J., Le Gorju E., Haji L., Guendouz M. Optical Waveguides Fabricated from Oxidised Porous Silicon. Journal of Porous Materials 2000; 7(1-3) 243-246.

[35] HadjersiT., GabouzeN. Photodetectors Based on Porous Silicon Produced by Ag-Assisted Electroless Etching. Optical Materials 2008; 30(6) 865–869.

[36] Saarinen J., Weiss S., Fauchet P., Sipe J.E. Optical Sensor Based on Resonant Porous Silicon Structures. Optics Express 2005; 13(10) 3754-3764.

[37] Pap A.E., Kordas K., Vahakangas J., Uusimaki A., Leppavuori S., Pilon L., Szatmari S. Optical Properties of Porous Silicon. Part III: Comparison of Experimental and Theoretical Results. Optical Materials 2006; 28 506–513.

[38] Kovalev D., Polisski G., Diener J., Heckler H., Künzner N., Timoshenko V.Yu., Koch F. Strong In-Plane Birefringence of Spatially Nanostructured Silicon. Applied Physics Letters 2001; 78, 916.

[39] Aktsipetrov O.A., Dolgova T.V., Soboleva I.V., Fedyanin A.A. Anisotropic Photonic Crystals and Microcavities Based on Mesoporous Silicon. Physics of the Solid State 2005; 47, 156-158.

[40] Ouchani N., Bria D., Djafari-Rouhani B., Nougaoui A. Defect Modes in One-Dimensional Anisotropic Photonic Crystal. Journal of Applied Physics 2009; 106 113107(1)-113107(8).

Chapter 5

Threshold Mode Structure of Square and Triangular Lattice Gain and Index Coupled Photonic Crystal Lasers

Marcin Koba

Additional information is available at the end of the chapter

http://dx.doi.org/10.5772/53712

1. Introduction

Photonic crystals (PC) are periodic structures with variation of the refractive index in one, two or three spatial dimensions. The dynamic development of experimental and theoretical works on photonic crystals has been launched by Yablonovitch [1],[2] and John [3] publications, although the idea of periodic structures had been known since Rayleigh [4].

The main properties of photonic crystals stem from the existence of frequency ranges for which the propagation of electromagnetic waves in the medium is not permitted. These frequency ranges are commonly known as photonic band gaps and give the ability to modify the structure parameters, e.g., group velocity, coherence length, gain, and spontaneous emission. Photonic crystals' properties are beneficial for both passive and active devices. This Chapter is devoted especially to the latter.

1.1. Two-dimensional photonic crystal lasers

Photonic structures are becoming more and more important component of light generating devices. They are used in lasers as mirrors [5],[6], active waveguides [7], coupled cavities [8], defect microcavities [9],[10], and the laser active region [11].

Lasers with defects within two-dimensional photonic crystals are known for their high finesse [12] and very low threshold [13].

Photonic crystal band-edge lasers allow to obtain edge [11] and surface emission [14],[15] of coherent light from large cavity area. These devices are able to emit single mode, high-power electromagnetic radiation by utilizing the presence of band-edge in the photonic band structure [16],[17]. They also allow to control the output beam pattern by manipulation of the structure geometry [18],[19], provide low threshold [20], and beams which have small divergence angle and can be focused to a size less than the wavelength [21]. Recently, the operation of PC lasers as an on-chip dynamical control of the emitted beam direction have been demonstrated [19],[22].

The photonic crystal structures lasing wavelengths span from terahertz [23]-[25], through infrared [26],[27] to visible [21],[28],[29].

1.2. Modeling of photonic crystal lasers

Laser action in photonic crystal structures has been theoretically studied and centered on the estimation of the output parameters e.g., [30],[31], and models describing light generation processes e.g.,[32]-[34]. The most general semi-classical model of light generation in photonic structures is presented in [34], where the description of one-, two-, and three-dimensional structures is given.

Theoretical analysis of photonic crystal lasers based on two-dimensional plane wave expansion method (PWEM) [15],[35]-[37] and finite difference time domain method (FDTD) [35],[38]-[40] confirm experimental results. Nevertheless these methods suffer from important disadvantages, i.e., plane wave method gives a good approximation for infinite structures, whereas finite difference time domain method is suited for structures with only a few periods and consumes huge computer resources for the analysis of real photonic structures. Therefore these methods are not very convenient for design and optimization of actual photonic crystal lasers. Hence, different, less complicated methods of analysis of two-dimensional photonic crystal lasers are being developed. These methods are meant to effectively support the design process of such lasers. They are based on a coupled-wave theory (CWT) [15],[41] and focused on square and triangular lattice photonic crystals e.g., [32],[33],[42]-[48]. Most of the works e.g., [32],[42]-[46] contain a mathematical description and numerical results of the threshold analysis of two-dimensional (2-D) square and triangular lattice photonic crystal laser with TM and TE polarization. They introduce general coupled mode relations for a threshold gain, a Bragg frequency deviation and field distributions, and give calculation results for some specific values of coupling coefficients. Further, in [42] the effect of boundary reflections has been investigated, and it has been shown that the mode properties can be adjusted by changing refractive index or boundary conditions. In, [46], the achievements of these works were summarized and supplemented with the analysis for the wide range of coupling coefficient. These studies concerned structures which were infinite in the direction normal to the 2D PC plane. This approach was improved and presented in [47], where a three-dimensional (3D) couple wave model was shown. This theory addressed some key issues in a modeling of threshold operation of surface-emitting-type PC lasers, i.e. the surface emission and the in-plane higher-order coupling effects. It has also been further developed to incorporate finite-size effects, and presented in [48]. Some other works such as for example [33],[44],[45],[49] present an above threshold analysis of 2D PC lasers. They illustrate gain saturation effect and describe the impact of structure parameters on the system efficiency.

In all of the cited works non have given much attention to simultaneous index and gain coupling. Thus in addition to the works already mentioned, this Chapter aims to remind crucial points of CMT and to show 2D coupled-wave analysis for structures with gain and index coupling. The study includes square and triangular lattice structures with TE and TM polarization of light.

The subsequent parts of this chapter include structure definition (Section 2), threshold analysis (Section 3), where 2D coupled-wave theory is reminded (Section 3.1), coupled-wave equations are shown (Section 3.2), and numerical analysis is performed (Section 3.3). The perspectives are sketched in Section 4, and finally conclusions are given in Section 5.

2. Structure definition

The Chapter describes two-dimensional photonic crystals which are characterized by the relative permittivity ε and gain α. Both parameters depend on the two-dimensional spatial structure of the medium. The cross-sections of discussed structures are schematically shown in Figure 1.

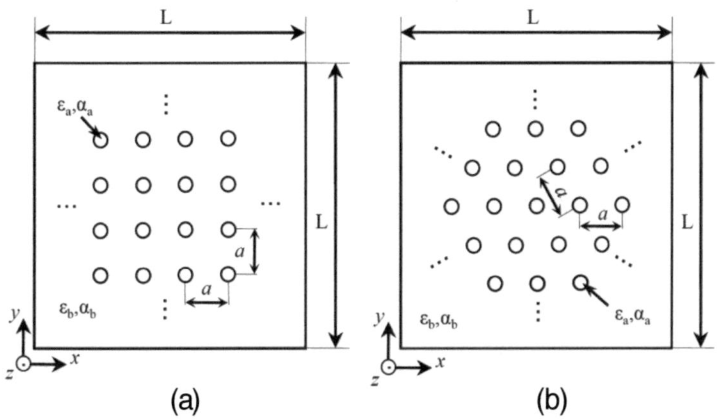

Figure 1. a) Square and b) triangular lattice photonic structures cross sections. (pairs ε_a, α_a and ε_b, α_b are relative permittivity and gain of rods and background material, respectively, *a* - lattice constant, L - cavity length).

From this point on, since photonic structures resemble the microscopic nature of crystals, a crystallography terminology will be used, see e.g., [49]. Throughout this Chapter only 2-D photonic crystals with a square, and hexagonal (also referred to as triangular) symmetry will be discussed, as it is depicted in Figure 1. The periodic pattern is created by cylinders called rods or holes. The structures in Figure 1 a) and b) are constrained in the xy plane by the square region of length L, and are assumed to be uniform and much larger than the wavelength in the z direction. The permittivity and gain of the rods and background material are represented by ε_a, α_a and ε_b, α_b, respectively. The number of periods in the xy plane is finite, but large enough to be expanded in Fourier series with small error. Schemes in Fig. 1 a) and 1 b) illustrate two spatial distributions of rods for two-dimensional photonic crystal, respectively, with square and triangular lattice.

Schemes in Figure 2 a) and Figure 3 a) show a view of photonic crystal cross sections in xy plane with cylinders arranged in square or triangular lattice with period a, and with depicted primitive vectors $\mathbf{a}_1$ and $\mathbf{a}_2$.

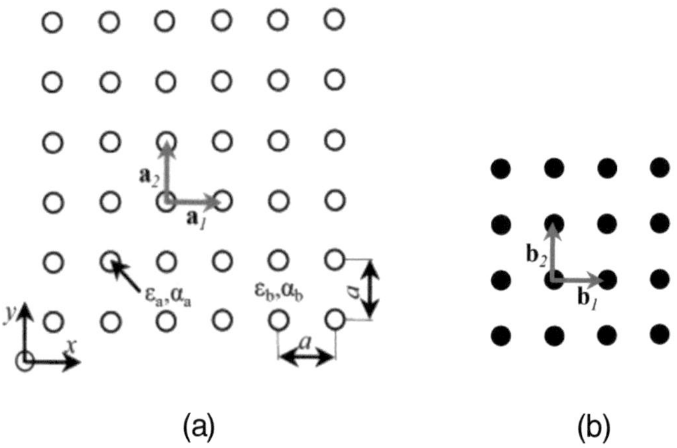

Figure 2. A schematic view of a) a square lattice photonic crystal with primitive vectors; and b) its representation in reciprocal space with reciprocal primitive vectors.

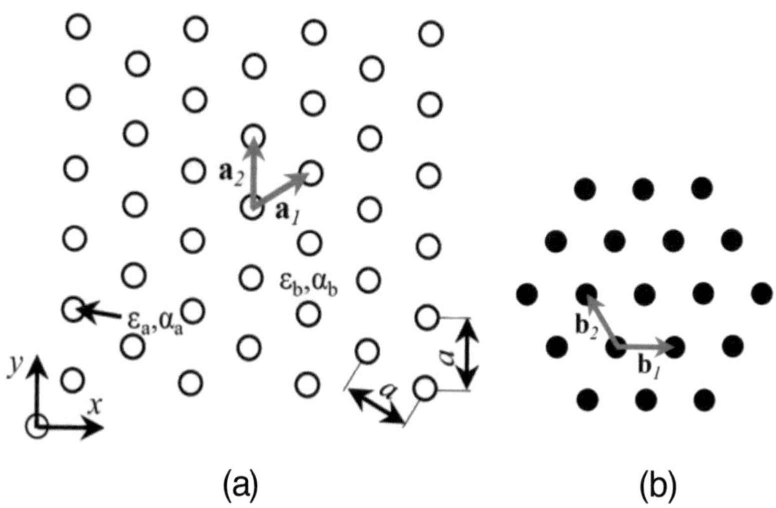

Figure 3. A schematic view of a) a triangular lattice photonic crystal with primitive vectors; and b) its representation in reciprocal space with reciprocal primitive vectors.

Figure 2 b) andFigure 3 b) show the reciprocal lattices corresponding, respectively, to the real square and triangular lattices. In the described case, the nodes of a two-dimensional structure can be expressed by (e.g., see [50])

$$\mathbf{x}_{||}(l)=l_1\mathbf{a}_1+l_2\mathbf{a}_2, \tag{1}$$

where $\mathbf{a}_1$ and $\mathbf{a}_2$ are primitive vectors, l_1 and l_2 are arbitrary integers, $x_{||}$ specifies the placement on the plane, $x_{||}=\hat{\mathbf{x}}x+\hat{y}y$, where $\hat{\mathbf{x}}$ and $\hat{\mathbf{y}}$ are unit vectors along x and y axis, respectively. Primitive vectors for square lattice are described by the expressions:$\mathbf{a}_1=(a,\ 0)$, $\mathbf{a}_2=(0,\ a)$ and for the triangular lattice: $\mathbf{a}_1=(\sqrt{3}a/2,\ a/2)$, $\mathbf{a}_2=(0,\ a)$.

In general, the reciprocal vectors can be written in the following form:

$$\mathbf{G}(h)=h_1\mathbf{b}_1+h_2\mathbf{b}_2 \tag{2}$$

where h_1 and h_2 are arbitrary integers, $\mathbf{b}_1$ and $\mathbf{b}_2$ are the primitive vectors of the two-dimensional reciprocal space, which are expressed by the following equations:

$$\mathbf{b}_1=\frac{2\pi}{a_c}\left(a_y^2,\ -a_x^2\right),\ \ \mathbf{b}_2=\frac{2\pi}{a_c}\left(-a_y^1,\ a_x^1\right) \tag{3}$$

where $a_j^{(i)}$ is the j-th Cartesian component (x or y) of the $\mathbf{a}_i$ vector ($i=1$ or 2) (e.g., see [32]).

The areas of primitive cells are $a_c=|\mathbf{a}_1\times\mathbf{a}_2|=a^2$ and $a_c=|\mathbf{a}_1\times\mathbf{a}_2|=\sqrt{3}a^2/2$ in case of square and triangular lattices, respectively.

Using Equations (3) and the expressions for square and triangular lattice primitive vectors and primitive cell areas the reciprocal primitive vectors are described by the following formulas:

$$\mathbf{b}_1=(2\pi/a,\ 0),\ \ \mathbf{b}_2=(0,2\pi/a)\ \text{– square lattice,} \tag{4}$$

and

$$\mathbf{b}_1=\left(4\pi/\sqrt{3}a,\ 0\right),\ \ \mathbf{b}_2=\left(-2\pi/\sqrt{3}a,\ 2\pi/a\right)\ \text{- triangular lattice.} \tag{5}$$

The spatial arrangement of periodic rods for the infinite two-dimensional square or triangular lattice can be expressed by the function:

$$\varepsilon(\mathbf{x}_{||})=\varepsilon_b+(\varepsilon_a-\varepsilon_b)\sum_l S(\mathbf{x}_{||}-\mathbf{x}_{||}(l)), \tag{6}$$

in terms of relative permittivity, and by

$$\alpha(\mathbf{x}_{||})=\alpha_b+(\alpha_a-\alpha_b)\sum_l S(\mathbf{x}_{||}-\mathbf{x}_{||}(l)), \tag{7}$$

in terms of gain. In Equations (6) and (7), function S

$$S(\mathbf{x}_{\|})=\begin{cases}1 & \text{for } \mathbf{x}_{\|}\in O\\ 0 & \text{for } \mathbf{x}_{\|}\notin O\end{cases}$$

specifies the location of rods in the structure, O is the area of the xy plane defined by the cross section of the rod, which symmetry axis intersects the plane at the point $\mathbf{x}_{\|}=0$.

In the next section an analysis based on the coupled mode theory is shown. It is conducted in the frequency domain, thus the relative permittivity as well as gain have to be Fourier transformed to fit reciprocal space [37],[51]. Functions (6) and (7) are now, respectively, written in the following form

$$\varepsilon(\mathbf{G})=\begin{cases}\varepsilon_a f+\varepsilon_b(1-f), & \mathbf{G}_{\|}=0\\ (\varepsilon_a-\varepsilon_b)f\dfrac{2J_1(\mathbf{G}_{\|}R)}{\mathbf{G}_{\|}R}, & \mathbf{G}_{\|}\neq 0\end{cases} \tag{8}$$

and

$$\alpha(\mathbf{G})=\begin{cases}\alpha_a f+\alpha_b(1-f), & \mathbf{G}_{\|}=0\\ (\alpha_a-\alpha_b)f\dfrac{2J_1(\mathbf{G}_{\|}R)}{\mathbf{G}_{\|}R}, & \mathbf{G}_{\|}\neq 0\end{cases} \tag{9}$$

where

$f=\pi r^2/a^2$- square lattice filling factor, $f=(2\pi/\sqrt{3})r^2/a^2$ - triangular lattice filling factor, r - rod radius, J_1 - Bessel function of the first kind.

In next parts of this Chapter four different cases are analyzed. Two of them are dedicated to the square lattice cavities with TE and TM polarization, and two remaining to the triangular lattice structures also with TE and TM polarization. For the purpose of this work, it is assumed that there is no gain in the background material, i.e., $\alpha_b=0$, but there is a gain in the rods $\alpha_a\neq 0$. The structure where $\alpha_a=\alpha_b$ will be referred to as Index Coupled, and where $\alpha_a\neq\alpha_b$ as Index and Gain Coupled.

The threshold analysis of the photonic crystal laser operation for the defined structures is shown in the next section.

3. A threshold analysis

3.1. 2D Coupled-wave model for 2D PC cavity

The electromagnetic wave behavior in the two-dimensional periodic system can be described by the set of scalar wave equations. Depending on the polarization of light it is easier to choose one specific field component, since then the set of equations may be reduced to a single one.

Thus, the scalar wave equations for the electric and magnetic fields E_z and H_z, respectively, are written in the following forms [37],[51]:

$$\frac{\partial^2 E_z}{\partial x^2} + \frac{\partial^2 E_z}{\partial y^2} + k^2 E_z = 0, \tag{10}$$

for TM polarization, and

$$\frac{\partial}{\partial x}\left\{\frac{1}{k^2}\frac{\partial}{\partial x}H_z\right\} + \frac{\partial}{\partial y}\left\{\frac{1}{k^2}\frac{\partial}{\partial y}H_z\right\} + H_z = 0, \tag{11}$$

for TE polarization.

In Equations (10) and (11) the constant k is given, correspondingly by [42]

$$k^2 = \beta^2 + 2i\alpha\beta + 2\beta\sum_{\mathbf{G}\neq 0}\kappa(\mathbf{G})\exp\left(i(\mathbf{G}\cdot\mathbf{r})\right), \tag{12}$$

and [32]

$$\frac{1}{k^2} = \frac{1}{\beta^4}\left(\beta^2 - i2\alpha\beta + 2\beta\sum_{\mathbf{G}\neq 0}\kappa(\mathbf{G})\exp\left(i(\mathbf{G}\cdot\mathbf{r})\right)\right). \tag{13}$$

In the expressions for k^2 and k^{-2}, β equals to $2\pi\varepsilon_0^{1/2}/\lambda$, where $\varepsilon_0=\varepsilon(\mathbf{G}=0)$ is the averaged dielectric permittivity ($\varepsilon_0^{1/2}$ corresponds to averaged refractive index n), α is an averaged gain in the medium, $\kappa(\mathbf{G})$ is the coupling constant, λ is the Bragg wavelength. Here, the reciprocal lattice vector (Equation (2)) is expressed by $\mathbf{G}=(m\mathbf{b}_1, n\mathbf{b}_2)$, where m and n are arbitrary integers, $\mathbf{b}_1$ and $\mathbf{b}_2$ depend on the structure symmetry and are written in the following forms $\mathbf{b}_1=\left(\beta_0^s, 0\right)$ and $\mathbf{b}_2=\left(0, \beta_0^s\right)$ for square lattice, and $\mathbf{b}_1=\left(\beta_0^t, 0\right)$ and $\mathbf{b}_2=\left(-\beta_0^t/2, \sqrt{3}\beta_0^t/2\right)$ for triangular lattice structure, $\beta_0^2=2\pi/a$ and $\beta_0^t=4\pi/\sqrt{3}a$. In the derivation of Equations (12) and (13) the following assumptions were set: $\alpha \ll \beta \equiv 2\pi\varepsilon_0^{1/2}/\lambda$, $\varepsilon_{\mathbf{G}\neq 0} \ll \varepsilon_0$, and $\varepsilon_{\mathbf{G}} \ll \beta$, e.g., see [42].

The periodic variation in the refractive index and gain is included as a small perturbation and appears in as the coupling constant $\kappa(\mathbf{G})$ of the form:

$$\kappa(\mathbf{G}) = -\frac{\pi}{\lambda\varepsilon_0^{1/2}}\varepsilon(\mathbf{G}) \pm i\frac{\alpha(\mathbf{G})}{2}. \tag{14}$$

In the above equation, plus sign refers to TM polarization (Equation (12)), while minus sign refers to TE polarization (Equation (13)). For the simplicity, it is set that $\eta(\mathbf{G}) = -\pi\varepsilon\left(\mathbf{G}\right)/\lambda\varepsilon_0^{1/2}$ and $\alpha(\mathbf{G})=\alpha(\mathbf{G})/2$, and Equation (14) is rewritten in the following form:

$$\kappa(\mathbf{G}) = \eta(\mathbf{G}) \pm i\alpha(\mathbf{G}). \tag{15}$$

In the two-dimensional system which is not confined in the third direction, in the vicinity of the Bragg wavelength only some of the diffraction orders contribute in a significant way. In general, a periodic perturbation produces an infinite set of diffraction orders. Keeping this in mind, the Bragg frequency orders have to be cautiously chosen. The Bragg frequency corresponding to the Γ (e.g., see [42]) is chosen for the purpose of this work, and the most significantly contributing coupling constants are expressed as follows:

$$\begin{aligned} \kappa_1 &= \eta(\mathbf{G}) \pm \alpha(\mathbf{G}) \Big|_{|\mathbf{G}|=\beta_0^{s,t}} \\ \kappa_2 &= \eta(\mathbf{G}) \pm \alpha(\mathbf{G}) \Big|_{|\mathbf{G}|=\sqrt{3}\beta_0^{s,t}} \\ \kappa_3 &= \eta(\mathbf{G}) \pm \alpha(\mathbf{G}) \Big|_{|\mathbf{G}|=2\beta_0^{s,t}} \end{aligned} \tag{16}$$

In Equations (10) and (11) electric and magnetic fields for the infinite periodic structure are given by the Bloch modes, [15],[37]:

$$E_z(\mathbf{r}) = \sum_{\mathbf{G}} e(\mathbf{G}) \exp\left(i(\mathbf{k}+\mathbf{G}) \cdot \mathbf{r}\right), \tag{17}$$

and

$$H_z(\mathbf{r}) = \sum_{\mathbf{G}} h(\mathbf{G}) \exp\left(i(\mathbf{k}+\mathbf{G}) \cdot \mathbf{r}\right), \tag{18}$$

where the functions $e(\mathbf{G})$ and $h(\mathbf{G})$ correspond to plane wave amplitudes, and the wave vector is denoted by k. In the first Brillouin zone at the Γ point the wave vector vanishes $\mathbf{k}=0$, see e.g., [41].

In a finite two-dimensional structure, the amplitude of each plane wave is not constant, so $e(\mathbf{G})$ and $h(\mathbf{G})$ become functions of space. At the Γ point, only the amplitudes ($e(\mathbf{G})$, $h(\mathbf{G})$) which are meant to be significant are considered, i.e., in most cases with $|\mathbf{G}|=\beta_0^{s,t}$, except for square lattice with TE polarization where additional $h(\mathbf{G})$ amplitudes with $|\mathbf{G}|=\sqrt{2}\beta_0^{s}$ have to be included [41]. The contributions of other waves of higher order in the Bloch mode are considered to be negligible. In general, where for example there is a three-dimensional confinement, this assumption have to be reconsidered.

3.2. Coupled-wave equations

3.2.1. Square lattice – TM polarization

For square lattice photonic crystal cavity in the case of TM polarization, it is assumed that at the center point of the Brillouin zone the most significant contribution to coupling is given by the electric waves which fulfill the condition $|\mathbf{G}|=\beta_0^{s}$. Therefore in the following derivation all higher order electric wave expansion coefficients ($|G|\geq\sqrt{2}\beta_0^{s}$) are neglected. Four basic waves most significantly contributing to coupling are depicted in Figure 4.

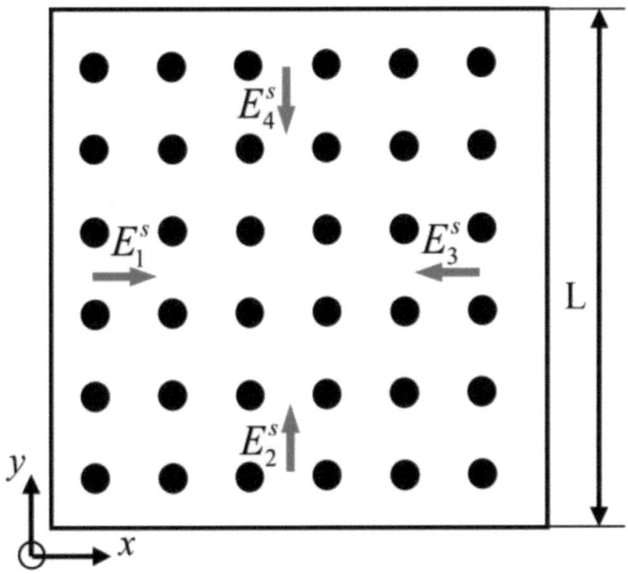

Figure 4. Schematic cross section of square lattice photonic crystal laser active region with the four basic waves involved in coupling for TM polarization.

Equation (17) in general describes infinite structures. It is possible to take into account the fact that the structure is finite by using the space dependent amplitudes, e.g., [42]. Thus, the electric field given by Equation (17) in the finite periodic structure can be expressed in the following way:

$$E_z = E_1^s(x, y)e^{-i\beta_0^s x} + E_3^s(x, y)e^{i\beta_0^s x} + E_2^s(x, y)e^{-i\beta_0^s y} + E_4^s(x, y)e^{i\beta_0^s y}. \tag{19}$$

In Equation (19) E_i^s, i=1..4 are the four basic electric field amplitudes propagating in four directions +x, -x, +y, -y. These amplitudes correspond to $e(\mathbf{G})$ (Equation (17)) satisfying the condition: $|\mathbf{G}| = \beta_0^s$. In further analysis, the space dependence is omitted for the simplicity of notation.

Using derived earlier reciprocal lattice vectors and Fourier expansions of spatial dependences of the square lattice PC with circular rods, the coupling coefficients $\kappa(\mathbf{G})$ (Equation (16)) can be written as:

$$\kappa_1 = \left(-\frac{\pi(\varepsilon_a - \varepsilon_b)}{a(\varepsilon_a f + \varepsilon_b(1 - f))} + i\frac{\alpha_a}{2}\right)\frac{2f\,J_1\left(2\sqrt{\pi f}\right)}{2\sqrt{\pi f}} = \eta_1 + i\alpha_1, \tag{20}$$

$$\kappa_2 = \left(-\frac{\pi(\varepsilon_a - \varepsilon_b)}{a(\varepsilon_a f + \varepsilon_b(1 - f))} + i\frac{\alpha_a}{2} \right) \frac{2f J_1\left(2\sqrt{2\pi f}\right)}{2\sqrt{2\pi f}} = \eta_2 + i\alpha_2, \tag{21}$$

$$\kappa_3 = \left(-\frac{\pi(\varepsilon_a - \varepsilon_b)}{a(\varepsilon_a f + \varepsilon_b(1 - f))} + i\frac{\alpha_a}{2} \right) \frac{2f J_1\left(4\sqrt{\pi f}\right)}{4\sqrt{\pi f}} = \eta_3 + i\alpha_3. \tag{22}$$

Combining Equations (12) and (19) with Equation (10), and assuming the slow varying electromagnetic field, one can get the set of coupled mode equations [42]:

$$-\frac{\partial}{\partial x} E_1^s + (\alpha - i\delta) E_1^s = (i\eta_3 - \alpha_3) E_3^s + (i\eta_2 - \alpha_2)\left(E_2^s + E_4^2\right), \tag{23}$$

$$\frac{\partial}{\partial x} E_3^s + (\alpha - i\delta) E_3^s = (i\eta_3 - \alpha_3) E_1^s + (i\eta_2 - \alpha_2)\left(E_2^s + E_4^s\right), \tag{24}$$

$$-\frac{\partial}{\partial y} E_2^s + (\alpha - i\delta) E_2^s = (i\eta_3 - \alpha_3) E_4^s + (i\eta_2 - \alpha_2)\left(E_1^s + E_3^s\right), \tag{25}$$

$$\frac{\partial}{\partial y} E_4^s + (\alpha - i\delta) E_4^s = (i\eta_3 - \alpha_3) E_2^s + (i\eta_2 - \alpha_2)\left(E_1^s + E_3^s\right), \tag{26}$$

where

$$\delta = \left(\beta^2 - \beta_0^{s2}\right) / 2\beta \approx \beta - \beta_0^s, \tag{27}$$

is the Bragg frequency deviation. Coupling coefficients κ_2 and κ_3 are expressed by Equations (21) and (22). The κ_2 coefficient is responsible for orthogonal coupling (e.g., the coupling of E_1^s to E_2^s and E_4^s), and κ_2 corresponds to backward coupling (e.g., the coupling of E_1^s to E_3^s). Solution of Equations (23)-(26) for the following boundary conditions:

$$E_1^s\left(-\frac{L}{2}, y\right) = E_3^s\left(\frac{L}{2}, y\right) = 0, \quad E_2^s\left(x, -\frac{L}{2}\right) = E_4^s\left(x, \frac{L}{2}\right) = 0, \tag{28}$$

defines eigenmodes of the photonic structure. The analysis of this solution is given in Section 3.3.

3.2.2. Square lattice – TE polarization

In the 2D square lattice PC-like resonator with TE polarization the coupling process in the most significant way involves magnetic waves satisfying following conditions: $|\mathbf{G}| = \beta_0$ and $|\mathbf{G}| = \sqrt{2}\beta_0$, [32]. In the presented analysis the higher order Bloch modes are neglected. Eight basic waves fulfilling the specified conditions are depicted in Figure 5.

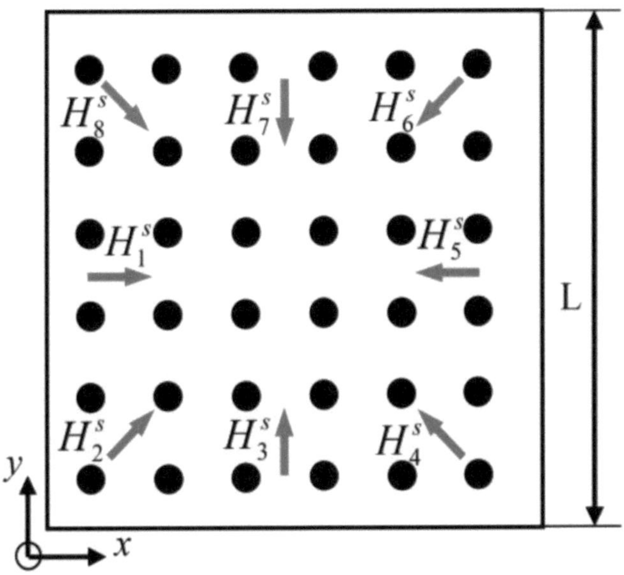

Figure 5. Schematic cross section of square lattice photonic crystal laser active region with the eight basic waves involved in coupling for TE polarization.

Similarly as it was stated in the case of TM polarization, the equation for magnetic field (18) describes modes for infinite structure, and the finite dimensions of the structure are introduced by spatial dependence of magnetic field amplitudes [32]. Thus, the magnetic field (18) is written in the following form:

$$\begin{aligned}H_z &= H_1^s(x,y)e^{-i\beta_0^s x} + H_5^s(x,y)e^{i\beta_0^s x} + H_3^s(x,y)e^{-i\beta_0^s y} + H_7^s(x,y)e^{i\beta_0^s y} \\ &+H_2^s(x,y)e^{-i\beta_0^s x - i\beta_0^s y} + H_4^s(x,y)e^{i\beta_0^s x - i\beta_0^s y} + H_6^s(x,y)e^{i\beta_0^s x + i\beta_0^s y} \\ &+H_8^s(x,y)e^{(-i\beta_0^s x + i\beta_0^s y)}.\end{aligned} \quad (29)$$

In Equation (29) H_i^s, i=1..8 are the basic magnetic field amplitudes of waves propagating in directions schematically shown in Figure 5. These amplitudes correspond to h (**G**) in Equation (18), where $|\mathbf{G}| = \beta_0$ and $|\mathbf{G}| = \sqrt{2}\beta_0$. Joining Equations (13), (29), and (11), and assuming slowly varying amplitudes, the coupled wave equations for TE modes in square lattice PC are obtained [32]:

$$-\frac{\partial}{\partial x}H_1^s + (\alpha - i\delta)H_1^s = (i\eta_3 + \alpha_3)H_5^s + i\frac{2(\eta_1 - i\alpha_1)^2}{\beta_0^s}\left(2H_1^s + H_3^s + H_7^s\right), \quad (30)$$

$$\frac{\partial}{\partial x}H_5^s + (\alpha - i\delta)H_5^s = (i\eta_3 + \alpha_3)H_1^s + i\frac{2(\eta_1 - i\alpha_1)^2}{\beta_0^s}\left(2H_5^s + H_3^s + H_7^s\right), \tag{31}$$

$$-\frac{\partial}{\partial x}H_3^s + (\alpha - i\delta)H_3^s = (i\eta_3 + \alpha_3)H_7^s + i\frac{2(\eta_1 - i\alpha_1)^2}{\beta_0^s}\left(2H_3^s + H_1^s + H_5^s\right), \tag{32}$$

$$\frac{\partial}{\partial x}H_7^s + (\alpha - i\delta)H_7^s = (i\eta_3 + \alpha_3)H_3^s + i\frac{2(\eta_1 - i\alpha_1)^2}{\beta_0^s}\left(2H_7^s + H_1^s + H_5^s\right). \tag{33}$$

In Equations (30)-(33), the spatial dependence of four magnetic field components H_i^s, i=2,4,6,8 was neglected, and it was assumed that $\alpha \ll \delta$. These steps let to formulate not eight but four partial differential equations (for details see [32] or [46]). The Bragg frequency deviation δ is given by (27). The coupling coefficients κ_1, κ_2, and κ_3 defined by Equations (16) are expressed by [32],[43]:

$$\kappa_1 = \left(-\frac{\pi(\varepsilon_a - \varepsilon_b)}{a(\varepsilon_a f + \varepsilon_b(1 - f))} - i\frac{\alpha_a}{2}\right)\frac{2f\,J_1\left(2\sqrt{\pi f}\right)}{2\sqrt{\pi f}} = \eta_1 - i\alpha_1, \tag{34}$$

$$\kappa_2 = \left(-\frac{\pi(\varepsilon_a - \varepsilon_b)}{a(\varepsilon_a f + \varepsilon_b(1 - f))} - i\frac{\alpha_a}{2}\right)\frac{2f\,J_1\left(2\sqrt{2\pi f}\right)}{2\sqrt{2\pi f}} = \eta_2 - i\alpha_2, \tag{35}$$

$$\kappa_3 = \left(-\frac{\pi(\varepsilon_a - \varepsilon_b)}{a(\varepsilon_a f + \varepsilon_b(1 - f))} - i\frac{\alpha_a}{2}\right)\frac{2f\,J_1\left(4\sqrt{\pi f}\right)}{4\sqrt{\pi f}} = \eta_3 - i\alpha_3. \tag{36}$$

In contrast to TM polarization, in Equations (30)-(33), the coupling coefficient responsible for coupling in perpendicular direction κ_2 vanishes. The coupling coefficient κ_3 has the same meaning as described in the previous (TM) case, whereas the coupling coefficient κ_1 describes the coupling of e.g., waves H_1^s, H_2^s, and H_8^s. Solution of Equations (30)-(33) for the following boundary conditions:

$$H_1^s\left(-\frac{L}{2}, y\right) = H_5^s\left(\frac{L}{2}, y\right) = 0, \quad H_3^s\left(x, -\frac{L}{2}\right) = H_7^s\left(x, \frac{L}{2}\right) = 0, \tag{37}$$

defines structure eigenmodes at lasing threshold i.e. in the linear case.

3.2.3. Triangular lattice - TM polarization

In the simple approximation scenario the coupling process in the triangular lattice photonic crystal cavity with TM polarization involves waves satisfying following condition: $|\mathbf{G}| = \beta_0$, and neglects higher order Bloch modes [43],[44]. There are six waves satisfying this condition and simultaneously most significantly contributing to coupling, they are depicted in Figure 6.

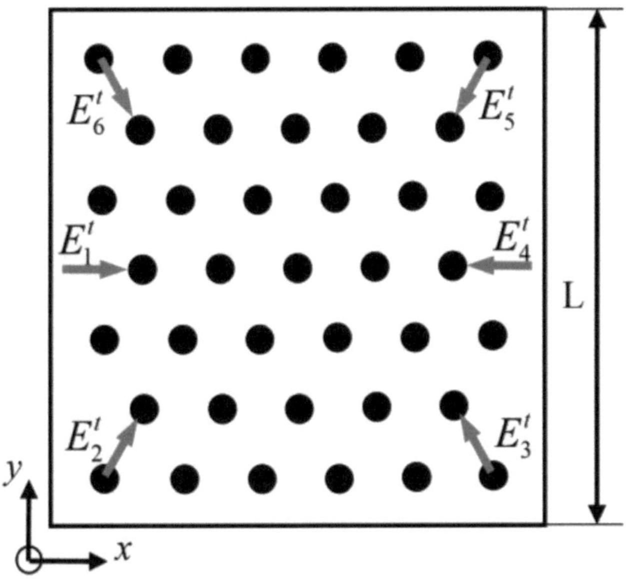

Figure 6. A schematic cross section of a triangular lattice photonic crystal laser active region with the six basic waves involved in the coupling for TM polarization.

The space dependent amplitudes for electric field $e(\mathbf{G})$ in triangular lattice PC cavity are written in the following form [44]:

$$E_z = E_1^t(x,y)e^{-i\beta_0^t x} + E_2^t(x,y)e^{-i\frac{\beta_0^t}{2}x - i\frac{\sqrt{3}\beta_0^t}{2}y} + E_3^t(x,y)e^{i\frac{\beta_0^t}{2}x - i\frac{\sqrt{3}\beta_0^t}{2}y}$$
$$+E_4^t(x,y)e^{i\beta_0^t x} + E_5^t(x,y)e^{i\frac{\beta_0^t}{2}x + i\frac{\sqrt{3}\beta_0^t}{2}y} + E_6^t(x,y)e^{-i\frac{\beta_0^t}{2}x + i\frac{\sqrt{3}\beta_0^t}{2}y} \quad (38)$$

In Equation (38), E_i^t, i=1..6, are the six electric field amplitudes propagating in the symmetry directions, Figure 6. Combining Equations (12), (38) with Equation (10), and assuming slowly varying amplitudes, the coupled wave equations for TM modes in triangular lattice PC are obtained:

$$-\frac{\partial}{\partial x}E_1^t + (\alpha - i\delta)E_1^t = (i\eta_1 - \alpha_1)(E_2^t + E_6^t) + (i\eta_2 - \alpha_2)(E_3^t + E_5^t) + (i\eta_3 - \alpha_3)E_4^t, \quad (39)$$

$$-\frac{1}{2}\frac{\partial}{\partial x}E_2^t - \frac{\sqrt{3}}{2}\frac{\partial}{\partial y}E_2^t + (\alpha - i\delta)E_2^t = (i\eta_1 - \alpha_1)(E_1^t + E_3^t) + (i\eta_2 - \alpha_2)(E_4^t + E_6^t) + (i\eta_3 - \alpha_3)E_5^t, \quad (40)$$

$$\frac{1}{2}\frac{\partial}{\partial x}E_3^t - \frac{\sqrt{3}}{2}\frac{\partial}{\partial y}E_3^t + (\alpha - i\delta)E_3^t = (i\eta_1 - \alpha_1)(E_2^t + E_4^t) + (i\eta_2 - \alpha_2)(E_1^t + E_5^t) + (i\eta_3 - \alpha_3)E_6^t, \tag{41}$$

$$\frac{\partial}{\partial x}E_4^t + (\alpha - i\delta)E_4^t = (i\eta_1 - \alpha_1)(E_3^t + E_5^t) + (i\eta_2 - \alpha_2)(E_2^t + E_6^t) + (i\eta_3 - \alpha_3)E_1^t, \tag{42}$$

$$\frac{1}{2}\frac{\partial}{\partial x}E_5^t + \frac{\sqrt{3}}{2}\frac{\partial}{\partial y}E_5^t + (\alpha - i\delta)E_5^t = (i\eta_1 - \alpha_1)(E_4^t + E_6^t) + (i\eta_2 - \alpha_2)(E_1^t + E_3^t) + (i\eta_3 - \alpha_3)E_2^t, \tag{43}$$

$$-\frac{1}{2}\frac{\partial}{\partial x}E_6^t + \frac{\sqrt{3}}{2}\frac{\partial}{\partial y}E_6^t + (\alpha - i\delta)E_6^t = (i\eta_1 - \alpha_1)(E_1^t + E_5^t) + (i\eta_2 - \alpha_2)(E_2^t + E_4^t) + (i\eta_3 - \alpha_3)E_3^t, \tag{44}$$

In Equations (39)-(44), like in the case of square lattice, δ is the Bragg frequency deviation (17), while κ_1, κ_2, and κ_3 are the coupling coefficients, which are defined by the following relations [44]:

$$\kappa_1 = \left(-\frac{\pi(\varepsilon_a - \varepsilon_b)}{a(\varepsilon_a f + \varepsilon_b(1 - f))} + i\frac{\alpha_a}{2}\right)\frac{2f\,J_1\left(\sqrt{8\pi f / \sqrt{3}}\right)}{\sqrt{8\pi f / \sqrt{3}}} = \eta_1 + i\alpha_1, \tag{45}$$

$$\kappa_2 = \left(-\frac{\pi(\varepsilon_a - \varepsilon_b)}{a(\varepsilon_a f + \varepsilon_b(1 - f))} + i\frac{\alpha_a}{2}\right)\frac{2f\,J_1\left(\sqrt{8\sqrt{3}\pi f}\right)}{\sqrt{8\sqrt{3}\pi f}} = \eta_2 + i\alpha_2, \tag{46}$$

$$\kappa_2 = \left(-\frac{\pi(\varepsilon_a - \varepsilon_b)}{a(\varepsilon_a f + \varepsilon_b(1 - f))} + i\frac{\alpha_a}{2}\right)\frac{2f\,J_1\left(2\sqrt{8\pi f / \sqrt{3}}\right)}{2\sqrt{8\pi f / \sqrt{3}}} = \eta_3 + i\alpha_3. \tag{47}$$

These coefficients describe strength and direction of the coupling of the waves, e.g., the coupling of E_1^t and E_4^t is described by κ_3, the coupling of E_1^t, E_2^t, and E_6^t by κ_1, and the coupling of E_1^t, E_3^t, and E_5^t by κ_2. Solution of Equations (39)-(44) for the boundary conditions:

$$\begin{aligned}
&E_1^t\left(-\frac{L}{2}, y\right) = 0,\; E_2^t\left(-\frac{L}{2}, y\right) = E_2^t\left(x, -\frac{L}{2}\right) = 0,\; E_3^t\left(\frac{L}{2}, y\right) = E_3^t\left(x, -\frac{L}{2}\right) = 0,\\
&E_4^t\left(\frac{L}{2}, y\right) = 0,\; E_5^t\left(\frac{L}{2}, y\right) = E_5^t\left(x, \frac{L}{2}\right) = 0,\; E_6^t\left(-\frac{L}{2}, y\right) = E_6^t\left(x, \frac{L}{2}\right) = 0
\end{aligned} \tag{48}$$

defines structure eigenmodes at lasing threshold.

3.2.4. Triangular lattice – TE polarization

The simple approximation of coupling process in 2D triangular lattice PC with TE polarization includes waves satisfying the same condition as it was shown for TM polarization, i.e., $|\mathbf{G}| = \beta_0$, [43]. Six waves satisfying this condition are depicted in Figure 7.

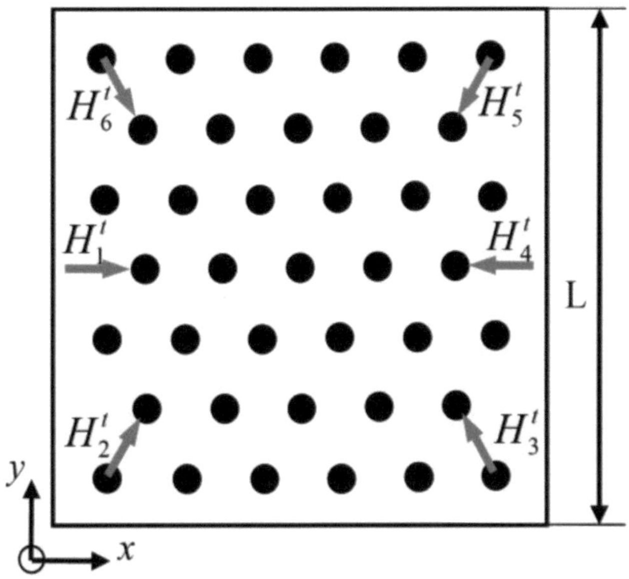

Figure 7. A schematic cross section of a triangular lattice photonic crystal laser active region with the six basic waves involved in the coupling for TE polarization are shown.

The magnetic field amplitudes h (**G**) in the triangular lattice PC cavity are written as follows [43]:

$$\begin{aligned} H_z = {} & H_1^t(x,y)e^{-i\beta_0^t x} + H_2^t(x,y)e^{-i\frac{\beta_0^t}{2}x - i\frac{\sqrt{3}\beta_0^t}{2}y} + H_3^t(x,y)e^{i\frac{\beta_0^t}{2}x - i\frac{\sqrt{3}\beta_0^t}{2}y} \\ & + H_4^t(x,y)e^{i\beta_0^t x} + H_5^t(x,y)e^{i\frac{\beta_0^t}{2}x + i\frac{\sqrt{3}\beta_0^t}{2}y} + H_6^t(x,y)e^{-i\frac{\beta_0^t}{2}x + i\frac{\sqrt{3}\beta_0^t}{2}y} \end{aligned} \tag{49}$$

In Equation (49), H_i^t, i=1..6, are the six magnetic field amplitudes propagating in the symmetry directions, Figure 7. Combining Equations (13), (49) and (11), and assuming slowly varying magnetic field amplitudes, the coupled wave equations for TE modes in triangular lattice PC are obtained:

$$-\frac{\partial}{\partial x}H_1^t + (\alpha - i\delta)H_1^t = -\frac{i\eta_1 + \alpha_1}{2}\left(H_2^t + H_6^t\right) + \frac{i\eta_2 + \alpha_2}{2}\left(H_3^t + H_5^t\right) + \frac{i\eta_3 + \alpha_3}{2}H_4^t \tag{50}$$

$$-\frac{1}{2}\frac{\partial}{\partial x}H_2^t - \frac{\sqrt{3}}{2}\frac{\partial}{\partial y}H_2^t + (\alpha - i\delta)H_2^t = -\frac{i\eta_1 + \alpha_1}{2}\left(H_1^t + H_3^t\right) + \frac{i\eta_2 + \alpha_2}{2}\left(H_4^t + H_6^t\right) + \frac{i\eta_3 + \alpha_3}{2}H_5^t \tag{51}$$

$$\frac{1}{2}\frac{\partial}{\partial x}H_3^t - \frac{\sqrt{3}}{2}\frac{\partial}{\partial y}H_3^t + (\alpha - i\delta)H_3^t = -\frac{i\eta_1 + \alpha_1}{2}\left(H_2^t + H_4^t\right) + \frac{i\eta_2 + \alpha_2}{2}\left(H_1^t + H_5^t\right) + \frac{i\eta_3 + \alpha_3}{2}H_6^t \tag{52}$$

$$\frac{\partial}{\partial x}H_4^t + (\alpha - i\delta)H_4^t = -\frac{i\eta_1 + \alpha_1}{2}\left(H_3^t + H_5^t\right) + \frac{i\eta_2 + \alpha_2}{2}\left(H_2^t + H_6^t\right) + \frac{i\eta_3 + \alpha_3}{2}H_1^t \tag{53}$$

$$\frac{1}{2}\frac{\partial}{\partial x}H_5^t + \frac{\sqrt{3}}{2}\frac{\partial}{\partial y}H_5^t + (\alpha - i\delta)H_5^t = -\frac{i\eta_1 + \alpha_1}{2}\left(H_4^t + H_6^t\right) + \frac{i\eta_2 + \alpha_2}{2}\left(H_1^t + H_3^t\right) + \frac{i\eta_3 + \alpha_3}{2}H_2^t \tag{54}$$

$$-\frac{1}{2}\frac{\partial}{\partial x}H_6^t + \frac{\sqrt{3}}{2}\frac{\partial}{\partial y}H_6^t + (\alpha - i\delta)H_6^t = -\frac{i\eta_1 + \alpha_1}{2}\left(H_1^t + H_5^t\right) + \frac{i\eta_2 + \alpha_2}{2}\left(H_2^t + H_4^t\right) + \frac{i\eta_3 + \alpha_3}{2}H_3^t \tag{55}$$

where the coupling coefficients κ_1, κ_2, and κ_3 are described by

$$\kappa_1 = \left(-\frac{\pi(\varepsilon_a - \varepsilon_b)}{a(\varepsilon_a f + \varepsilon_b(1 - f))} - i\frac{\alpha_a}{2}\right)\frac{2f\,J_1\left(\sqrt{8\pi f / \sqrt{3}}\right)}{\sqrt{8\pi f / \sqrt{3}}} = \eta_1 - i\alpha_1 \tag{56}$$

$$\kappa_2 = \left(-\frac{\pi(\varepsilon_a - \varepsilon_b)}{a(\varepsilon_a f + \varepsilon_b(1 - f))} - i\frac{\alpha_a}{2}\right)\frac{2f\,J_1\left(\sqrt{8\sqrt{3}\pi f}\right)}{\sqrt{8\sqrt{3}\pi f}} = \eta_2 - i\alpha_2 \tag{57}$$

$$\kappa_2 = \left(-\frac{\pi(\varepsilon_a - \varepsilon_b)}{a(\varepsilon_a f + \varepsilon_b(1 - f))} - i\frac{\alpha_a}{2}\right)\frac{2f\,J_1\left(2\sqrt{8\pi f / \sqrt{3}}\right)}{2\sqrt{8\pi f / \sqrt{3}}} = \eta_3 - i\alpha_3 \tag{58}$$

and have the same physical meaning like it was described in the TM polarization case. The boundary conditions for the square region of PC with triangular symmetry are written as:

$$\begin{aligned} &H_1^t\left(-\frac{L}{2}, y\right) = 0,\ H_2^t\left(-\frac{L}{2}, y\right) = H_2^t\left(x, -\frac{L}{2}\right) = 0,\ H_3^t\left(\frac{L}{2}, y\right) = H_3^t\left(x, -\frac{L}{2}\right) = 0, \\ &H_4^t\left(\frac{L}{2}, y\right) = 0,\ H_5^t\left(\frac{L}{2}, y\right) = H_5^t\left(x, \frac{L}{2}\right) = 0,\ H_6^t\left(-\frac{L}{2}, y\right) = H_6^t\left(x, \frac{L}{2}\right) = 0. \end{aligned} \tag{59}$$

3.3. Numerical analysis of the PC laser threshold operation

3.3.1. *Square lattice – TM and TE polarization*

Figure 8 shows enlarged areas of square lattice photonic crystal dispersion characteristics for the first four modes (A,B,C,D) in the vicinity of Γpoint (where the cavity finesse increases, and the active medium is used more efficiently). The dispersion curves are plotted for a) TM polarization and b) TE polarization. They have been obtained by using Plane Wave Method (PWM) [52], and they describe the infinite two-dimensional PC structures with circular holes ε_b=9.8 arranged in square lattice with background material permittivity: ε_a=12.0. The rods radius to lattice constant ratio amounts to 0.24.

In each plot, i.e., Figure 8 a) and Figure 8 b), the pairs of degenerate modes: B,C for TM polarization and C,D for TE polarization are marked. These modes have the same frequency at the Γpoint. Modes marked as A have the lowest frequency.

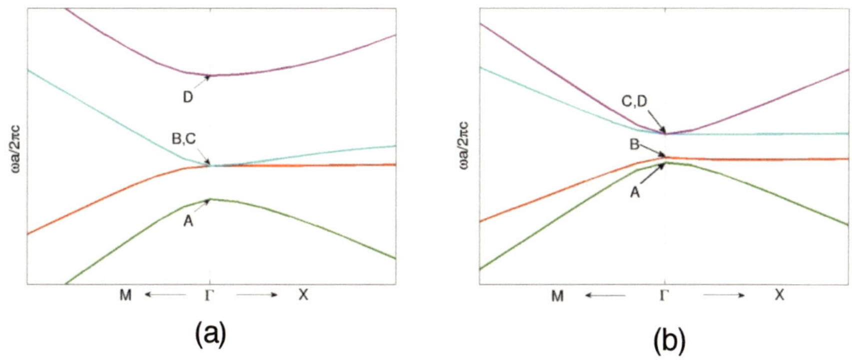

Figure 8. An enlarged area of a square lattice photonic crystal dispersion curves for the first four modes in the vicinity of Γ point. Square lattice, a) TM polarization, and b) TE polarization.

In Figure 8 each of the marked points (A,B,C,D) represents a mode, which is characterized by: Bragg frequency deviationδ, threshold gainα, and threshold field distribution. These characteristic quantities were calculated by the numerical solution of Equations (23)-(26) for TM polarization and Equations (30)-(33) for TE polarization. The similar description of modes, shown in Figure 8, where no gain coupling is considered has already been presented in [46] or [49].

In order to assign appropriate points A,B,C,D to the obtained numerical values, it was necessary to use the analytic expressions for the Bragg frequency deviation. These expressions are not affected by the gain modulation, and have the following form:

$$\delta_A = -2\kappa_2 - \kappa_3, \quad \delta_{B,C} = \kappa_3, \quad \delta_D = 2\kappa_2 - \kappa_3 \tag{60}$$

in case of TM polarization, and

$$\delta_A = -8\kappa_1^2 / \beta_0 - \kappa_3, \quad \delta_B = -\kappa_3, \quad \delta_{C,D} = -4\kappa_1^2 / \beta_0 + \kappa_3 \tag{61}$$

in case of TE polarization.

The numerical solution of Equations (23)-(26) and (30)-(33) for the wide range of coupling coefficient is divided into two stages. In the first phase the gain expansion coefficients α_1, α_2, and α_3 are neglected, and the equations and their solutions are reduced to known forms, e.g., see [46]. The second step uses the solutions obtained in first stage and iteratively solves Equations (23)-(26) and (30)-(33) for $\alpha_1 \neq 0$, $\alpha_2 \neq 0$, and $\alpha_3 \neq 0$, using the relations (20)-(22) and (34)-(36).

The obtained solutions were grouped: $\left(\left(\delta,\ \alpha,\ E_m^s\right)^j\right)_{\kappa_{3i}}$ or $\left(\left(\delta,\ \alpha,\ H_m^s\right)^j\right)_{\kappa_{3i}}$, where κ_{3i} corresponds to subsequent values of coupling coefficient for different modes *j*=A,B,C,D; m=1..4; s - denotes square lattice symmetry (here: *square*); δ and α are values of simultaneously index and gain coupled structure. Assigning numerical values of δ_j to analytical solutions (60) and (61), the mode structure of 2-D square lattice index and gain coupled PC laser with TM and TE polarization was obtained.

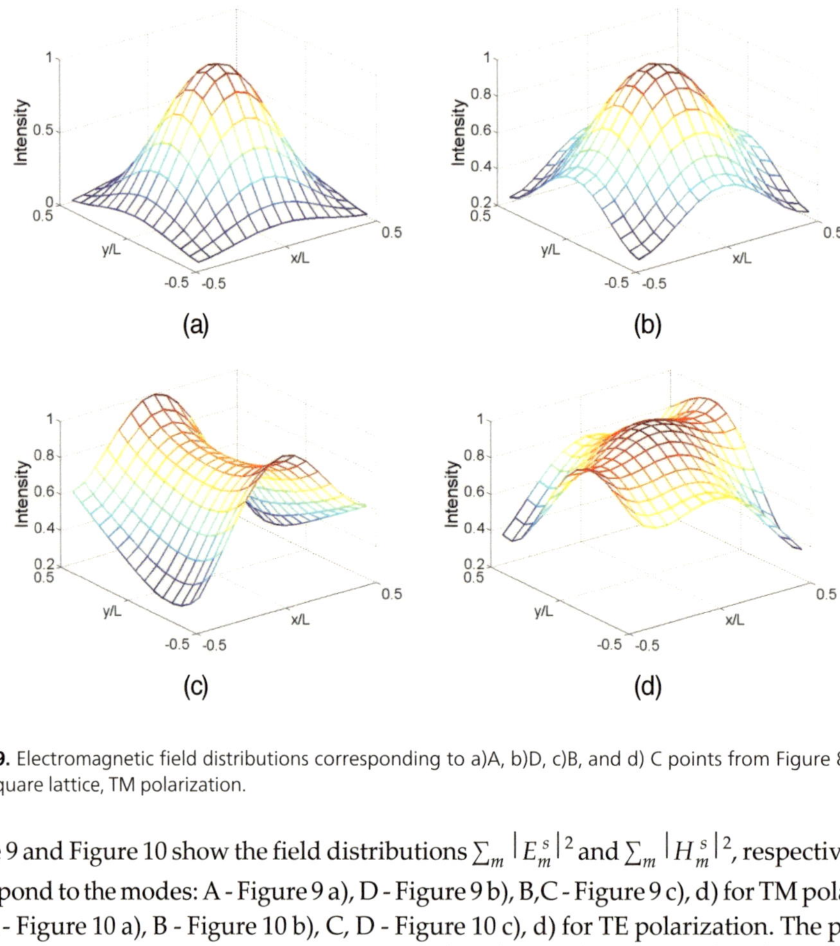

Figure 9. Electromagnetic field distributions corresponding to a)A, b)D, c)B, and d) C points from Figure 8 a), respectively. Square lattice, TM polarization.

Figure 9 and Figure 10 show the field distributions $\sum_m |E_m^s|^2$ and $\sum_m |H_m^s|^2$, respectively. They correspond to the modes: A - Figure 9 a), D - Figure 9 b), B,C - Figure 9 c), d) for TM polarization, and A - Figure 10 a), B - Figure 10 b), C, D - Figure 10 c), d) for TE polarization. The plots were made for the normalized coupling coefficients $|\kappa_1 L| = 5.5$, $|\kappa_2 L| = 4.1$, $|\kappa_3 L| = 2$, and filling factor f = 0.16.

In each case (TM and TE polarization), the doubly degenerate modes are orthogonal and show saddle-shaped patterns. The slight discrepancies arise from numerical inaccuracy. All non-degenerate modes are similar and exhibit Gaussian-like pattern, and this suggests that these modes should more efficiently use the photonic cavity. These modes (A) also have lower threshold, Figure 11.

http://dx.doi.org/10.5772/53712

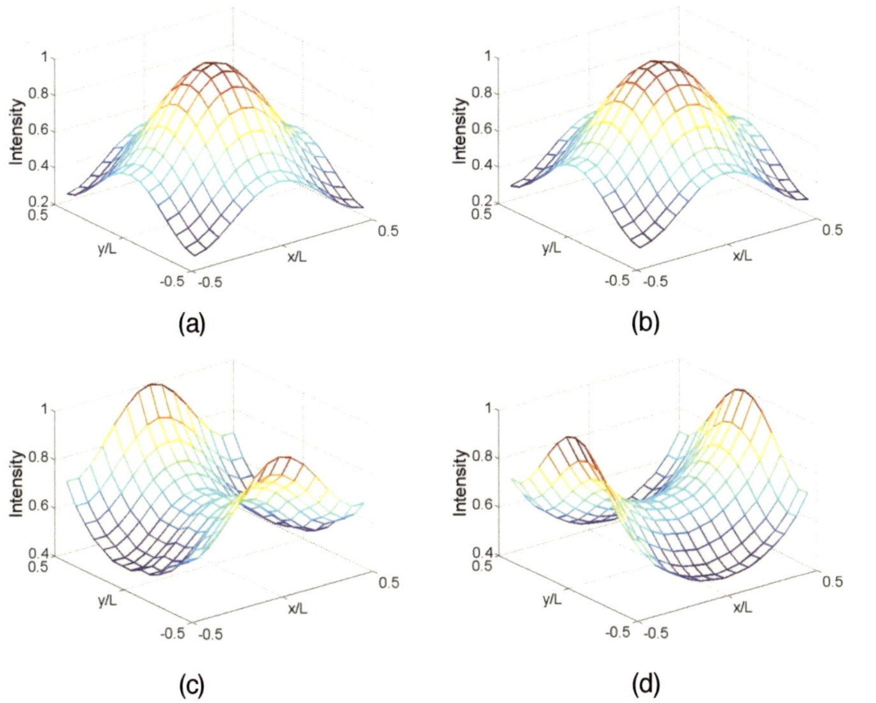

Figure 10. Electromagnetic field distributions corresponding to a) A, b) B, c) C, and d) D points from Figure 8 b), respectively. Square lattice, TE polarization.

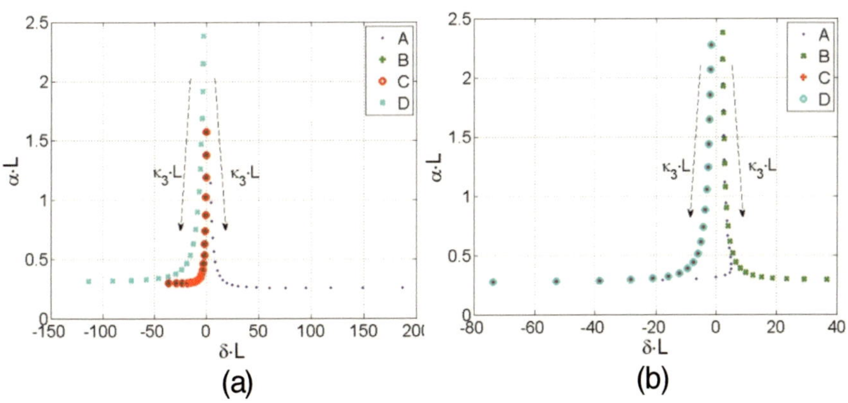

Figure 11. The dependence of threshold gain versus Bragg frequency deviation. Square lattice, a) TM polarization and b) TE polarization.

In Figure 11 a) and Figure 11 b), the normalized threshold gain αL was plotted as a function of Bragg frequency deviation δL, for various values of the normalized coupling coefficient $|\kappa_3 L|$ (which takes values from 0.5 to 40). The characteristics in the figures show that by increasing the value of coupling coefficient the Bragg frequency deviation increases and the threshold gain decreases. Simultaneously, the value of threshold gain saturates for all modes and eventually tends to similar values. This tendency is a consequence of growing field confinement in the cavity (for high index contrast all modes become Gaussian-like). For this reason the mode designation for higher values of coupling coefficients is difficult and only possible by the careful comparison of frequency deviation δ, and threshold gain α values. It is also worth noting that the threshold gain values for mode A are the lowest in wide range of coupling coefficient.

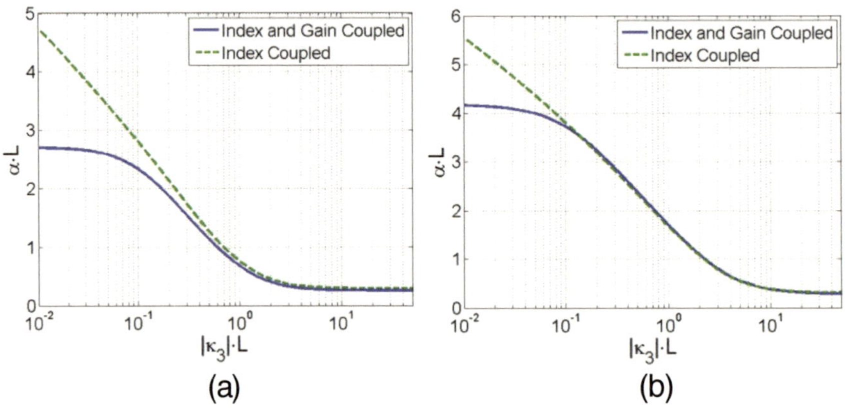

Figure 12. The dependence of normalized threshold gain versus normalized coupling coefficient for mode A for Index and Gain Coupled (solid line) and Index Coupled (dashed line) structures. Square lattice, a) TM polarization and b) TE polarization.

The impact of simultaneous gain and index coupling is depicted in Figure 12, where threshold gain for mode A is plotted as a function of coupling coefficient $|\kappa_3 L| \in (0.01;50)$. The characteristics compare the structure with gain (solid line) and without gain coupling (dashed line). It can be easily observed that the nonuniformity of the gain in the low index contrast structures has a strong impact on the threshold gain and cannot be disregarded. Therefore, by inducing gain coupling in the index coupled structure it is possible to lower threshold gain particularly for low index contrast photonic crystals.

3.3.2. Triangular lattice — TM and TE polarization

By repeating all the calculations shown for square lattice structures, threshold characteristics for triangular lattice structures are obtained. In Figure 13 enlarged areas of triangular lattice photonic crystals dispersion curves for the first six modes (A,B,C,D,E,F) in the vicinity of Γ point are shown. Figure 13 a) corresponds to TM polarization, and Figure 13 b) refers to TE polarization. For the calculations the circular holes $\varepsilon_b = 9.8$ arranged in triangular lattice with background material $\varepsilon_a = 12.0$ were assumed. The rods radius to lattice constant ratio was set to 0.24. In each plot, i.e., Figure 13 a) and Figure 13 b), there are two pairs of doubly degenerate modes (i.e., they have the same frequency at the Γ point): B,C and D,E for TM polarization, and B,C and E,F for TE polarization. Modes A have the lowest frequency.

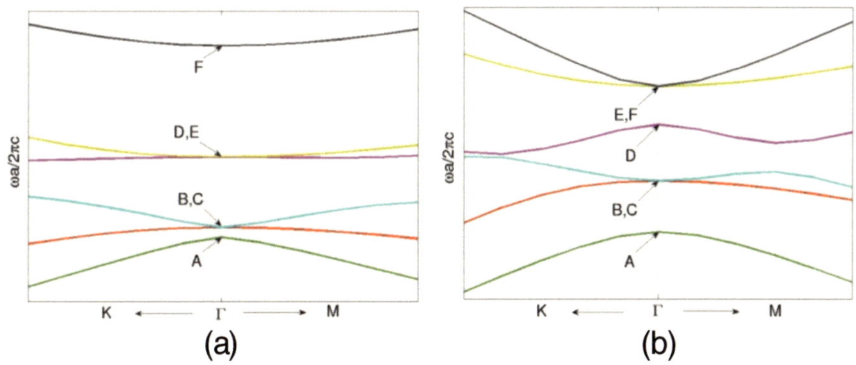

Figure 13. An enlarged area of dispersion curves of photonic crystal for the first four modes in the vicinity of Γ point. Triangular lattice, a) TM polarization, and b) TE polarization.

Bragg frequency deviation (for points marked as A,B,C,D,E,F in Figure 13) depending on coupling coefficient is analytically expressed in the following form for the TM polarization:

$$\delta_A = -2\kappa_1 - 2\kappa_2 - \kappa_3, \quad \delta_{B,C} = -\kappa_1 + \kappa_2 + \kappa_3, \quad \delta_{D,E} = \kappa_1 + \kappa_2 - \kappa_3, \quad \delta_F = 2\kappa_1 - 2\kappa_2 + \kappa_3, \qquad (62)$$

and for TE polarization:

$$\delta_A = -\kappa_1 - \kappa_2 + \kappa_3, \quad \delta_{B,C} = -\frac{\kappa_1}{2} + \frac{\kappa_2}{2} - \kappa_3, \quad \delta_{D,E} = \kappa_1 - \kappa_2 - \kappa_3, \quad \delta_F = \frac{\kappa_1}{2} + \frac{\kappa_2}{2} + \kappa_{3.}\,. \qquad (63)$$

Figure 14 shows the field distributions $\sum_m |E_m^t|^2$, m=1..6 corresponding to the modes: A - Figure 14 a), F - Figure 14 b), B,C - Figure 14 c), d), D,E - Figure 14 e), f).

Figure 15 shows the field distributions $\sum_m |H_m^t|^2$, m=1..6 corresponding to the modes: A - Figure 15 a), D - Figure 15 b), B,C - Figure 15 c), d), E,F - Figure 15 e), f).

The values of the normalized coupling coefficients for TM and TE polarization are set as follows: $|\kappa_1 L| = 7.0$, $|\kappa_2 L| = 3.3$, $|\kappa_3 L| = 2$, and the value of the filling factor f=0.16.

In both discussed cases, all degenerate modes are orthogonal and show similar patterns. For TM polarization, Figure 14, modes B,C are very similar to the non-degenerate mode A. This

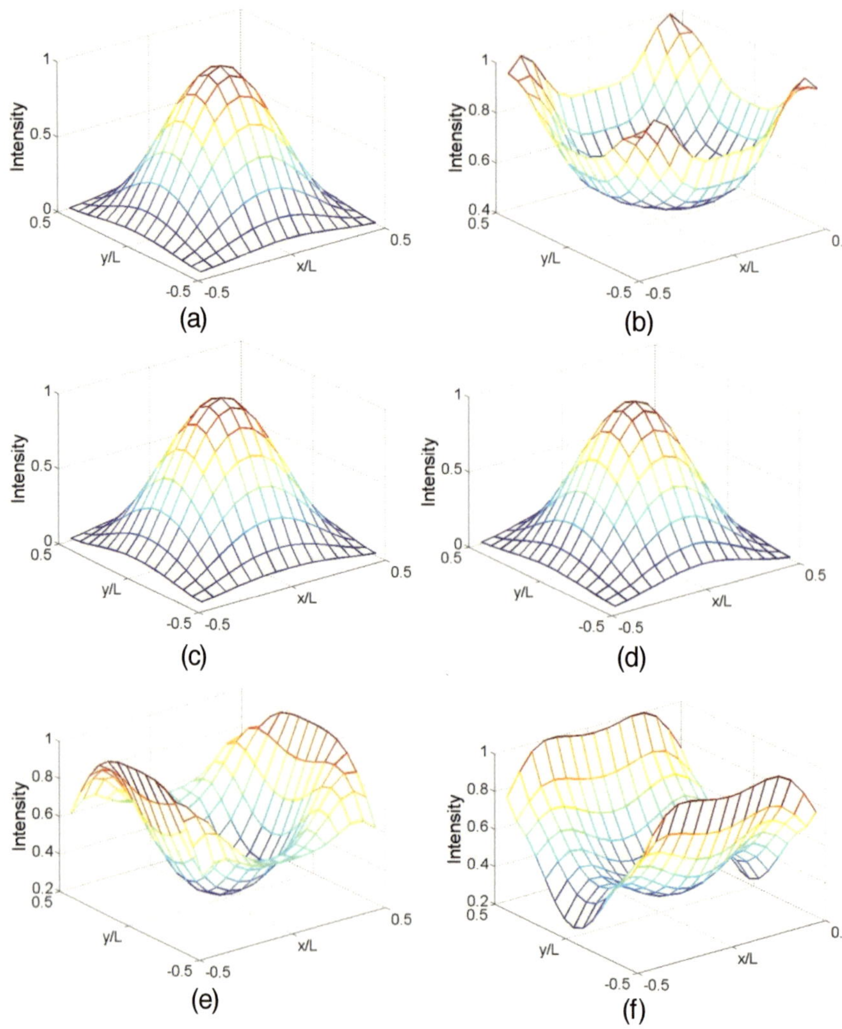

Figure 14. Electromagnetic field distributions corresponding to a)A, b)F, c)B, d)C, e)D, and f)E points from Figure 13 a), respectively. Triangular lattice, TM polarization.

means that the coupling coefficients, for which they are plotted, have high enough values to achieve strong field confinement. Similar situation is shown for TE polarization, Figure 15, where two pairs of doubly-degenerate modes are comparable to non-degenerate mode. Likewise, it is due to relatively high values of coupling coefficients and mode confinement.

In Figure 16 a), and Figure 16 b) the normalized threshold gain αL was plotted as a function of Bragg frequency deviation δL, for various values of the normalized coupling coefficient $|\kappa_3 L| \in (1;40)$.

Figure 16 shows similar tendency as in earlier examples of square lattice, i.e., by increasing the values of coupling coefficient the Bragg frequency deviation increases and the threshold gain

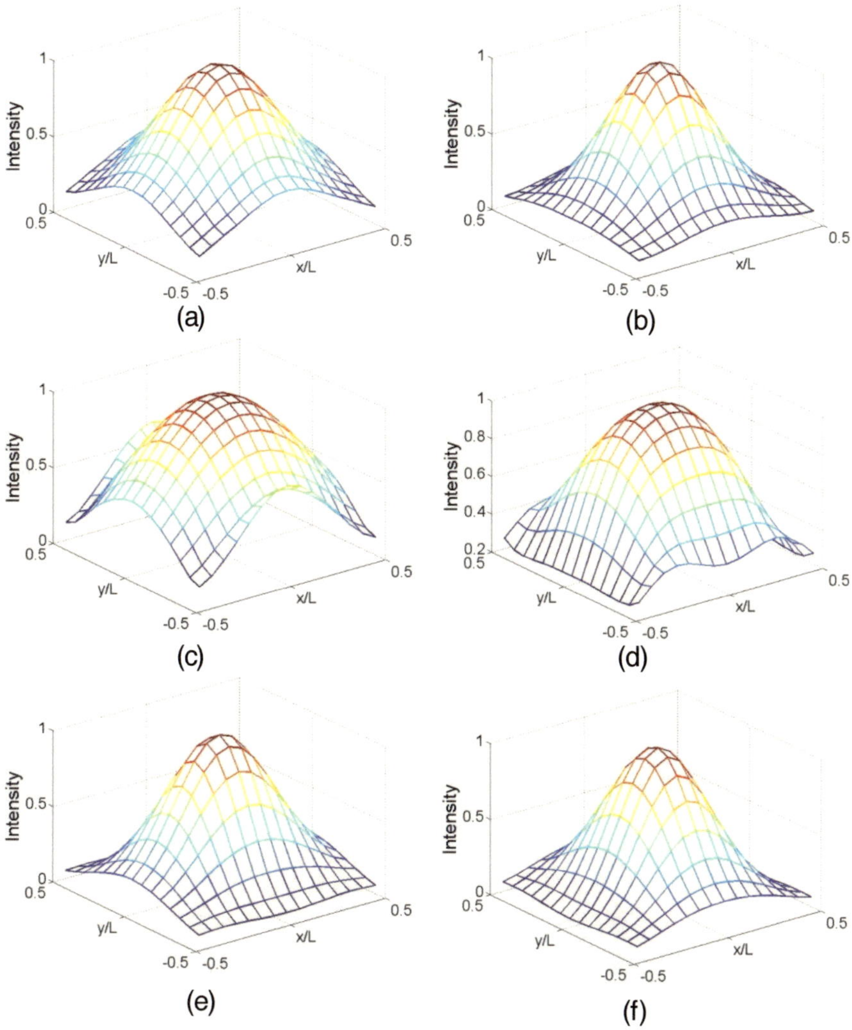

Figure 15. Electromagnetic field distributions corresponding to a)A, b)D, c)B, d)C, e)E, and f)F points from Figure 13 b), respectively. Triangular lattice, TE polarization.

decreases. Simultaneously, for larger values of coupling coefficient the threshold gain tends to alike values. This fact is due to the growing field confinement in the cavity (all modes become Gaussian-like, e.g., Figure 14 and Figure 15. The difference in the threshold gain values of degenerate modes stems from numerical inaccuracy, and the degenerate modes' threshold gain values should be averaged.

Figure 17 depicts the impact of simultaneous gain and index coupling. Here, the threshold gain for mode A is plotted as a function of coupling coefficient $|\kappa_3 L| \in (0.01;50)$. The characteristics compare the structure with gain and without gain coupling. Similarly as it is shown for square lattice structures, it is clearly seen that the incorporation of gain modulation in the

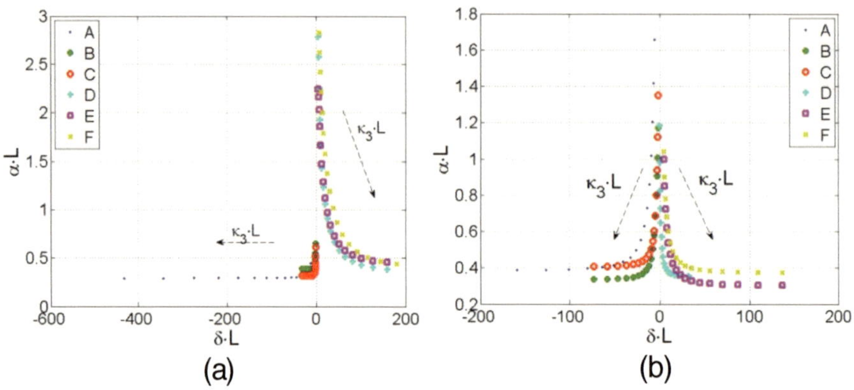

Figure 16. The dependence of threshold gain versus Bragg frequency deviation. Triangular lattice, a) TM polarization, and b) TE polarization.

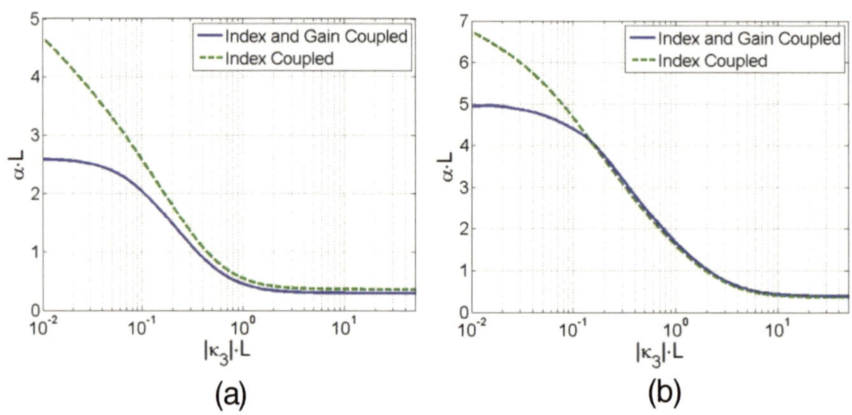

Figure 17. The dependence of normalized threshold gain versus normalized coupling coefficient for mode A for Index and Gain Coupled (solid line) and Index Coupled (dashed line) structures. Triangle lattice, a) TM polarization and b) TE polarization.

structure has a significant effect on the threshold gain characteristics. They are substantially changed in the lower range of coupling coefficient values and slightly lower the threshold gain in the entire range. This shows that the index and gain coupled structures can lower the threshold gain, especially in low index contrast photonic crystals.

Now, as an example of the model utilization let us consider the square lattice PC structure, which schematic cross-section is shown in Figure 1 a). For this structure following parameters are assumed: cavity length L = 50 nm, lattice constant a = 290 nm, and filling factor f = 0.16. The background material has higher permittivity than the rods $\varepsilon_a < \varepsilon_b$, and the active material is situated in the rods $\alpha_a \neq 0$, $\alpha_b = 0$. As schematically shown in Figure 12 a) for lower values of the coupling coefficient, i.e., a low refractive index difference, (e.g., $\kappa_3 L \in (0.01,\ \ 0.1)$) this structure has lower lasing threshold than it would have if the gain was uniformly distributed in the

http://dx.doi.org/10.5772/53712

medium, i.e., $\alpha_a = \alpha_b$. Thus for example: $\varepsilon_b = 12.00$ and $\varepsilon_a = 11.94$, then $\kappa_3 L = 0.09$ and the threshold gain drops by a factor of ~1.5. The supported lasing wavelength in such a cavity amounts to $\lambda = a\varepsilon_0^{1/2}$, where $\varepsilon_0^{1/2} = \sqrt{\varepsilon_a f + \varepsilon_b(1 - f)}$, that is $\lambda \sim 1\mu$m.

4. Perspectives

This Chapter discusses only some problems in threshold operation of 2D PC lasers. Thus, future work should be devoted to further investigation of gain coupling in photonic crystal cavities, e.g., such as comparison of solely index and solely gain coupled structures. Moreover, an above threshold analysis for gain coupled PC laser may apply as well as it did to index coupled structures, e.g., see [44],[45]. Finally, since more and more works on three-dimensional structures are published, it seems interesting to develop coupled wave models for threshold analysis of different symmetries incorporating gain and index modulation.

5. Conclusions

This work presents the systematic studies on the threshold operation of two-dimensional photonic crystal laser. It gives the comprehensive coupled mode description of gain and index coupled photonic crystal laser threshold operation. The calculations are conducted in the wide range of coupling coefficient for all four cases (square and triangular lattice with TM and TE polarization). It has been shown that the nonuniformity of the gain in the low index contrast structures has a strong impact on the threshold gain, by lowering it. Consequently, by inducing gain coupling in the index coupled structure it is possible to lower threshold gain particularly for low index contrast photonic crystals. This outcome helps understand the principles of PC band-edge laser operation and it may be useful in supporting the design process of PC laser structures.

Acknowledgements

The author would like to thank prof. P. Szczepanski for fruitful discussions and comments. This work was partly supported by the Polish MNiSW project IP2011 024771.

Author details

Marcin Koba

National Institute of Telecommunications, University of Warsaw, Warsaw University of Technology, Warsaw, Poland

References

[1] Yablonovitch E. Inhibited Spontaneous Emission in Solid-State Physics And Electronics, Physical Review Letters 1987;58(20) 2059-2062.

[2] Yablonovitch E. Photonic Band-Gap Crystals, Journal of Physics: Condensed Matter 1993;5(16) 2443-2460.

[3] John S. Strong Localization of Photons in Certain Disordered Dielectric Superlattices, Physical Review Letters 1987;58(23) 2486-2489.

[4] Strutt J. (Lord Rayleigh), Philosophical Magazine 1887;24(147) 145-159.

[5] Scherer H, Gollub D, Kamp M, Forchel A. Tunable GaInNAs Lasers with Photonic Crystal Mirrors, IEEE Photonics Technology Letters 2005;17(11) 2247-2249.

[6] Dunbar L, Moreau V, Ferrini R, Houdré R, Sirigu L, Scalari G, Giovannini M, Hoyler N, Faist J. Design, Fabrication and Optical Characterization of Quantum Cascade Lasers at Terahertz Frequencies Using Photonic Crystal Reflectors, Optics Express 2005;13(22) 8960-8968.

[7] Watanabe H, Baba T. Active/Passive-Integrated Photonic Crystal Slab μ-laser, Electronics Letters 2006;42(12) 695-696.

[8] Steinberg B, Boag A. Propagation in Photonic Crystal Coupled-Cavity Waveguides with Discontinuities in Their Optical Properties, Journal of Optical Society of America B 2006;23(7) 1442-1450.

[9] Asano T, Song B.-S, Noda S. Analysis of the Experimental Q Factors (~1 million) of Photonic Crystal Nanocavities, Optics Express 2006;14(5) 1996--2002.

[10] Lee K.-H, Baek J.-H, Hwang I.-K, Lee Y.-H, Lee G.-H, Ser J.-H, Kim H.-D, Shin H.-E. Square-lattice Photonic-Crystal Vertical-Cavity Surface-Emitting Lasers, Optics Express 2004;12(17) 4136-4143.

[11] Cojocaru C, Raineri F, Raj R, Monnier P, Drisse O, Legouezigou L, Chandouineau J.-P, Pommereau F, Duan G.-H, Levenson A. Room-Temperature Simultaneous In-Plane and Vertical Laser Operation in a Deep-Etched InP-based Two-Dimensional (2D) Photonic Crystal, IEE Optoelectronics Proceedings 2005;152(2) 86-89.

[12] Monat C, Seassal C, Letartre X, Viktorovitch P, Regreny P, Gendry M, Rojo-Romeo P, Hollinger G, Jalaguier E, Pocas S, Aspar B. InP 2D Photonic Crystal Microlasers on Silicon Wafer: Room Temperature Operation at 1.55 μm, Electronics Letters 2001;37(12) 764-766.

[13] Nomura M, Iwamoto S, Kumagai N, Arakawa Y. Ultra-Low Threshold Photonic Crystal Nanocavity Laser, Physica E: Low-dimensional Systems and Nanostructures 2008;40(6) 1800-1803.

[14] Turnbull G, Andrew P, Barnes W, Samuel I. Operating Characteristics of a Semiconducting Polymer Laser Pumped by a Microchip Laser, Applied Physics Letters 2003;82(3) 313-315.

[15] Vurgaftman I, Meyer J. Design Optimization for High-Brightness Surface-Emitting Photonic-Crystal Distributed-Feedback Lasers, IEEE Journal of Quantum Electronics 2003;39(6) 689-700.

[16] Imada M, Noda S, Chutinan A, Tokuda T, Murata M, Sasaki G. Coherent Two-Dimensional Lasing Action in Surface-Emitting Laser with Triangular-Lattice Photonic Crystal Structure, Applied Physics Letters 1999;75(3) 316-318.

[17] Ohnishi D, Okano T, Imada M, Noda S. Room Temperature Continuous Wave Operation of a Surface-Emitting Two-Dimensional Photonic Crystal Diode Laser, Optics Express 2004;12(8) 1562-1568.

[18] Miyai E, Sakai K, Okano T, Kunishi W, Ohnishi D, Noda S. Photonics: Lasers Producing Tailored Beams, Nature 2006;441 946-946.

[19] Kurosaka Y, Iwahashi S, Liang Y, Sakai K, Miyai E, Kunishi W, Ohnishi D, Noda S. On-Chip Beam-Steering Photonic-Crystal Lasers, Nature Photonics 2010;4(7) 447-450.

[20] Susa N. Threshold Gain and Gain-Enhancement Due to Distributed-Feedback in Two-Dimensional Photonic-Crystal Lasers, Journal of Applied Physics 2001;89(2) 815-823.

[21] Matsubara H, Yoshimoto S, Saito H, Jianglin Y, Tanaka Y, Noda S. GaN Photonic Crystal Surface-Emitting Laser at Blue-Violet Wavelengths, Science 2008;319(5862) 445-447.

[22] Kamp M. Photonic Crystal Lasers: On-Chip Beam Steering, Nature Photonics, News and Views, 2010:4 412-413.

[23] Sirigu L, Terazzi R, Amanti M, Giovannini M, Faist J, Dunbar L, Houdré R. Terahertz Quantum Cascade Lasers Based on Two-Dimensional Photonic-Crystal Resonators, Optics Express 2008;16(8) 5206-5217.

[24] Chassagneux Y, Colombelli R, Maineult W, Barbieri S, Beere H, Ritchie D, Khanna S, Linfield E, Davies A. Electrically Pumped Photonic-Crystal Terahertz Lasers Controlled by Boundary Conditions, Nature 2009;457 174-178.

[25] Mahler L, Tredicucci A. Photonic Engineering of Surface-Emitting Terahertz Quantum Cascade Lasers, Laser & Photonics Review 2011;5(5) 647-658.

[26] Kim M, Kim C, Bewley W, Lindle J, Canedy C, Vurgaftman I, Meyer J. Surface-Emitting Photonic-Crystal Distributed-Feedback Laser for the Midinfrared, Applied Physics Letters 2006;88(19) 191105 -191105-3.

[27] Xu G, Chassagneux Y, Colombelli R, Beaudoin G, Sagnes I. Polarized Single-Lobed Surface Emission in Mid-Infrared, Photonic-Crystal, Quantum-Cascade Lasers, Optics Letters 2010;35(6) 859-861.

[28] Zhang Z, Yoshie T, Zhu X, Xu J, Scherer A. Visible Two-Dimensional Photonic Crystal Slab Laser, Applied Physics Letters 2006;89(7) 071102-071102-3.

[29] Lu T, Chen S, Lin L, Kao T, Kao C, Yu P, Kuo H, Wang S, Fan S. GaN-Based Two-Dimensional Surface-Emitting Photonic Crystal Lasers with AlN/GaN Distributed Bragg Reflector, Applied Physics Letters 2008;92(1) 011129 1-3.

[30] Czuma P, Szczepanski P. Analytical Model of One-Dimensional SiO_2:Er-Doped Photonic Crystal Fabry-Perot Laser: Semiclassical Approach, Proceedings of the SPIE 2005:5723 307-315.

[31] Lesniewska-Matys K, Mossakowska-Wyszynska A, Szczepanski P. Nonlinear Operation of Planar Waveguide Laser with Photonic Crystal, Physica Scripta 2005;T118 107.

[32] Sakai K, Miyai E, Noda S. Coupled-Wave Theory for Square-Lattice Photonic Crystal Lasers with TE Polarization, IEEE Journal of Quantum Electronics 2010;46(5) 788-795.

[33] Koba M, Szczepanski P. Approximate Analysis of Nonlinear Operation of Square Lattice Photonic Crystal Laser, IEEE Journal of Quantum Electronics 2010;46(6) 1003-1008.

[34] Florescu L, Busch K, John S. Semiclassical Theory of Lasing in Photonic Crystals, Journal of Optical Society of America B 2002;19(9) 2215-2223.

[35] Imada M, Chutinan A, Noda S, Mochizuki M. Multidirectionally Distributed Feedback Photonic Crystal Lasers, Physical Review B 2002;65(19) 195306 1-8.

[36] Sakai K, Miyai E, Sakaguchi T, Ohnishi D, Okano T, Noda S. Lasing band-edge identification for a surface-emitting photonic crystal laser, IEEE Journal on Selected Areas in Communications, 2005;23(7) 1335-1340.

[37] Plihal M, Maradudin A. Photonic Band Structure of Two-Dimensional Systems: The Triangular Lattice, Physical Review B 1991;44(16) 8565-8571.

[38] Yokoyama M, Noda S. Finite-Difference Time-Domain Simulation of Two-Dimensional Photonic Crystal Surface-Emitting Laser, Optics Express 2005;13(8) 2869-2880.

[39] Fan S, Joannopoulos J. D. Analysis of Guided Resonances in Photonic Crystal Slabs, Physical Review B 2002;65(23) 235112 1-8.

[40] Ryu H. Y, Notomi M, Lee Y. H. Finite-Difference Time-Domain Investigation of Band-Edge Resonant Modes in Finite-Size Two-Dimensional Photonic Crystal Slab, Physical Review B 2003;68(40) 045209 1-8.

[41] Sakai K, Miyai E, Noda S. Coupled-Wave Model for Square-Lattice Two-Dimensional Photonic Crystal with Transverse-Electric-Like Mode, Applied Physics Letters 2006;89(2) 021101 1-3.

[42] Sakai K, Miyai E, Noda S. Two-Dimensional Coupled Wave Theory for Square-Lattice Photonic-Crystal Lasers with TM-Polarization, Optics Express 2007;15(7) 3981-3990.

[43] Sakai, K, Yue J, Noda S. Coupled-Wave Model for Triangular-Lattice Photonic Crystal with Transverse Electric Polarization, Optics Express 2008;16(9) 6033-6040.

[44] Koba M, Szczepanski P, Kossek T. Nonlinear Operation of a 2-D Triangular Lattice Photonic Crystal Laser, IEEE Journal of Quantum Electronics 2011;47(1) 13-19.

[45] Koba M, Szczepanski P, Osuch T. Nonlinear Analysis of Photonic Crystal Laser, Journal of Modern Optics 2011;58(17) 1538-1550.

[46] Koba M, Szczepanski P. The Threshold Mode Structure Analysis of The Two-Dimensional Photonic Crystal Lasers, Progress In Electromagnetics Research 2012;125 365-389.

[47] Liang Y, Peng C, Sakai K, Iwahashi S, Noda S. Three-Dimensional Coupled-Wave Model for Square Lattice Photonic-Crystal Lasers with Transverse Electric Polarization: A General Approach, Physical Review B 2011;84(19) 195119 1-11.

[48] Liang Y, Peng C, Sakai K, Iwahashi S, Noda S. Three-Dimensional Coupled-Wave Analysis for Square-Lattice Photonic Crystal Surface Emitting Lasers with Transverse-Electric Polarization: Finite-Size Effects, Optics Express 2012;20(14) 15945-15961.

[49] Koba M., Szczepański P. Coupled Mode Theory of Photonic Crystal Lasers. In: Massaro A. (ed.) Photonic Crystals - Introduction, Applications and Theory. Rijeka: InTech; 2012. p291-318.

[50] Kittel C. Introduction to Solid State Physics. Hoboken, NJ: Wiley; 2005.

[51] Plihal M, Shambrook A, Maradudin A, Sheng P. Two-Dimensional Photonic Band Structures, Optics Communications 1991;80(3-4) 199-204.

[52] Johnson S, Joannopoulos J. Block-Iterative Frequency-Domain Methods for Maxwell's Equations in a Planewave Basis, Optics Express 2001;8(3) 173-190.

Chapter 6

Two-Component Gap Solitons in Self-Defocusing Photonic Crystals

Thawatchai Mayteevarunyoo,
Athikom Roeksabutr and Boris A. Malomed

Additional information is available at the end of the chapter

http://dx.doi.org/10.5772/53605

1. Introduction

Studies of solitons in spatially periodic (lattice) potentials have grown into a vast area of research, with profoundly important applications to nonlinear optics, plasmonics, and matter waves in quantum gases, as outlined in recent reviews [1]-[4]. In ultracold bosonic and fermionic gases, periodic potentials can be created, in the form of optical lattices (OLs), by coherent laser beams illuminating the gas in opposite directions [5]-[7]. Effective lattice potentials for optical waves are induced by photonic crystals (PhCs), which are built as permanent structures by means of various techniques [2, 8, 9], or as laser-induced virtual structures in photorefractive crystals [10]. Reconfigurable PhCs can be also based on liquid crystals [11]. Parallel to the progress in the experiments, the study of the interplay between the nonlinearity and periodic potentials has been an incentive for the rapid developments of theoretical methods [12, 13]. Both the experimental and theoretical results reveal that solitons can be created in lattice potentials, if they do not exist in the uniform space [this is the case of gap solitons (GSs) supported by the self-defocusing nonlinearity, see original works [14]-[17] and reviews [7, 18]], and solitons may be stabilized, if they are unstable without the lattice (multidimensional solitons in the case of self-focusing, as shown in Refs. [17], [19]-[25], see also reviews [1, 3, 4]). The stability of GSs has been studied in detail too—chiefly, close to edges of the corresponding bandgaps—in one [27]-[29] and two [30] dimensions alike.

In fundamental and applied optics, PhCs provide ways to tailor the effective dispersion and diffraction of the medium, and control the transmission and routing of electromagnetic waves [8]. Fundamental characteristics of the PhCs are the band diagrams, which reveal gaps where linear (Bloch) waves cannot propagate. In PhCs made of nonlinear materials, GSs may self-trap as a result of the interplay of the Kerr-type nonlinearity and bandgap structures [18, 31–33]. Unlike ordinary bright solitons supported by the balance between the self-focusing nonlinearity and diffraction in uniform media [34], the dispersion relation induced by the PhC makes it possible to create GSs in focusing and defocusing materials alike. Combining assets of PhCs and regular solitons, GSs offer a considerable potential for applications to nonlinear photonics. On the other hand, the use of GSs is limited by the modulational [27] and oscillatory instabilities [35, 36]. One of solutions of this problem is the enhancement of the stability (and also mobility of the GSs) in nonlocal nonlinear media [37–41].

In addition to optics, GSs of matter waves have also been theoretically studied [42] and experimentally created [43] in Bose-Einstein condensates formed by atoms with repulsive interactions, loaded into OL potentials. In fact, the OLs controling the dynamics of matter waves may be considered as counterparts of PhCs for coherent atomic wave.

An essential extension of the theme is the study of two-component solitons in lattice potentials. In particular, if both the self-phase-modulation and cross-phase-modulation (SPM and XPM) nonlinearities, i.e., intra- and inter-species interactions, are repulsive, one can construct two-component GSs of *intra-gap* and *inter-gap* types, with chemical potentials of the components (or propagation constants, in terms of optical media) falling, respectively, into the same or different bandgaps of the underlying linear spectrum [44, 45]. In the case of the attractive SPM, a family of stable *semi-gap solitons* was found too, with one component residing in the infinite gap, while the other stays in a finite bandgap [45]. The GSs supported by the XPM repulsion dominating over the intrinsic (SPM-mediated) attraction may be regarded as an example of *symbiotic solitons*. In the free space (without the lattice potential), symbiotic solitons are supported by the XPM attraction between their two components, despite the action of the repulsive SPM in each one [46]-[48]. This mechanism may be additionally enhanced by the linear coupling (interconversion) between the components [49]. Another case of the "symbiosis" was reported in Ref. [50], where the action of the lattice potential on a single component was sufficient for the stabilization of two-dimensional (2D) two-component solitons against the collapse, the stabilizing effect of the lattice on the second component being mediated by the XPM interaction. In addition, the attraction between the components, competing with the intrinsic repulsion, may cause spatial splitting between two components of the GS, as for these components, whose effective masses are negative [16], the attractive interaction potential gives rise to a repulsion force [45, 51].

The ultimate form of the model which gives rise to two-component GSs of the symbiotic type is the one with no intra-species nonlinearity, the formation of the GSs being accounted for by the interplay of the repulsion between the components and the lattice potential acting on both of them. In optics, the setting with the XPM-only interactions is known in the form of the "holographic nonlinearity", which can be induced in photorefractive crystals for a pair of coherent beams with a small angle between their wave vectors, giving rise to single- [52, 53] and double-peak [54] solitons. Both beams are made by splitting a single laser signal, hence the power ratio between them (which is essential for the analysis reported below) can be varied by changing the splitting conditions. The creation of 2D spatial "holographic solitons" in a photorefractive-photovoltaic crystal with the self-focusing nonlinearity was demonstrated in Ref. [55] (such solitons are stable, as the collapse is arrested by the saturation of the self-focusing). To implement the situation considered here, the sign of the nonlinearity may be switched to self-defocusing by the reversal of the bias voltage, and the effective lattice potential may be induced by implanting appropriate dopants, with the concentration periodically modulated in one direction, which will render the setting quasi-one-dimensional.

In binary bosonic gases, a similar setting may be realized by switching off the SPM nonlinearity with the help of the Feshbach resonance, although one may need to apply

http://dx.doi.org/10.5772/53605

two different spatially uniform control fields (one magnetic and one optical) to do it simultaneously in both components. On the other hand, the same setting is natural for a mixture of two fermionic components with the repulsive interaction between them, which may represent two states of the same atomic species, with different values of the total atomic spin (F). If spins of both components are polarized by an external magnetic field, the SPM nonlinearity will be completely suppressed by the Pauli blockade while the inter-component interaction remains active [6], hence the setting may be described by a pair of Schrödinger equations for the two wave functions, coupled by XPM terms.

The objective of this work is to present basic families of one-dimensional symbiotic GSs, supported solely by the repulsive XPM nonlinearity in the combination with the lattice potential, and analyze their stability, via the computation of eigenvalues for small perturbations and direct simulations of the perturbed evolution. The difference from the previously studied models of symbiotic solitons [44, 45] is that the solitons where created there by the SPM nonlinearity separately in each component, while the XPM interaction determined the interaction between them and a possibility of creating two-component bound states. Here, the two-component GSs may exist solely due to the repulsive XPM interactions between the components.

We conclude that the symmetric solitons, built of equal components, are destabilized by symmetry-breaking perturbations above a certain critical value of the soliton's power. The analysis is chiefly focused on asymmetric symbiotic GSs, and on breathers into which unstable solitons are transformed. The model is introduced in Section II, which is followed by the analytical approximation presented in Section III. It is an extended version of the Thomas-Fermi approximation, TFA, which may be applied to other models too. In Section IV, we report systematic numerical results obtained for fundamental solitons of both the intra- and inter-gap types, hosted by the first two finite bandgaps of the system's spectrum. The most essential findings are summarized in the form of plots showing the change of the GS stability region with the variation of the degree of asymmetry of the two-component symbiotic solitons, which is a new feature exhibited by the present system. In particular, the stability area of intra-gap solitons shrinks with the increase of the asymmetry, while inter-gap solitons may be stable only if the asymmetry is large enough, in favor of the first-bandgap component, and intra-gap solitons in the second bandgap are completely unstable. The paper is concluded by Section V.

2. The model

The model outlined above is represented by the system of XPM-coupled Schrödinger equations for local amplitudes of co-propagating electromagnetic waves in the planar optical waveguide, $u(x,z)$ and $v(x,z)$, where x and z are the transverse coordinate and propagation distance, without the SPM terms, and with the lattice potential of depth $2\varepsilon > 0$ acting on both components:

$$i\frac{\partial u}{\partial z} + \frac{1}{2}\frac{\partial^2 u}{\partial x^2} - |v|^2 u + \varepsilon\cos(2x)u = 0, \tag{1}$$

$$i\frac{\partial v}{\partial z} + \frac{1}{2}\frac{\partial^2 v}{\partial x^2} - |u|^2 v + \varepsilon\cos(2x)v = 0. \tag{2}$$

The variables are scaled so as to make the lattice period equal to π, and the coefficients in front of the diffraction and XPM terms equal to 1. In the case of matter waves, u and v are wave functions of the two components, and z is replaced by time t. Direct simulations of Eqs. (1) and (2) were performed with the help of the split-step Fourier-transform technique.

Stationary solutions to Eqs. (1), (2) are looked as

$$u\,(x,z) = e^{ikz}U(x),\ v\,(x,z) = e^{iqz}V(x), \tag{3}$$

where the real propagation constants, k and q, are different, in the general case, and real functions $U(x)$ and $V(x)$ obey equations

$$-kU + \frac{1}{2}U'' - V^2U + \varepsilon\cos(2x)U = 0, \tag{4}$$

$$-qV + \frac{1}{2}V'' - U^2V + \varepsilon\cos(2x)V = 0, \tag{5}$$

with the prime standing for d/dx. Numerical solutions to Eqs. (4) and (5) were obtained by means of the Newton's method.

Solitons are characterized by the total power,

$$P = \int_{-\infty}^{+\infty}(|U|^2 + |V|^2)dx \equiv P_u + P_v, \tag{6}$$

with both P_u and P_v being dynamical invariants of Eqs. (1), (2), and by the *asymmetry ratio*,

$$R = (P_u - P_v)\,/\,(P_u + P_v)\,. \tag{7}$$

The total power and asymmetry may be naturally considered as functions of the propagation constants, k and q.

The well-known bandgap spectrum of the linearized version of Eqs. (4), (5) (see, e.g., book [12]) is displayed in Fig. 1, the right edge of the first finite bandgap being

$$k_{\max}\,(\varepsilon = 6) \approx 3.75. \tag{8}$$

The location of GSs is identified with respect to bandgaps of the spectrum. In this work, results are reported for composite GSs whose two components belong to the first and second finite bandgaps.

Stability of the stationary solutions can be investigated by means of the linearization against small perturbations [26]-[28]. To this end, perturbed solutions of Eqs. (1) and (2) are looked for as

$$u\,(x,z) = e^{ikz}\left[U(x) + u_1(x)e^{-i\lambda z} + u_2^*(x)e^{i\lambda^* z}\right],$$

$$v\,(x,z) = e^{iqz}\left[V(x) + v_1(x)e^{-i\lambda z} + v_2^*(x)e^{i\lambda^* z}\right], \tag{9}$$

http://dx.doi.org/10.5772/53605

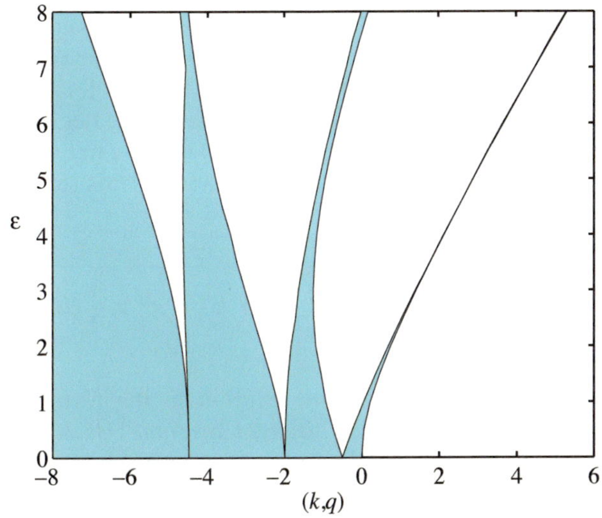

Figure 1. (Color online) The identical bandgap structures produced by the linearization of Eqs. (4) and (5) for $\varepsilon = 6$. Shaded areas are occupied by the Bloch bands, where gap solitons do not exist.

where $u_{1,2}$ and $v_{1,2}$ are wave functions of infinitesimal perturbations, and λ is the respective instability growth rate, which may be complex (the asterisk stands for the complex conjugate). The instability takes place if there is at least one eigenvalue with $\mathrm{Im}(\lambda) > 0$. The substitution of ansatz (9) into Eqs. (1), (2) and the linearization with respect to the small perturbations leads to the eigenvalue problem based on the following equations:

$$qv_1 - \frac{1}{2}v_1'' + U^2(x)v_1 + U(x)V(x)(u_1 + u_2) - \varepsilon\cos(2x)v_1 = \lambda v_1, \tag{10}$$

$$-qv_2 + \frac{1}{2}v_2'' - U^2(x)v_2 - U(x)V(x)(u_1 + u_2) + \varepsilon\cos(2x)v_2 = \lambda v_2, \tag{11}$$

$$ku_1 - \frac{1}{2}u_1'' + V^2(x)u_1 + U(x)V(x)(v_1 + v_2) - \varepsilon\cos(2x)u_1 = \lambda u_1, \tag{12}$$

$$-ku_2 + \frac{1}{2}u_2'' - V^2(x)u_2 - U(x)V(x)(v_1 + v_2) + \varepsilon\cos(2x)u_2 = \lambda u_2. \tag{13}$$

These equations were solved by means of the fourth-order center-difference numerical scheme.

Results for the shape and stability of GSs of different types are presented below for lattice strength $\varepsilon = 6$, which adequately represents the generic case. Note, in particular, that Fig. 1 was plotted for this value of the lattice-potential's strength.

3. The extended Thomas-Fermi approximation

It is well known that, close to edges of the bandgap, GSs feature an undulating shape, which may be approximated by a Bloch wave function modulated by a slowly varying envelope [14, 16]. On the other hand, deeper inside the bandgap, the GSs are strongly localized (see, e.g., Figs. 4 and 7 below), which suggests to approximate them by means of the variational method based on the Gaussian ansatz [45, 56]. This approximation was quite efficient for the description of GSs in single-component models [56], while for two-component systems it becomes cumbersome [44, 45]. Explicit analytical results for well-localized patterns can be obtained by means of the TFA [5], which, in the simplest case, neglects the kinetic-energy terms, U'' and V'', in Eqs. (4), (5). Assuming, for the sake of the definiteness, $q < k$ and also $|k| < \varepsilon$ (the TFA is irrelevant for $|k| > \varepsilon$), the approximation yields the fields inside the *inner layer* of the solution:

$$\left\{ \begin{array}{c} U^2(x) \\ V^2(x) \end{array} \right\}_{\text{inner}} = \left\{ \begin{array}{c} \varepsilon \cos(2x) - q \\ \varepsilon \cos(2x) - k \end{array} \right\}, \text{ at } |x| < x_0 \equiv \frac{1}{2} \cos^{-1}\left(\frac{k}{\varepsilon}\right). \tag{14}$$

Thus, the TFA predicts the core part of the solution in the form of peaks in the two components with the same width, $2x_0$, but different heights, $\{U^2, V^2\}_{\max} = \{\varepsilon - q, \varepsilon - k\}$. This structure complies with numerically generated examples of asymmetric solitons displayed in Figs. 7(a,b).

Further, the expansion of expressions (14) around the soliton's center ($x = 0$), yields

$$\left\{ \begin{array}{c} U(x) \\ V(x) \end{array} \right\} \approx \left\{ \begin{array}{c} \sqrt{\varepsilon - q} - \left(\varepsilon / \sqrt{\varepsilon - q}\right) x^2, \\ \sqrt{\varepsilon - k} - \left(\varepsilon / \sqrt{\varepsilon - k}\right) x^2. \end{array} \right\} \tag{15}$$

The substitution of the second derivatives of the fields at $x = 0$, calculated as per Eq. (15), into Eqs. (4), (5) yields a corrected expression for the soliton's amplitudes:

$$\left\{ \begin{array}{c} U(x=0) \\ V(x=0) \end{array} \right\} \approx \left\{ \begin{array}{c} \sqrt{\varepsilon - q} - \varepsilon \left[2(\varepsilon - k)\sqrt{\varepsilon - q}\right]^{-1}, \\ \sqrt{\varepsilon - k} - \varepsilon \left[2(\varepsilon - q)\sqrt{\varepsilon - k}\right]^{-1}, \end{array} \right\} \tag{16}$$

along with the condition for the applicability of the TFA:

$$\varepsilon \ll (\varepsilon - k)(\varepsilon - q). \tag{17}$$

For the symmetric-GS families in the two first finite bandgaps, the amplitude predicted by the improved TFA in the form of Eq. (16) is displayed, as a function of $k = q$, in Fig. 2(a) and compared to its numerically found counterpart. It is worthy to note that the correction terms in Eq. (16) essentially improve the agreement of the TFA prediction with the numerical findings.

At $|x| > x_0$, the TFA gives $V(x) = 0$, which is a continuous extension of the respective expression (14) in the V component, while the continuity of fields $U(x)$ and $U'(x)$ makes it necessary to match the respective expression (14) to "tails", which, in the lowest approximation, satisfy equation $U'' = 0$. The continuity is provided by the following tail solution:

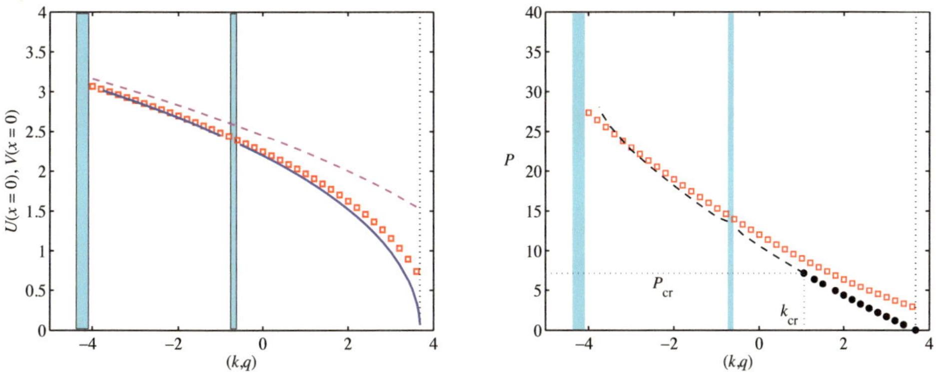

(a) (b)

Figure 2. (Color online) (a) The continuous (blue) curves show the numerically found amplitude of the fundamental symmetric gap solitons (with equal components), versus propagation constant $k = q$, in the first and second bandgaps at $\varepsilon = 6.0$. The chain of symbols is the analytical approximation for the same dependence, as produced by the improved TFA in the form of Eq. (16). The dashed curve is the result of the usual TFA, which corresponds to Eq. (16) without the correction (second) terms. (b) Total power P for the same soliton families, whose stable and unstable portions are designated by the bold dotted and dashed lines, respectively. The latter one is destabilized by symmetry-breaking perturbations, while the entire family is stable in the framework of the single-component model. Coordinates of the stability/instability border are given by Eq. (21). The chain of squares shows the analytical dependence produced by the TFA, see Eq. (20).

$$\left\{U^2(x)\right\}_{\text{outer}} = \left\{ \begin{array}{c} \left[\sqrt{k-q} - \sqrt{(\varepsilon^2 - k^2)\,/\,(k-q)}\,(|x| - x_0)\right]^2, \\ \text{at } 0 < |x| - x_0 < (k-q)\,/\sqrt{\varepsilon^2 - k^2}; \\ 0, \text{ at } |x| - x_0 > (k-q)\,/\sqrt{\varepsilon^2 - k^2}. \end{array} \right\} \tag{18}$$

The integration of expressions (14) and (18) yields the following approximation for the powers of the two components:

$$\left\{ \begin{array}{c} P_u \\ P_v \end{array} \right\}_{\text{TFA}} = \left\{ \begin{array}{c} \sqrt{\varepsilon^2 - k^2} - q\cos^{-1}(k/\varepsilon) \\ +(2/3)\,(k-q)^2\,/\sqrt{\varepsilon^2 - k^2}; \\ \sqrt{\varepsilon^2 - k^2} - k\cos^{-1}(k/\varepsilon)\,. \end{array} \right\} \tag{19}$$

The substitution of approximation (19) into definitions (6) and (7) of the total power and asymmetry demonstrates an agreement with numerical results. For instance, the slope of the curve $R(q)$ for the intra-gap GSs at fixed k (see Fig. 8 below) at the symmetry point ($k = q$), as predicted by Eq. (19) for $\varepsilon = 6$ and $k = 1$, is $(\partial R/\partial q)\,|_{q=k} \approx -0.155$, while its numerically found counterpart is ≈ -0.160. Further, the analysis of Eq. (19) readily demonstrates that the strongly asymmetric solitons may exist up to the limit of $R \to 1$, which is corroborated by the existence area for the intra-gap solitons shown below in Fig. 11(b).

Another corollary of Eqs. (19) is the prediction for the total power for the symmetric solitons,

$$P(k = q) = 2\left[\sqrt{\varepsilon^2 - k^2} - k\cos^{-1}(k/\varepsilon)\right], \tag{20}$$

which is plotted in Fig. 2(b), along with its numerically found counterpart. Although the TFA does not predict edges of the bandgaps, the overall analytical prediction for $P(k)$, as well as the soliton's amplitude shown in Fig. 2(a), run quite close to the numerical curves, except for near the right edge, where, indeed, condition (17) does not hold for $\varepsilon = 6$ and $k = 3.75$, see Eq. (8). Note that these simple analytical approximations were not derived before in numerous works dealing with single-component GSs.

4. Results of the numerical analysis

4.1. Symmetric solitons

Obviously, the shape of symmetric solitons, built of two equal components [with $k = q$ and $U(x) = V(x)$, see Eq. (3)], is identical to that of GSs in the single-component model. However, there is a drastic difference in the stability of the symmetric GSs between the single- and two-component systems. Almost the entire symmetric family is stable against symmetric perturbations, i.e., it is stable in the framework of the single-component equation (in agreement with previously known results [12]), except for a weak oscillatory instability, accounted for by quartets of complex-conjugate eigenvalues, in the form of $\lambda = \pm i\mathrm{Im}\,(\lambda) \pm \mathrm{Re}\,(\lambda)$ (with two mutually independent signs $\pm$), which appears near the left edge of the second bandgap—namely, at $k < k_{\min} \approx -3.45$. An example of the development of the latter instability is displayed in Fig. 3.

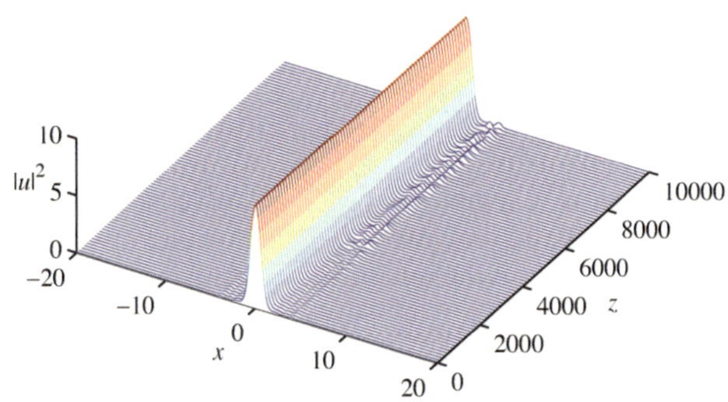

Figure 3. (Color online) The evolution of a weakly unstable single-component fundamental soliton at $k = -3.7$.

On the other hand, Fig. 2 demonstrates that a considerable part of the family in the first finite bandgap, and the entire family in the second bandgap are unstable against symmetry-breaking perturbations in the two-component system. The boundary separating the stable and unstable subfamilies of the fundamental symmetric GSs in the first finite bandgap corresponds to the power and propagation constants is found at

$$P_{\mathrm{cr}} \approx 7.19,\ k_{\mathrm{cr}} \approx 1.05, \tag{21}$$

the symmetric solitons being stable in the intervals of $0 < P < 7.19$, $1.05 < k < k_{\max} \approx 3.75$ [see Eq. (8)]. These results were produced by a numerical solution of Eqs. (10)-(13) (the instability is oscillatory, characterized by complex eigenvalues).

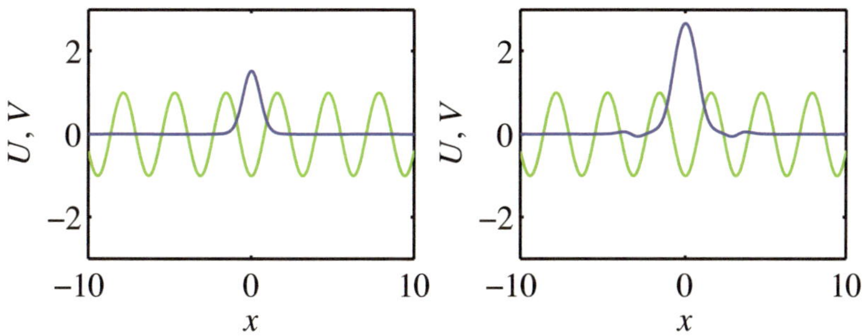

Figure 4. (Color online) Examples of fundamental symmetric gap solitons found in the first and second bandgaps, for $k = q = 2.0$ and $k = q = -2.0$ (left and right panels, respectively). Here and in similar figures below, the background pattern (green sinusoid) represents the underlying periodic potential. Both solitons are stable as solutions of the single-component model, but only the one corresponding to $k = q = 2.0$ remains stable in the two-component system, while its counterpart pertaining to $k = q = -2$ is destabilized by symmetry-breaking perturbations.

Typical examples of stable and unstable fundamental symmetric GSs, found in the first and second bandgaps (not too close to their edges), are displayed in Fig. 4. Further, direct simulations demonstrate that the evolution transforms the unstable symmetric solitons into persistent localized breathers, as shown in Figs. 5 and 6, in accordance with the fact that the corresponding instability eigenvalues are complex. Although the emerging breather keeps the value of $R = 0$, see Eq. (7), the u- and v- components of the breather generated by the symmetry-breaking instability are no longer mutually identical. This manifestation of the symmetry-breaking instability is illustrated by Fig. 5(c), which displays the evolution of the difference between the peak powers of the two components, and the separation between their centers. The latter is defined as

$$X_u - X_v \equiv \frac{1}{P_u}\int_{-\infty}^{+\infty} |u(x,z)|^2 x dx - \frac{1}{P_v}\int_{-\infty}^{+\infty} |v(x,z)|^2 x dx. \tag{22}$$

It is relevant to mention that the second finite bandgap also contains a branch of the so-called subfundamental solitons, whose power is smaller than that of the fundamental GSs [29], [57], [58]. These are odd modes, squeezed, essentially, into a single cell of the underlying lattice potential. The subfundamental solitons are unstable, tending to rearrange themselves into fundamental ones belonging to the first finite bandgap, therefore they are not considered below.

4.2. Asymmetric solitons of the intra-gap type

As said above, two-component asymmetric fundamental GSs, with different propagation constants, $k \neq q$, may be naturally classified as solitons of the intra- and inter-gap types if k and q belong to the same or different finite bandgaps [44]. In this subsection, we report results for asymmetric intra-gap solitons with both k and q falling into the first finite bandgap, as well as for asymmetric breathers developing from such solitons when they are unstable.

Examples of stable asymmetric GSs of the intra-gap type are displayed in Fig. 7, for a fixed asymmetry ratio, $R = -0.5$, defined as per Eq. (7). The GS family, along with the family of persistent breathers into which unstable solitons are spontaneously transformed, is

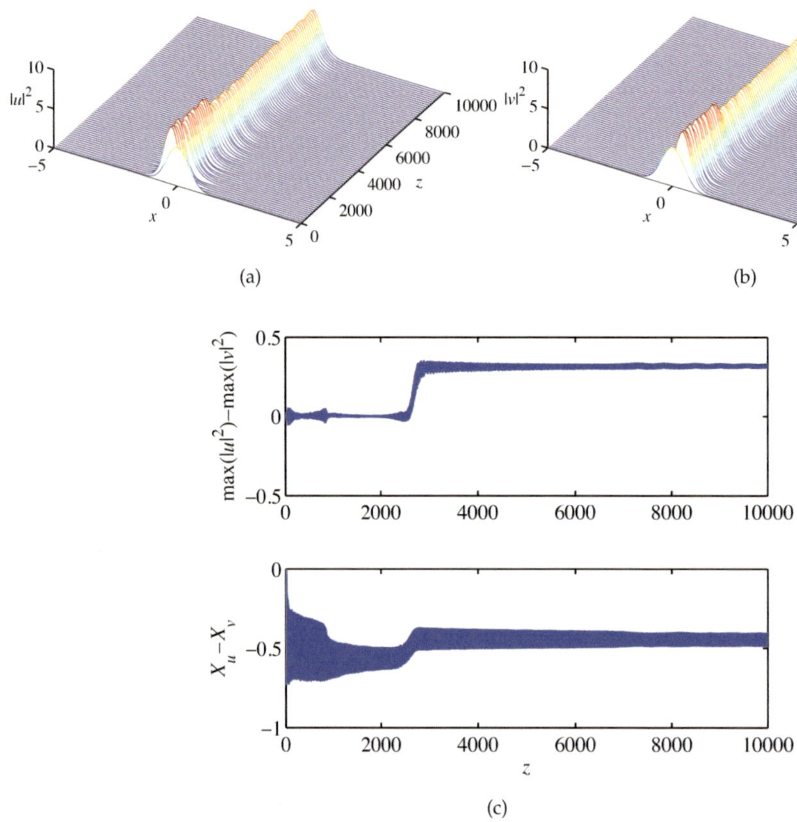

Figure 5. (Color online) (a) and (b) The spontaneous transformation of an unstable symmetric fundamental soliton for u- and v-component, in the first bandgap, with $k = q = 0$, into a stable asymmetric breather. (c) The top and bottom plots display, respectively, the evolution of the peak-power difference, $\max(|u(x,z)|^2) - \max(|u(x,z)|^2)$, and the separation between centers of the two components, X_u and X_v, which is defined as per Eq. (22).

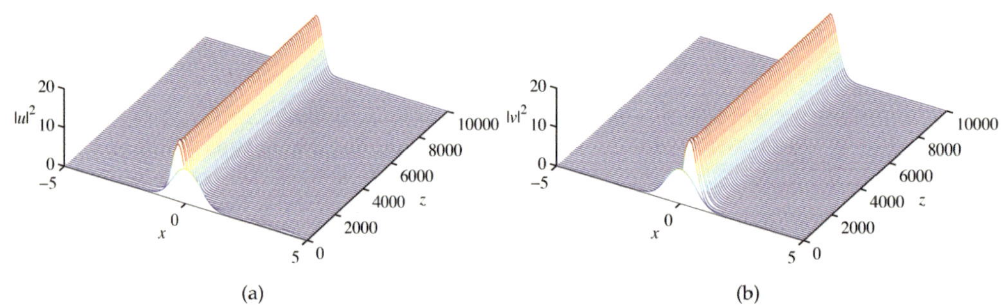

Figure 6. (Color online) (a,b) The same as in Fig. 5(a,b), but for an unstable soliton in the second bandgap, with $k = q = -3.5$.

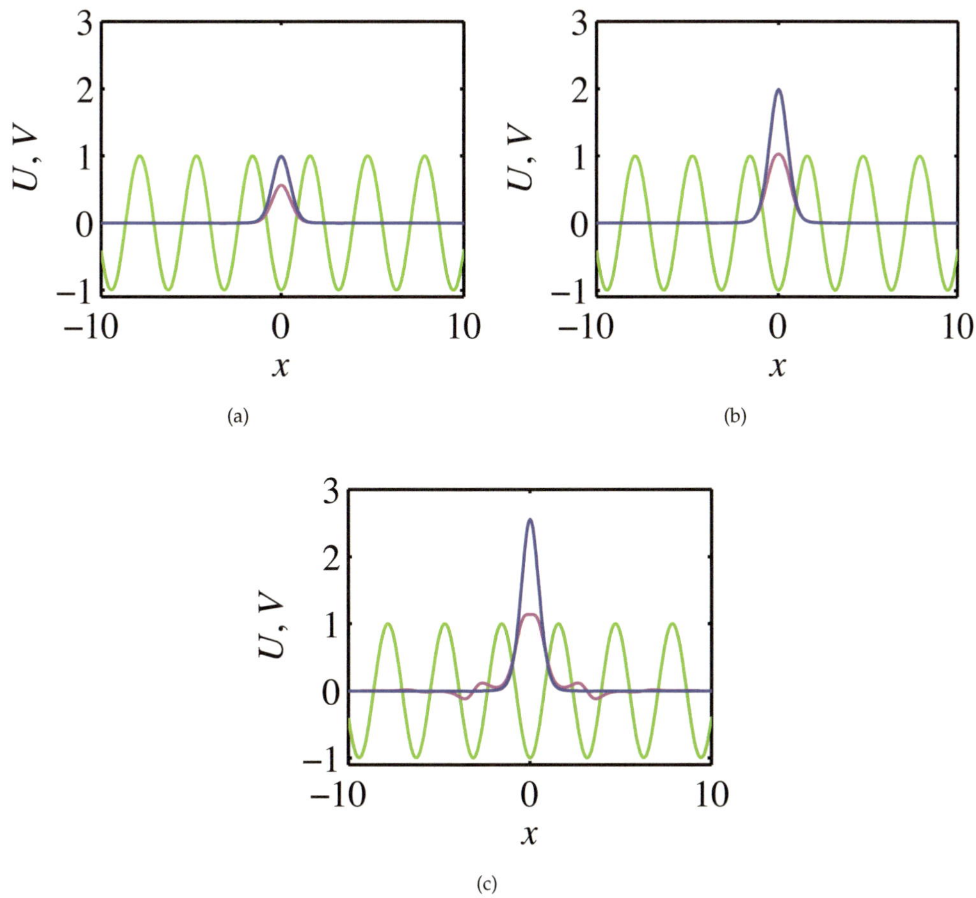

Figure 7. (Color online) Examples of stable solitons of the intra-gap type found in the first finite bandgap, with fixed asymmetry $R = -0.5$: (a) $k = 3$ and $q = 3.4601$; (b) $k = 1$ and $q = 2.843$; (c) $k = -0.5$ and $q = 2.5116$. Fields $U(x)$ and $V(x)$, which pertain to propagation constants k and q, are shown, respectively, by the magenta (lower) and blue (higher) profiles.

represented in Fig. 8 by dependences $R(q)$ at different fixed values of the other propagation constant, k.

It is possible to explain the fact that all the $R(q)$ curves converge to $R = -1$, as q approaches the right edge of the bandgap in Fig. 8. In this case, the V component turns into the delocalized Bloch wave function with a diverging power, P_v, that corresponds to $P_u/P_v \to 0$ [it is tantamount to $R \to -1$, as per Eq. (7)]. An example of a stable GS, close to this limit, with $k = -0.5$, $q = 3.65$ and $R = -0.9811$, is shown in Fig. 9. The central core of the V-component is described by the TFA, based on Eq. (14), as the corresponding necessary condition (17) holds in this case, while the TFA does not apply to the U-component. The presence of undulating tails, which are close to the Bloch functions, rather than the simple approximation (18), which is valid far from the edge of the bandgap, is also visible in Fig. 8.

Those asymmetric intra-gap GSs which form unstable subfamilies in Fig. 8 are destabilized by oscillatory perturbations. The instability transform the solitons into breathers, see a typical

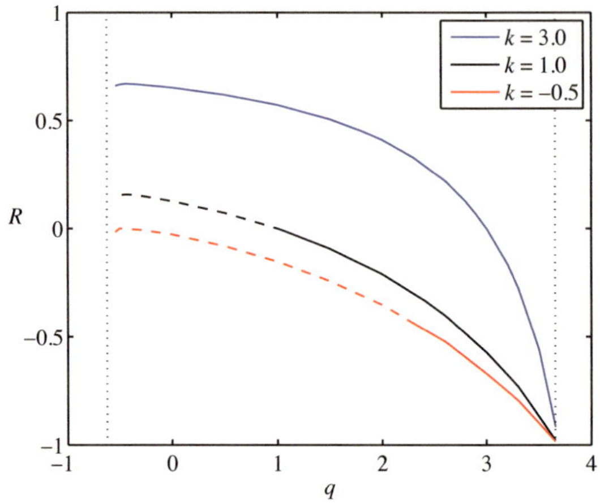

Figure 8. (Color online) The asymmetry ratio, R [defined as per Eq. (7)], versus propagation constant q, at fixed values of $k = 3.0$, 1.0, and $k = -0.5$ (the top, middle, and bottom curves, respectively), for asymmetric fundamental solitons of the intra-gap type. Stable and unstable branches are shown by solid and dashed lines, respectively.

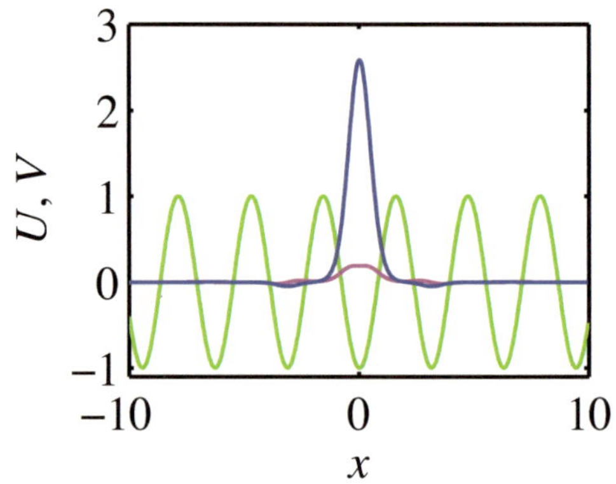

Figure 9. (Color online) An example of a stable strongly asymmetric soliton with $k = -0.5$ and $q = 3.65$. Fields $U(x)$ and $V(x)$, which pertain to propagation constants k and q, are shown, respectively, by the magenta (lower) and blue (taller) profiles.

example in Fig. 10 [cf. the examples of the destabilization of the symmetric GSs shown in Fig. 5(b,c)]. The emerging breathers keep values of the asymmetry ratio (7) almost identical to those of their parent GSs; for instance, in the case displayed in Fig. 10, the unstable soliton with $R_{\text{initial}} = -0.3617$ evolves into the breather with $R_{\text{final}} = -0.3623$.

It is relevant to stress that the transformation of unstable stationary GSs into the breathers gives rise to little radiation loss of the total power, P. On the other hand, in the general case a given unstable gap soliton does not have a stable counterpart with a close value of P, hence this unstable soliton cannot transform itself into a slightly excited state of another stable GS. Thus, the breathers represent a distinct species of localized modes.

The most essential results of the stability analysis for the asymmetric solitons of the intra-gap type, and for breathers replacing unstable solitons, are summarized by diagrams in the planes

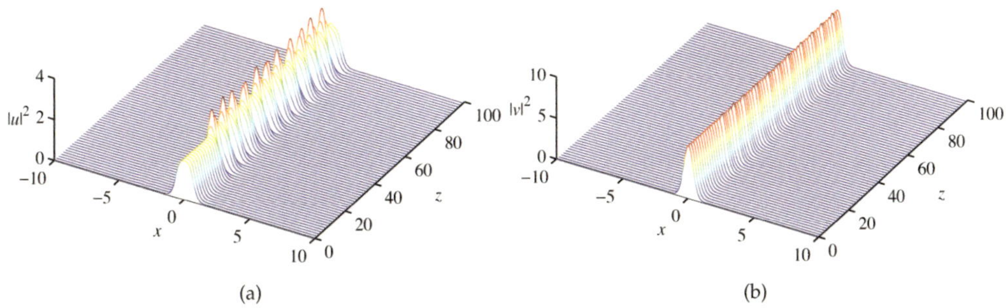

Figure 10. (Color online) A typical example of the transformation of the unstable asymmetric gap solitons into a breather, for $k = -0.5$ and $q = 2.0$.

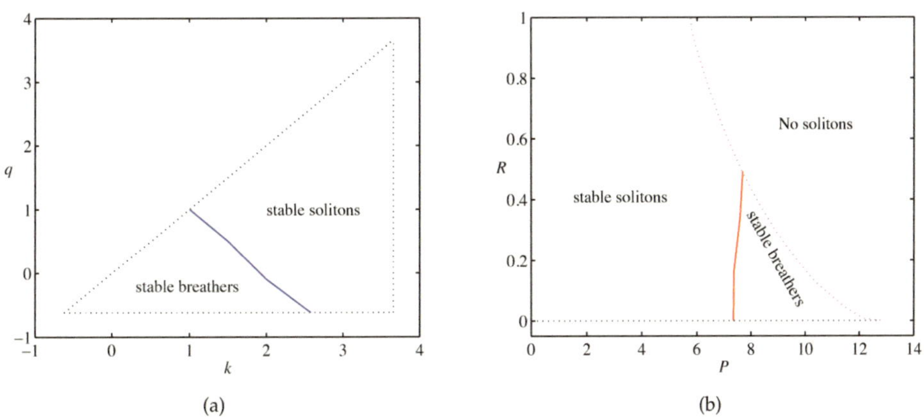

Figure 11. (Color online) (a) The stability border in the plane of the propagation constants, (k, q), for asymmetric solitons of the intra-gap type. Only half of the plane is shown, delineated by the dotted triangle, within which wavenumbers k and q belong to the first finite bandgap, as the other half is a mirror image of the displayed one. (b) The same in the plane of the total power and asymmetry ratio, (P, R), defined as per Eqs. (6) and (7). Localized modes do not exist above the right boundary of the stability regions in panel (b). The diagram at $R < 0$ is a mirror image of the one displayed here for $R > 0$.

of (k, q) and (P, R), which are displayed in Fig. 11. The predictions of the analysis based on the computation of the stability eigenvalues for the stationary solitons, as per Eqs. (10)-(13), always comply with stability tests provided by direct simulations of Eqs. (1) and (2).

As mentioned above, the instability of a part of the branch of the symmetric solitons along the line of $R = 0$ in Fig. 11(b) implies that the symmetry-breaking perturbations destabilize the symmetric solitons in the first finite bandgap at $P > P_{\text{cr}}$, see Eq. (21), while their counterparts are stable in the single-component system. Another clear conclusion is that the stability region gradually shrinks with the increase of the asymmetry.

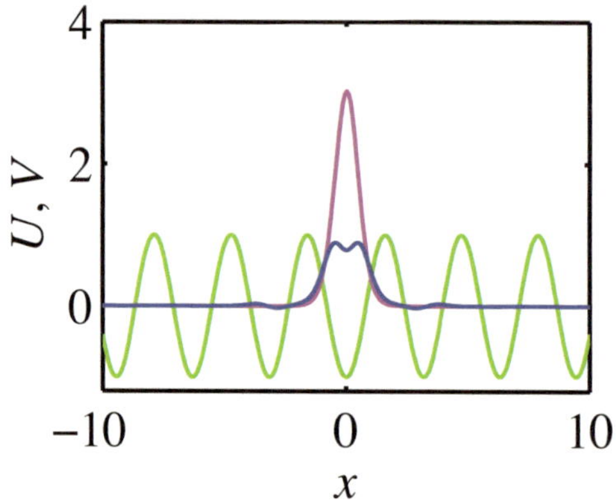

Figure 12. (Color online) An example of a stable inter-gap soliton, for $k = 3$ and $q = -1.65$. The single-peak and split-peak profiles, $U(x)$ and $V(x)$, represent, respectively, the components in the first and second finite bandgaps.

4.3. Solitons of the inter-gap type

All the GSs of the inter-gap type, with two propagation constants belonging to the two different finite bandgaps, are naturally asymmetric, even if their components have equal powers. Examples of stable and unstable inter-gap solitons are displayed in Figs. 12 and 13, respectively. A noteworthy feature exhibited by these examples is a split-peak structure of the component belonging to the second finite bandgap.

The asymmetry measure, $R(q)$, for families of the inter-gap solitons is plotted in Fig. 14(a) versus the propagation constant q in the second bandgap, at fixed values of k (the propagation constant in the first bandgap). The stability of the respective GS families is also shown in Fig. 14.

The (in)stability of the inter-gap solitons is summarized by the diagrams in the planes of (k, q) and (P, R) presented in Fig. 15, cf. similar diagrams for intra-gap solitons shown above in Fig. 11. As well as in that case, unstable inter-gap solitons are spontaneously replaced by robust localized breathers. The spontaneous transformation increases the initial degree of the asymmetry: For instance, an unstable inter-gap soliton with $R = 0.019$ is converted into a breather with $R = 0.028$.

In the present case too, the existence region of stable modes shrinks with the increase of the asymmetry; note also that the stationary inter-gap solitons may be stable solely at sufficiently large values of the asymmetry, $R \geq R_{\text{min}} \approx 0.5$. The asymmetric shape of the stability diagram in Fig. 15(b) with respect to $R > 0$ and $R < 0$ [unlike the symmetry of the diagram for the intra-gap solitons implied in Fig. 11(b)] is explained by the fact that, in definition (7), $R > 0$ implies that the dominant component resides in the first finite bandgap, where it is more robust than in the second bandgap.

Finally, out additional analysis has demonstrated that all the stationary GSs—not only the symmetric ones [see Fig. 2], but also all the asymmetric solitons of the intra-gap type—are completely unstable in the second finite bandgap. They too tend to spontaneously rearrange themselves into breathers, which is not shown here in detail.

http://dx.doi.org/10.5772/53605

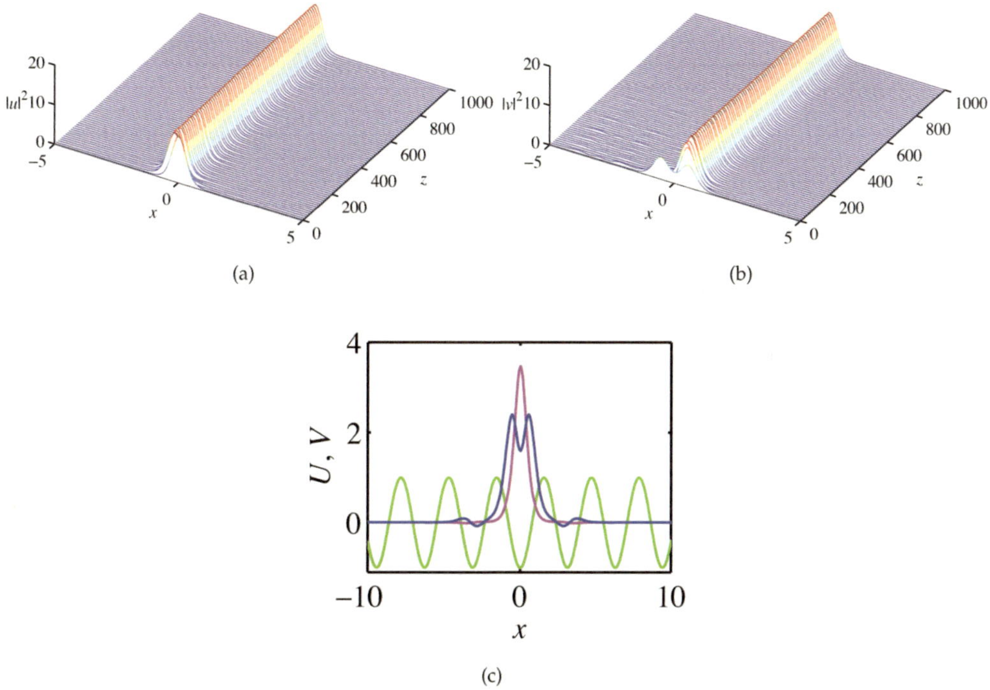

Figure 13. (Color online) A typical example of the transformation of an unstable intergap soliton into a stable breather, at $k = 0$ and $q = -2$. (a,b) The evolution of $|u|^2$ and $|v|^2$. (c) The initial profiles of $U(x)$ and $V(x)$ (single-peak and split-peak shapes, respectively).

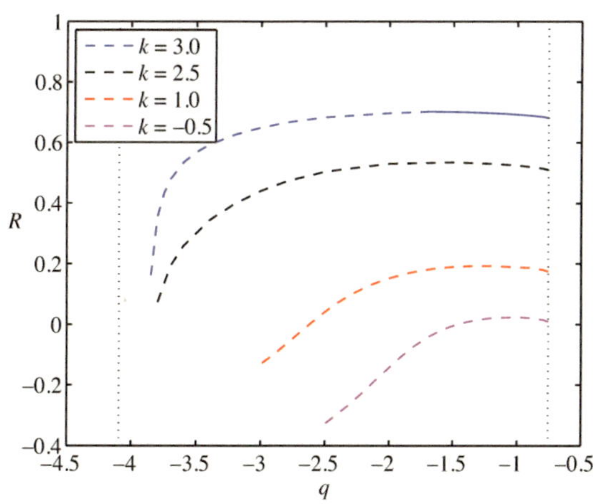

Figure 14. (Color online) Asymmetry ratio R for the inter-gap solitons versus propagation constant q in the second finite bandgap, at fixed values $k = 3.0$, 2.5, 1.0, and -0.5 (from the top to the bottom) of wavenumber k in the first bandgap. Solid and dashed lines designate stable stationary solitons and breathers, respectively.

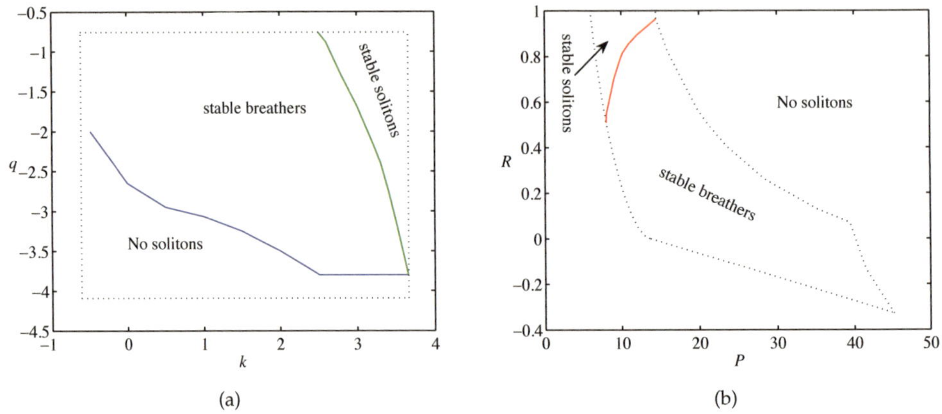

Figure 15. (Color online) The same as in Fig. 11, but for inter-gap solitons. In (a), the dotted rectangle delineates the region occupied by wavenumbers k and q belonging to the first and second finite bandgaps, respectively.

5. Conclusion

We have introduced the model of symbiotic two-component GSs (gap solitons), based on two nonlinear Schrödinger equations coupled by the repulsive XPM terms and including the lattice potential acting on both components, in the absence of the SPM nonlinearity. The model has a realization in optics, in terms of "holographic solitons" in photonic crystals, and as a model of binary quantum gases (in particular, a fully polarized fermionic one) loaded into the optical-lattice potential. Families of fundamental asymmetric GSs have been constructed in the two lowest finite bandgaps, including the modes of both the intra-gap and inter-gap types, i.e., those with the propagation constants of the two components belonging to the same or different bandgaps, respectively. The existence and stability regions of the symbiotic GSs and breathers, into which unstable solitons are transformed, have been identified. A noteworthy finding is that symmetry-breaking perturbations destabilize the symmetric GSs in the first finite bandgap, if their total power exceeds the critical value given by Eq. (21), along with all the symmetric solitons in the second bandgaps. It was demonstrated too that the stability area for the intra-gap GSs shrinks with the increase of the asymmetry ratio, R. On the other hand, inter-gap GSs may be stable only for sufficiently large ratio, $R > 0.5$. The intra-gap solitons are completely unstable in the second bandgap. Some features of the GS families were explained by means of the extended TFA (Thomas-Fermi approximation), augmented by the tails attached to the taller component, in the case of asymmetric solitons.

A natural extension of the analysis may deal with 2D symbiotic gap solitons, supported by the square- or radial-lattice potentials. In that case, it may be interesting to consider two-component solitary vortices too.

Acknowledgment

The work of T.M. was supported by the Thailand Research Fund through grant RMU5380005. B.A.M. appreciates hospitality of the Mahanakorn University of Technology (Bangkok, Thailand).

http://dx.doi.org/10.5772/53605

Author details

Thawatchai Mayteevarunyoo[1,*],
Athikom Roeksabutr[1] and Boris A. Malomed[2]

* Address all correspondence to: thawatch@mut.ac.th

1 Department of Telecommunication Engineering, Mahanakorn University of Technology, Bangkok, Thailand
2 Department of Physical Electronics, School of Electrical Engineering, Faculty of Engineering, Tel Aviv University, Tel Aviv, Israel

References

[1] B. A. Malomed, D. Mihalache, F. Wise, and L. Torner, "Spatiotemporal optical solitons", J. Opt. B: Quant. Semicl. Opt. 7, R53–R72 (2005).

[2] F. Lederer, G. I. Stegeman, D. N. Christodoulides, G. Assanto, M. Segev, and Y. Silberberg, Phys. Rep. 463, 1 (2008).

[3] Kartashov, V. A. Vysloukh, and L. Torner, "Soliton shape and mobility control in optical lattices", Progr. Opt. 52, 63-148 (2009).

[4] Y. V. Kartashov, B. A. Malomed, and L. Torner, "Solitons in nonlinear lattice," Rev. Mod. Phys. 83, 247-306 (2011).

[5] L. Pitaevskii and S. Stringari, *Bose-Einstein Condensate* (Clarendon Press: Oxford, 2003).

[6] H. T. C. Stoof, K. B. Gubbels, and D. B. M. Dickerscheid, *Ultracold Quantum Fields* (Springer: Dordrecht, 2009).

[7] O. Morsch and M. Oberthaler, "Dynamics of Bose-Einstein condensates in optical lattices," Rev. Mod. Phys. 78, 179-215 (2006).

[8] J. D. Joannopoulos, S. G. Johnson, J. N. Winn, and R. D. Meade, *Photonic Crystals: Molding the Flow of Light* (Princeton University Press: Princeton, 2008).

[9] A. Szameit, D. Blömer, J. Burghoff, T. Schreiber, T. Pertsch, S. Nolte, and A. Tüunnermann, 2005, "Discrete nonlinear localization in femtosecond laser written waveguides in fused silica," Opt. Express 13, 10552-10557 (2005).

[10] N. K. Efremidis, S. Sears, D. N. Christodoulides, J. W. Fleischer, and M. Segev, "Discrete solitons in photorefractive optically induced photonic lattices," Phys. Rev. E 66, 046602 (2002).

[11] M. Peccianti, K.A. Brzdkiewicz, and G. Assanto, *Opt. Lett.* 27, 1460 (2002).

[12] J. Yang, *Nonlinear Waves in Integrable and Nonintegrable Systems* (SIAM: Philadelphia, 2010).

[13] D. E. Pelinovsky, *Localization in Periodic Potentials* (Cambridge University Press: Cambridge, UK, 2011).

[14] B. B. Baizakov, V. V. Konotop, and M. Salerno, "Regular spatial structures in arrays of Bose–Einstein condensates induced by modulational instability", J. Phys. B: At. Mol. Opt. Phys. 35, 5105–5119 (2002).

[15] P. J. Y. Louis, E. A. Ostrovskaya, C. M. Savage, and Y. S. Kivshar, "Bose-Einstein condensates in optical lattices: Band-gap structure and solitons", Phys. Rev. A 67, 013602 (2003).

[16] H. Sakaguchi and B. A. Malomed, "Dynamics of positive- and negative-mass solitons in optical lattices and inverted traps", J. Phys. B 37, 1443-1459 (2004).

[17] B. Baizakov, B. A. Malomed, and M. Salerno, "Matter-wave solitons in radially periodic potentials," Phys. Rev. E 74, 066615 (2006).

[18] V. A. Brazhnyi and V. V. Konotop, "Theory of nonlinear matter waves in optical lattices," Mod. Phys. Lett. B 18, 627-651 (2004).

[19] B. B. Baizakov, B. A. Malomed and M. Salerno, "Multidimensional solitons in periodic potentials", Europhys. Lett. 63, 642–648 (2003).

[20] J. Yang, and Z. H. Musslimani, "Fundamental and vortex solitons in a two-dimensional optical lattice," Opt. Lett. 28, 2094-2096 (2003).

[21] Z. H. Musslimani and J. Yang, "Self-trapping of light in a two-dimensional photonic lattice", J. Opt. Soc. Am. B 21, 973-981 (2004).

[22] B. B. Baizakov, B. A. Malomed and M. Salerno, "Multidimensional solitons in a low-dimensional periodic potential", Phys. Rev. A 70, 053613 (2004).

[23] D. Mihalache, D. Mazilu, F. Lederer, Y. V. Kartashov, L.-C. Crasovan, and L. Torner, "Stable three-dimensional spatiotemporal solitons in a two-dimensional photonic lattice", Phys. Rev. E 70, 055603(R) (2004).

[24] B. B. Baizakov, B. A. Malomed, and M. Salerno, "Multidimensional semi-gap solitons in a periodic potential", Eur. Phys. J. 38, 367-374 (2006).

[25] T. Mayteevarunyoo, B. A. Malomed, B. B. Baizakov, and M. Salerno, "Matter-wave vortices and solitons in anisotropic optical lattices", Physica D 238, 1439-1448 (2009).

[26] K. M. Hilligsoe, M. K. Oberthaler, and K. P. Marzlin, "Stability of gap solitons in a Bose-Einstein condensate", Phys. Rev. A 66, 063605 (2002).

[27] D. E. Pelinovsky, A. A. Sukhorukov, and Y. S. Kivshar, "Bifurcations and stability of gap solitons in periodic potentials", Phys. Rev. E 70, 036618 (2004).

[28] G. Hwanga, T. R. Akylas, and J. Yang, "Gap solitons and their linear stability in one-dimensional periodic media", Physica D 240, 1055–1068 (2011).

[29] J. Cuevas, B. A. Malomed, P. G. Kevrekidis, and D. J. Frantzeskakis, "Solitons in quasi-one-dimensional Bose-Einstein condensates with competing dipolar and local interactions", Phys. Rev. A 79, 053608 (2009).

[30] Z. Shi, J. Wang, Z. Chen, and J. Yang, "Linear instability of two-dimensional low-amplitude gap solitons near band edges in periodic media", Phys. Rev. A 78, 063812 (2008).

[31] "Nonlinear Photonic Crystals", edited by R.E. Slusher and B.J. Eggleton (Springer-Verlag, Berlin, 2003).

[32] B. J. Eggleton, R. E. Slusher, C. M. de Sterke, P. A. Krug, and J. E. Sipe, "Bragg grating solitons", Phys. Rev. Lett. 76, 1627 (1996).

[33] F. Kh. Abdullaev, B. B. Baizakov, S. A. Darmanyan, V. V. Konotop, and M. Salerno, "Nonlinear excitations in arrays of Bose-Einstein condensates", Phys. Rev. A 64, 043606 (2001); I. Carusotto, D. Embriaco, and G. C. La Rocca, "Nonlinear atom optics and bright-gap-soliton generation in finite optical lattices", *ibid.* 65, 053611 (2002); E. A. Ostrovskaya and Yu. S. Kivshar, "Matter-wave gap solitons in atomic band-gap structures", Phys. Rev. Lett. 90, 160407 (2003).

[34] Yu.S. Kivshar and G.P. Agrawal, "Optical Solitons: from Fibers to Photonic Crystals" (Academic Press, San Diego, 2003).

[35] B. A. Malomed and R. S. Tasgal, "Vibration modes of a gap soliton in a nonlinear optical medium", Phys. Rev. E 49, 5787 (1994).

[36] I. V. Barashenkov, D. E. Pelinovsky, and E. V. Zemlyanaya, "Vibrations and Oscillatory Instabilities of Gap Solitons", Phys. Rev. Lett. 80, 5117 (1998).

[37] Z. Xu, Y.V. Kartashov, and L. Torner, "Soliton Mobility in Nonlocal Optical Lattices", Phys. Rev. Lett. 95, 113901 (2005).

[38] Y. Y. Lin, I.-H. Chen, and R.-K. Lee, "Breather-like collision of gap solitons in Bragg gap regions within nonlocal nonlinear photonic crystals", J. Opt. A: Pure Appl. Opt. 10, 044017 (2008).

[39] Y. Y. Lin, R.-K. Lee, and B. A. Malomed, "Bragg solitons in nonlocal nonlinear media", Phys. Rev. A 80, 013838 (2009).

[40] Y. Y. Lin, C. P. Jisha, C. J. Jen, R.-K. Lee, and B. A. Malomed, "Gap solitons in optical lattices embedded into nonlocal media", Phys. Rev. A 81, 063803 (2010).

[41] K.-H. Kuo, Y. Y. Lin, R.-K. Lee, and B. A. Malomed, "Gap solitons under competing local and nonlocal nonlinearities", Phys. Rev. A 83, 053838 (2011).

[42] T. J. Alexander, E. A. Ostrovskaya, and Y. S. Kivshar, "Self-Trapped Nonlinear Matter Waves in Periodic Potentials", Phys. Rev. Lett. 96, 040401 (2006).

[43] B. Eiermann, Th. Anker, M. Albiez, M. Taglieber, P. Treutlein, K.-P. Marzlin, and M. K. Oberthaler, "Bright Bose-Einstein gap solitons of atoms with repulsive interaction," Phys. Rev. Lett. 92, 230401 (2004).

[44] A. Gubeskys, B. A. Malomed, and I. M. Merhasin, "Two-component gap solitons in two- and one-dimensional Bose-Einstein condensate", Phys. Rev. A 73, 023607 (2006).

[45] S. K. Adhikari and B. A. Malomed, "Symbiotic gap and semigap solitons in Bose-Einstein condensates", Phys. Rev. A 77, 023607 (2008).

[46] V. M. Pérez-García and J. B. Beitia, "Symbiotic solitons in heteronuclear multicomponent Bose-Einstein condensates," Phys. Rev. A 72, 033620 (2005).

[47] S. K. Adhikari, "Bright solitons in coupled defocusing NLS equation supported by coupling: Application to Bose-Einstein condensation," Phys. Lett. A 346, 179-185 (2005).

[48] S. K. Adhikari, "Fermionic bright soliton in a boson-fermion mixture," Phys. Rev. A 72, 053608 (2005).

[49] S. K. Adhikari and B. A. Malomed, "Two-component gap solitons with linear interconversion", Phys. Rev. A 79, 015602 (2009).

[50] O. V. Borovkova, B. A. Malomed and Y. V. Kartashov, "Two-dimensional vector solitons stabilized by a linear or nonlinear lattice acting in one component", EPL 92, 64001 (2010).

[51] M. Matuszewski, B. A. Malomed, and M. Trippenbach, "Competition between attractive and repulsive interactions in two-component Bose-Einstein condensates trapped in an optical lattice", Phys. Rev. A 76, 043826 (2007).

[52] O. Cohen, T. Carmon, M. Segev, and S. Odoulov, "Holographic solitons," Opt. Lett. 27, 2031-2033 (2002).

[53] O. Cohen, M. M. Murnane, H. C. Kapteyn, and M. Segev, "Cross-phase-modulation nonlinearities and holographic solitons in periodically poled photovoltaic photorefractives," Opt. Lett. 31, 954-956 (2006).

[54] J. R. Salgueiro, A. A. Sukhorukov, and Y. S. Kivshar, "Spatial optical solitons supported by mutual focusing," Opt. Lett. 28, 1457-1459 (2003).

[55] J. Liu, S. Liu, G. Zhang, and C. Wang, "Observation of two-dimensional holographic photovoltaic bright solitons in a photorefractive-photovoltaic crystal", Appl. Phys. Lett. 91, 111113 (2007).

[56] S. Adhikari and B. A. Malomed, "Gap solitons in a model of a superfluid fermion gas in optical lattices", Physica D 238, 1402-1412 (2009).

[57] N. K. Efremidis and D. N. Christodoulides, "Lattice solitons in Bose-Einstein condensates", Phys. Rev. A 67, 063608 (2003).

[58] T. Mayteevarunyoo and B. A. Malomed, "Stability limits for gap solitons in a Bose-Einstein condensate trapped in a time-modulated optical lattice", Phys. Rev. A 74, 033616 (2006).

Chapter 7

Very Long Photon-Lifetimes Achieved by Photonic Atolls

S. Nojima

Additional information is available at the end of the chapter

http://dx.doi.org/10.5772/53435

1. Introduction

It is one of the primary interests in recent nano- and micro-photonics to achieve a strong confinement of light in a small region, because it finds a variety of applications in optical physics and engineering where it is exploited in low-threshold lasers [1], nonlinear optical devices [2], and cavity quantum electrodynamics devices [3]. Extensive efforts have therefore been devoted to developing a cavity that can confine light efficiently—a high-quality optical resonator. The quality of resonators is described here by the photon lifetime τ which is the time that elapses before a photon trapped in the resonator escapes from it, or by the quality factor defined by $Q=\omega\tau$ where ω is the angular frequency of light [4].

In order to achieve high-quality optical resonators, the two directions seem to have been explored so far: one is the use of the extended waves and another is the use of the localized waves. The photonic crystals (PCs) may be the first candidate high-quality resonators, the Q factors for which have been found to be increased by the slowed-down light (the extended waves, or the Bloch waves in this case) near the photonic band edge [5-7]. The typical example for the exploitation of the localized waves can be found in the defect mode that is localized around a disorder in the PC [8-13], which provides more pronounced light-confinement than the band edge modes in the PCs. Although the defect itself generally occupies a very small region, this confinement requires the presence of a large periodic medium around it in order for the defect mode to be sufficiently isolated from its environment. Light can also be localized in the central part of a three-dimensional (3D) fractal structure (Menger sponge) made up of cubes that need not have high Q factors [14]. A single microstructure with a variety of forms [15-19] also creates high-Q modes called whispering gallery modes (WGMs) that occur when the light waves circulate within the microstructure because they undergo total internal reflection at its boundaries. This could be regarded as intermediate between the two directions mentioned before, since the waves for WGMs are propagating extended waves but confined within the microstructure.

In the context mentioned above, we describe in Sec. 3 an entirely different type of resonator: a closed chain array made up of dielectric microstructures arranged periodically in the background material (e.g., the air). We call it a photonic atoll (PA) resonator because it resembles an atoll in the ocean. This PA resonator is thought to have a prominent function to confine light very strongly for the following reasons: (1) the multiple scattering of light by the periodic quasi-one-dimensional (q1D) array causes a slowing down of extended light-waves and (2) the closed optical path forces a photon once trapped in the array to keep circulating in the loop, both of which would undoubtedly increase the photon lifetime. Factor (1) is the same as the factor responsible for the lifetime enhancement at the band edge of the PCs (see the preceding paragraph) while factor (2) reminds us of the analogy to the ring accelerator for elementary particles. Because of the features mentioned above, this PA structure could also be called a distributed feedback ring-resonator. The above concept was previously [20] applied to the PA resonator of the two-dimensional (2D) circular array consisting of the fifty rods. This resonator was actually found to create an extremely high radiative Q factor of the order of 10^{15} and the resultant very long lifetime of the order of one second for visible light at the modes near the photonic band edges created in this q1D closed PC. The idea of PA was conceived during the investigation of circulating modes in a two-dimensional PC [6, 21] and a microdisk [22], so we believe that the investigation of it and its structure effects will also help us understand the behavior of light in those structures.

Since we have confirmed that the PA structure has a potential to achieve very long lifetimes, our next step of research is to investigate what kind of PA shapes would provide the most efficient optical resonator (Sec. 4). This is because the PA has the degree of freedom that permits it to have an arbitrary loop form: note that the first work (Sec. 3) has focused on a circular PA. In the process of investigations to pursue the optimum PA structure that maximizes the Q value, we observed the remarkable metamorphoses of eigen modes whose degeneracy has been lifted in the modified PAs. This kind of phenomena has so far not been observed in the PA, but it could be considered in the context of the phenomena such as the Stark effect [23] and the Zeeman effect [23] for the electronic energy and the Sagnac effect [22, 24] for the optical energy. This is because the mode splittings in all of these phenomena are caused by some perturbations applied to the system: the Stark effect is caused by the modification of the electronic potential by the electric field, the Zeeman effect by the magnetic field, the Sagnac effect by the mechanical rotation, and our case by the modification of the optical potential by the rearrangement of rods.

Finally, we describe in Sec. 5 the laser actions in the PA resonators with extremely high Q values as an example of their application to a practical optical device. The threshold amplitude gain for laser oscillation is calculated together with the lifetimes. We find that these values are well correlated, in particular that the threshold gain is inversely proportional to the lifetime obtained for the same PA resonators. Although other possible losses of light remain to be considered before this structure is put to practical use, the results obtained here suggest that it would be an excellent structure for confining light. In particular, the fact that it does not require a large size to achieve a strong light confinement will prove a great advantage over other ways of light confinement when it is incorporated into optical integrated circuits.

2. Theory

2.1. Multiple scattering of light

The analytic multiple-scattering theory is used here to evaluate the light confinement effects. Since the general theory is described in the reports [6, 7], here we briefly outline the framework of the calculation. We consider a 2D array consisting of a finite number N of cylindrical rods (made of material A) with radius d placed at arbitrary points in the background material (material B, usually air). Here we focus on the polarization for which the electric field is parallel to the rod axis (E-polarization). By considering the scattering of the incident plane-wave with the unit amplitude by these rods, we obtain the electric field of the total scattered wave $E^s(\boldsymbol{r})$:

$$E^s(\boldsymbol{r}) = \sum_{n=1}^{N} \sum_{l=-\infty}^{+\infty} b_{nl} H_l^{(1)}(Kr_n) e^{il\theta_n} \equiv \boldsymbol{b} \cdot \boldsymbol{\varphi}(\boldsymbol{r}), \tag{1}$$

where $\boldsymbol{r}$ is the generic coordinate, (r_n, θ_n) is the polar coordinates of the center of the nth rod, and $K = \omega / c$ is the wave number of light in the air. Here, the second equality in Eq. (1) implies the inner product of vectors $\boldsymbol{b} = (b_{nl})$ and $\boldsymbol{\varphi}(\boldsymbol{r}) = \left(H_l^{(1)}(Kr_n)e^{il\theta_n}\right)$ where $H_l^{(1)}(x)$ is the Hankel function [25] of the first kind. Vector $\boldsymbol{b}$ is calculated from the relation $T\boldsymbol{b} = \boldsymbol{q}$, where $\boldsymbol{q}$ is a vector, the size of which is proportional to the amplitude of the incident wave, and T is a matrix:

$$T_{nl,n'l'} = \delta_{nn'}\delta_{ll'} - (1 - \delta_{nn'}) e^{i(l'-l)\phi_{n'n}} H_{l-l'}^{(1)}(KR_{nn'}) s_{l'}, \tag{2}$$

where δ is Kronecker's delta, $R_{nn'}$ is the distance between the centers of the nth and n'th rods, and $\phi_{n'n}$ is the angle that indicates the direction of the n'th rod center as viewed from the nth rod center. Here, s_l is a parameter related to the boundary conditions at the rod surface: see the previous report [6] for its details.

2.2. Modes and lifetimes

To determine the photon lifetime in the photonic atoll, we assume real dielectric constants (i.e., no optical gain) and a complex photon frequency $\omega = \omega' - i\omega''$. Since the frequency dependence of the amplitude of the resonance scattered-wave follows the Breit-Wigner formula [23], the first-order pole of the scattered-wave amplitude gives the complex frequency $\omega_m = \omega'_m - i\omega''_m$ of the resonance mode. Hereafter, we use the subscript m to indicate the specific mode obtained. Since the electric-field amplitude has to diverge at $\omega = \omega_m$ irrespective of position $\boldsymbol{r}$ in $\varphi(\boldsymbol{r})$, this divergence must occur in vector $\boldsymbol{b}$. This implies that the condition $det(T) = 0$ determines the complex frequency ω_m. The photon lifetime of the relevant mode is given by $\tau_m = 1/2\omega''_m$ and the Q factor is given by $\omega'_m \tau_m$.

Here, we refer to the physical meaning of the above method for determining the photon lifetimes. The imaginary part ω''_m of the complex frequency thus determined must be positive since the lifetime is positive. The positive ω''_m means that k''_m is positive due to the relation $\omega_m = ck_m$ in the air, i.e., $\omega'_m - i\omega''_m = c(k'_m - ik''_m)$, where $k_m = k'_m - ik''_m$ is the complex wave number and c is the light velocity (the positive value). Since the 2D scattered wave behaves like $\exp(ik_m r)/\sqrt{r} = \exp(ik'_m r) \cdot \exp(k''_m r)/\sqrt{r}$ at large r, we find that it diverges at the limit of $r \to \infty$ because $k''_m > 0$. This may appear to be unusual, because it is as if light be amplified despite the absence of optical gain in the present physical system. Note, however, that this is true. This actually occurs because the resonance state decays exactly at this resonance frequency to magnify the light intensity outside the PC (not due to gain). In this consideration, the temporal variation of the field should be taken into account at the same time: the light field decreases with the factor $|\exp(-i\omega_m t)| = \exp(-\omega''_m t)$ since $\omega''_m > 0$. The overall behavior of the light field is described by the product of the two factors: the increasing spatial part and the decreasing temporal part. The total light field is thus known to remain unchanged at the simultaneous limits of $r \to \infty$ and $t \to \infty$. We find that the light field energy is conserved during the whole decaying process of the resonance states. This is in marked contrast to the case where the PC has optical gain and therefore the light field energy in the total system is amplified.

2.3. Threshold amplitude-gain for laser oscillation

In the calculation of lasing thresholds [6] in the photonic atoll, we assume that every rod has the same optical amplitude gain K''_a that is the negative imaginary part of the complex wavenumber ($K_a \equiv K'_a - iK''_a$) of light propagating in material A. We introduce the complex dielectric function in order to describe the light amplification in the rod:

$$\varepsilon_a(\omega) = \varepsilon_{a0} - i2c\sqrt{\varepsilon_{a0}}K''_a/\omega, \tag{3}$$

where the photon frequency ω is a real value. Here, ε_{a0} is the dielectric constant of material A in the absence of gain. Taking this complex dielectric function into account in the calculation of the scattered waves, the expansion coefficients $\boldsymbol{b}$ in Eq. (1) can be uniquely determined as $\boldsymbol{b} = T^{-1}\boldsymbol{q}$ when the inverse matrix T^{-1} exists. When there is no incident light wave, we know that $\boldsymbol{q} = 0$, hence $\boldsymbol{b} = 0$, and so we obtain no scattered wave: $E^s(\boldsymbol{r}) = 0$. Note, however, that there is an exception: if the inverse matrix of T (i.e., T^{-1}) does not exist, we can observe a finite intensity of light even for no light-wave incidence. This is nothing other than a laser oscillation, if it exists. The condition for the nonexistence of T^{-1}, i.e., det(T)=0 can therefore be regarded as the laser oscillation condition. Since matrix T is a complex function of both the photon frequency ω and the amplitude gain K_a'', we can obtain the mode frequency ω_m as well as the threshold amplitude gain K_{am}'' for laser oscillation by searching for the pair of variables (ω, K_a'') at which the determinant for T vanishes. The mode frequency values ω_m thus determined must coincide with those obtained in Sec. 2.2. The easily-oscillating modes have relatively low K_{am}'' values, while those which do not oscillate have higher K_{am}'' values. Therefore, we call the modes the *unlasing modes*, which do not laser-oscillate even under very high K_a'' values that exceed 1.0 (in the units of $2\pi / L$, where L is the period of the rod array: see the first paragraph in Sec. 5 for this normalization).

3. Modes and lifetimes in photonic atolls

The schematic photonic-atoll structure is shown by the inset in Fig. 1. It consists of periodically arranged 50 GaAs rods (with the dielectric constant ε_a=13.18) in the background material air (ε_b=1.0). The typical atoll shape is a perfect circle with the filling factor $f = d / L$ (d: rod radius, L : array period) of 0.45. The expansion up to $|l| \le l_{\max}$=12 was used in Eq. (1) on the basis of the detailed study of its convergence. By numerically solving the equation det(T)=0, we obtain the root ω_m with a sufficient accuracy even for very high Q. Because of the scaling rule that holds in our calculation in a similar manner to in the PCs, the ω and τ values normalized in the units of $2\pi c / L$ and L / c respectively are determined by f (neither d nor L). Hence, the Q factor obtained is independent of the choice of L . Here, we simply use ω instead of ω', the real part of the complex angular frequency, to represent the mode frequency in the description of the results.

3.1. Mode distributions

Figure 1 shows the distribution of optical modes and Q factors for a circular photonic atoll with the filling factor f=0.45. These modes seem to be grouped into several bundles separated by regions with no optical modes. As will be made clearer (see Figs. 2 and 3), the mode bundles and hiatuses seen in Fig. 1 are respectively thought to be photonic bands and band gaps created by the periodic loop array of microstructures, i.e., by the q1D PC.

Let us call these bands #1, #2... etc. from lower to higher frequencies. The #2 and #4 bands are very narrow, but the others are so wide they can be regarded as real bands. These narrow bands are not localized modes, however, because this structure does not contain disorders causing light localization. Actually, these modes have extended (unlocalized) distributions of the light intensity [see Fig. 3(b)]. Although these Q factors have been calcu-

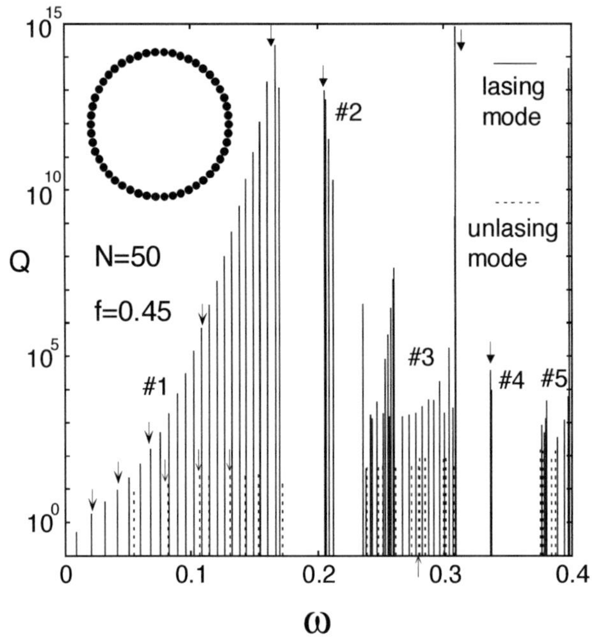

Figure 1. Distribution of optical modes on the frequency axis for a closed circular array of 50 GaAsmicrorods (see the inset). The heights of the columns indicate the values of the Q factors for the modes, ω is normalized in the units of $2\pi c / L$, and the filling factor $f = d / L$ (d: rod radius, L: array period) is 0.45. Here, no optical gain is considered in the calculation of the Q factors (see Sec. 2.2). Modes shown by solid and dotted lines respectively indicate those which lase and unlase (see Sec. 2.3 for the definition of *unlase*). See also Fig. 3 for the modes denoted by arrows.

lated assuming no optical gain, another examination assuming optical gain in every rod revealed that these modes can be classified into two types depending on whether they did lase (solid lines) or did not lase even by giving very high optical gains (dotted lines, see also §2.3). As shown in Fig. 1, the Q factors (i.e., those obtained assuming no gain) for the lasing modes are very high whereas those for unlasing modes are relatively low (10 to 100). What causes this difference will be clarified later (see Sec. 3.3). In the first band #1, the Q factor for lasing modes increases rapidly toward the band edge and reaches a maximum near it. Although the Q-factor variation for lasing modes in other bands is more complicated, the Q factor there also tends to become higher at the band edges. At the top edge of band #3 (ω_m=0.3097), the Q factor reaches the extremely high value of 0.8×10^{15}. This high Q would allow light to stay in the photonic atoll for about 1 second, a surprisingly long time for the visible light, if we assume ideal circumstances that let us neglect other losses. Although one may think that these high-Q modes have been created fortuitously, we have confirmed that they are always observed at the same band edges in this kind of structure. These results indicate that the atoll structure has a potential to confine light very strongly: its geometry inherently involves the high-Q effect.

The significance of the above results can be better understood by comparing them with the results obtained with other resonators. The Q factors of modes at the band edge of a 2D PC with 53 rods are ~100 [7]. The Q factors for modes in a photonic atoll, in contrast, are more than 12 orders of magnitude larger (~10^{15}) despite the fact that both stem from the extended modes in similar-sized resonators. The high-Q localization also occurs in a defect mode in the PC with lateral Q factors of the order of 10^4 [8, 9], which are still lower than the present Q. This localization occurs around a defect that is sufficiently separated from the outer environment so that the coupling between them is cut off. This implies that the defect has to be surrounded by a considerable volume of the periodic medium (PC). It is actually reported that nearly 400 rods are needed to isolate the defect mode in a PC [8]. The presence of so many rods despite the smallness of the defect would be disadvantageous with regard to incorporating an optical cavity in the PC. The present photonic atoll permits us to obtain much higher Q factors by using a small number of rods with little deterioration of them by the presence of dielectric materials near it. This demonstrates the advantage of this resonator over the defect mode localized in a regular PC. These Q factors are also found to be higher than those for the WGMs in a rod (~10^{10}) [17]. Although the present model does not consider the vertical Q that comes to play a certain role in the slab, we find that the photonic atoll greatly strengthens the confinement of light propagating in the 2D space.

3.2. Filling-factor effects

The finding of bands and band gaps has given impetus to the study of filling-factor effects, as is often carried out in the ordinary PCs [26]. Figure 2 shows the variation of the positions of bands (shaded areas) and band gaps (blank areas) as a function of the f value. Here, we focused on bands and band gaps created by lasing modes (solid lines in Fig. 1), because they are verified later to be generated along the rod loop (see Fig. 3). The vertical broken line corresponds to Fig. 1. Let us scan the results from low to high f values. Since $f=0$ implies uniform air (no rods), the mode distribution is continuous (i.e., without bands and band gaps). With increasing f, the closed periodic array structure comes into existence and as a result the continuous free-space dispersion gradually splits into several bands: this occurs at $f<0.1$ (not explicitly shown in Fig. 2). We see that a large band gap is produced between bands #1 and #3. Further increase in f splits band #3 to create a new narrow band #2. This new band remains narrow until f reaches its maximum value. A similar phenomenon occurs in band #5, which splits to form a narrow band #4 around $f=0.4$. The first band gap formed between bands #1 and #3 ($f<0.35$) appears to be maximized at a certain filling factor that is intermediate between 0 and 0.25, because the band gap vanishing at $f=0$ undoubtedly increases for $f>0$ but decreases for $f>0.25$. The existence of an optimum filling factor for the large-gap generation is similar to what is seen in the ordinary PCs [26]. The formation of narrow bands for $f>0.35$ is not well understood, but some modes with nodes produced in the loop-radial direction may be involved in their formation because of the increased rod radius for higher f values. These results substantiates for the first time the creation of photonic bands and band gaps in a q1D looped array structure like the photonic-atoll resonator proposed here.

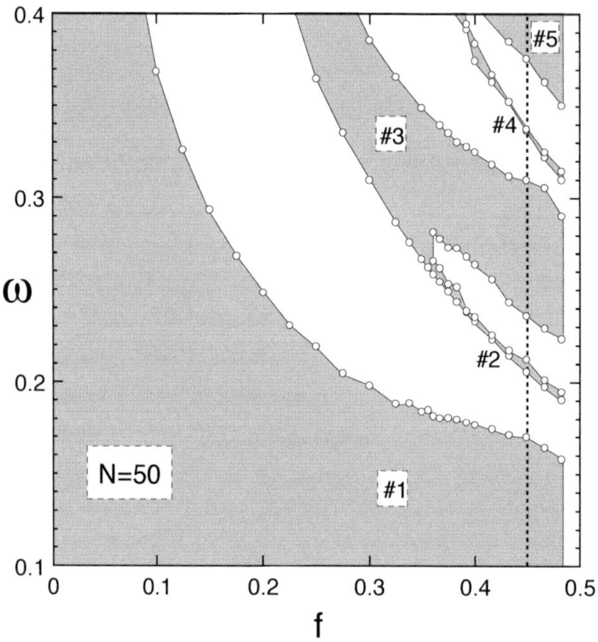

Figure 2. Variation of the bands (shaded areas) and band gaps (blank areas) formed by lasing modes (solid lines in Fig. 1) as a function of rod filling factor *f*. *f* =0 corresponds to only air (no rods), and *f* =0.5 corresponds to the configuration in which neighboring rods touch one another.

3.3. Light intensity distributions

In order to clarify what occurs for these modes, we next investigate the field intensity distributions in the photonic atoll. Note again that no gain is assumed for this calculation. We first select several lasing modes from band #1, because the modes in the first band with relatively long wavelengths are expected to provide a variety of clues to the understanding of the fundamental processes of light localization. Figure 3(a) shows the light-intensity distributions for the four lower lasing modes in band #1 (indicated by arrows in Fig. 1). In the colored figures, the intensity increases in the order blue, white, yellow, red, and black. The numerals in the figure are the mode frequency values and from left to right in Fig. 3(a), they respectively correspond to Q factors of 1.7, 9, 150, and 6.6×10^5. Although the light confinement is not very strong for these modes, we can clearly recognize the process in which light comes to be localized along the loop as Q increases. We also find a noticeable variation of the field distributions. First, the lowest-frequency mode (ω_m=0.0216) appears to have two loops and nodes of light waves along the array loop. This implies that the wavelength λ is comparable to the circumference D of the circular loop. If we assume the light propagation along the array loop with a wavenumber vector $\boldsymbol{K}=K\boldsymbol{e}$ ($\boldsymbol{e}$ is the unit vector along it) and a light velocity $v=c/n_{eff}$, this mode gives an effective refractive index n_{eff} of about 0.93. This value is close to the n_{eff} of 1 for air, which is reasonable since most light is leaked into the air

http://dx.doi.org/10.5772/53435

because of the small Q of this mode. The second mode (ω_m=0.0421) appears to have six loops and nodes giving $\lambda \sim D/3$, which leads to $n_{eff} \sim 1.4$. Further increase in the mode frequency enables us to observe distinct light localizations toward the rod array. The modes ω_m=0.0678 and ω_m=0.1086 respectively give wavelengths of $\lambda \sim D/6$ and $\sim D/12$. The similar estimation of n_{eff} leads to respective n_{eff} values of 1.8 and 2.2 for these modes. The gradual increase in the estimated n_{eff} toward $\sqrt{\varepsilon_a}$=3.6 for the rods reconfirms the increased light confinement in the rod array loop. The light confinement along the rod chain may also be construed by the coupling between a WGM mode in a rod and the one in its neighboring rod via their Fano resonances with the outer region [27].

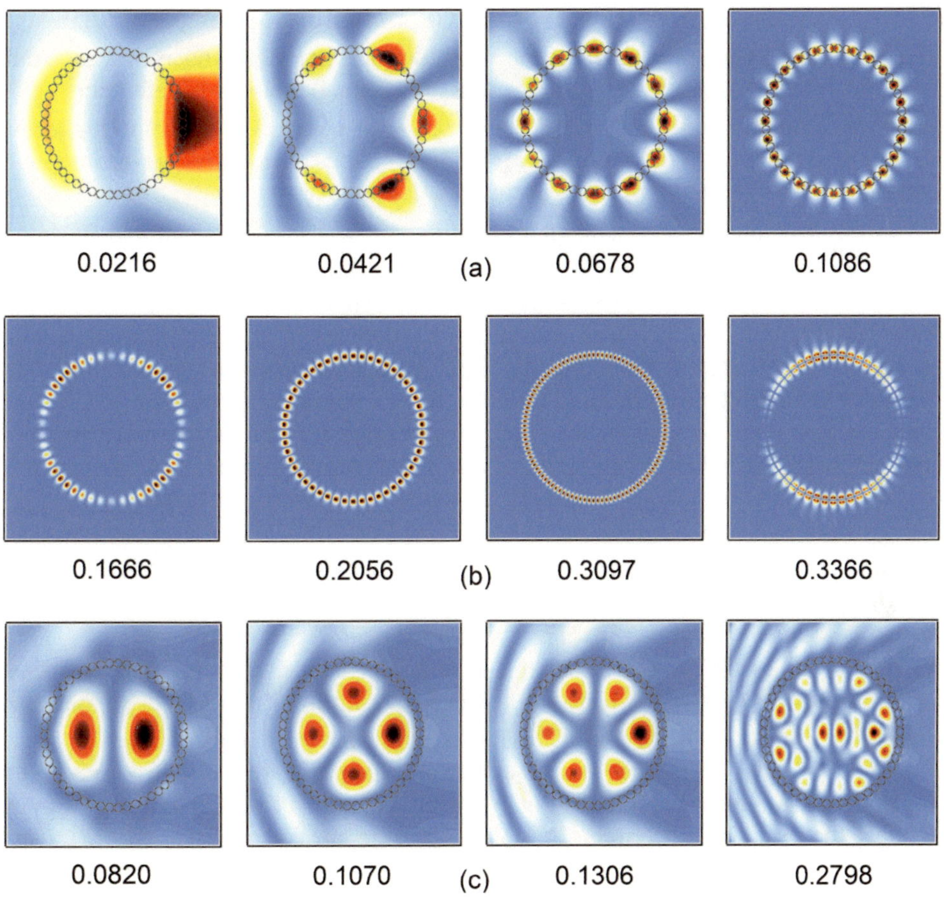

Figure 3. Color) Light intensity distributions for (a) lasing modes in the first band #1, (b) lasing modes with very high Q-factors, and (c) unlasing modes (see §2.3 for the definition of *unlasing modes*). Here, note that no optical gain is assumed for the calculation of these intensity distributions. The numeral shown under each figure is the normalized-mode frequency: see Fig. 1, where different symbols of arrows are used to distinguish the modes in Figs. 3(a), (b) and (c). The intensity increases in the following order: blue < white < yellow < red < black. Rod positions are indicated in Figs. 3(a) and 3(c) by black circles but for clarity they are not indicated in Fig. 3(b).

Several examples for the modes with very high Q-factors are shown in Fig. 3(b). These modes are selected as those having the maximum Q factor values in each band from #1 to #4 (see the arrows in Fig. 1). The strong light-confinement for these modes is confirmed by comparing the intensity (i.e., unity) of the incident plane-wave and the maximum light intensity in the rod array: the first three modes (ω_m=0.1666, 0.2056, and 0.3097) all have intensity of the order of 10^{11} and the last one (ω_m=0.3366) the order of 10^5. Note that this intense light is obtained in the array with no optical gain (no gain is assumed here!). It is entirely due to the extremely long photon lifetimes attained by the use of photonic-atoll resonators. Although light is focused on the rod array, its intensity distributions are not easy to construe. Let us focus on the bright regions, which are the loops of light waves. While the contour of bright regions for the mode in band #1 (ω_m=0.1666) is simple, the contour of bright regions for the mode in band #2 (ω_m=0.2056) contains two brightest points (not clearly seen for a small drawing). The mode in band #3 (ω_m=0.3097) has many bright regions (100, or twice the rod number of 50) though its contour is simple. The most extraordinary results are found in the mode in band #4 (ω_m=0.3366), for which the bright region consists of two small units separated in the radial direction of the loop and each small unit has two brightest points. The modes in higher-index bands thus tend to become complicated. This evidently stems from the higher-order Bragg reflections to create these bands, which occur in the q1D closed array.

Shown in Fig. 3(c) are for the modes that never lase even if they have very high gains (the modes shown by dotted lines in Fig. 1). Light for these modes is clearly confined in the inner region of the atoll but not along the rod array. We see an increase in the number of loops and nodes as the mode frequency increases, indicating that they are formed by light trapped in the inner region and reflected at the rod array of the atoll. The observed Q factors (10–100) are small despite the expectations for the WGMs to produce strong light confinement. The above results are reasonable, however, because both inner and outer regions are made from the air and hence the thin array loop does not serve as a solidly made pool for light. We also understand that these modes do not lase because light stays in the region with no gain.

4. Shape effects of photonic atolls

In this section, we assume the PA that consists of 20 GaAs rods (with the dielectric constant ε_a=13.18) in the air (ε_b=1.0). We consider a variety of elliptical PAs created by changing its eccentricity e from 0 to 0.968. For all the PAs studied here, however, the filling factor $f = a/L$ (d: rod radius, L : period) is fixed at 0.45 and the period of the rod chain is assumed to be the same. Moreover, we modify the PA form keeping its circumference fixed in order to facilitate the comparison between the PAs with different eccentricities. The angular frequency ω and the lifetime τ are expressed in the units of $2\pi c/L$ and L/c, respectively. Here, we again simply use ω instead of ω', the real part of the complex angular frequency, to represent the mode frequency in the description of the results.

4.1. Splitting of degenerate modes

Prior to showing the detailed properties of the PAs, we first present the basic results for the optical modes created in the PA. Figure 4 shows the angular frequency ω positions for the optical modes (denoted as n=0, 1, 2, …, 10 for low to high ω modes), where the height of the columns indicates the lifetime τ of these modes. The thin lines are the results for the circular PA (e=0) shown in the inset. In the limit of the infinite number of rods, these modes get accumulated densely to form the first photonic band of the q1D closed photonic crystal [20] and hence the shaded region in Fig. 4 can be regarded as the first photonic band gap. As shown in Fig. 4, the lifetime becomes longer with the mode frequency that is increasing and approaching the first band edge. Here, let us examine the specific numerical values. We take mode 10 as an example with the dimensionless values $\omega = 0.1691$ and $\tau = 6.30 \times 10^6$, and $L = 0.1 \mu m$ for the periodicity. These values give the actual frequency $\omega / 2\pi = 510\,THz$ (the visible light with wavelength $\lambda = 0.59\,\mu m$) and the actual lifetime $\tau = 2.1\ ns$. The above results demonstrate the presence of the enhanced confinement of light near the band edge [6, 7]. The thick lines indicate the results for an extremely deformed PA (e=0.866), the form of which is also shown in the inset. We see the lifetime remarkably decreased by the use of the elliptical PA, which occurs more pronouncedly for higher modes. The examination of a variety of elliptical PAs showed that any PA modifications caused the decrease in the lifetime. We have to admit that these results for the deformed PAs are discouraging from the viewpoint of the achievement of longer lifetimes. Here, we observe some notable phenomena, however: twin modes are isolated in the vicinity of the circular-PA modes with n=1, 2, …, 9. Note that no twin modes are created for n=0 and 10.

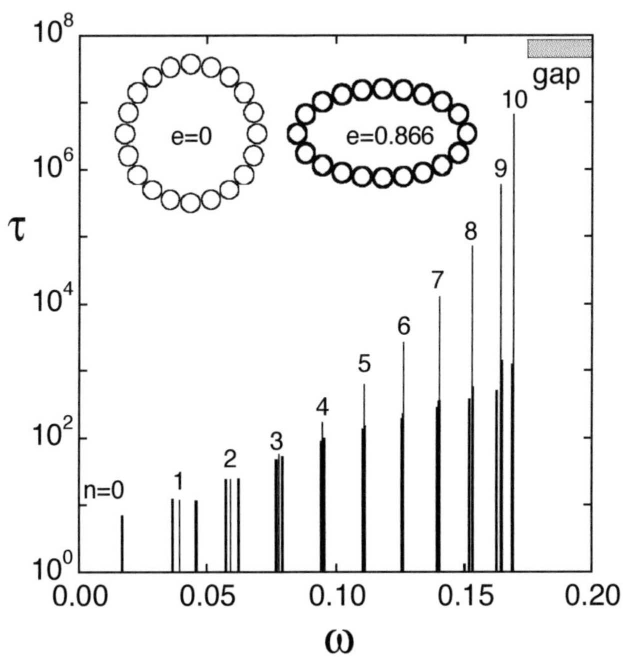

Figure 4. Optical modes distributed on the frequency axis, where the height of the columns indicates the lifetime of these modes. The thin and thick lines are the results for the PAs as shown in the insets: a perfect circle (e=0) and an ellipse (e=0.866), respectively, where e is the eccentricity of the elliptic PA. Here, several thin lines are hidden behind thick lines since the latter are superimposed onto the former. In this paper, the angular frequency $\omega$and the lifetime τ are normalized in the units of $2\pi c / L$ and L / c, respectively, where L is the period of the PA chain.

The similar studies have been carried out for elliptical PAs with a variety of eccentricities. The results are summarized in Fig. 5, which shows the variation of the mode frequency as a function of the eccentricity *e* (from 0 to 0.968). As clearly shown in Fig. 5, each of modes 1-9 is found to split into two with the increasing eccentricity. Those modes that are increasing and decreasing, respectively, with the growing eccentricity are denoted by open and close circles. While we succeeded in locating almost all split-modes very precisely, we failed in isolating several higher modes (open circles) split for mode 1 at e>0.954. Here, we can read in Fig. 5 what follows. First, the splitting width strongly depends upon the optical mode as well as the PA form. In fact, modes 1, 2, and 9 already exhibit slight but clear splittings even at relatively low *e* values (e<0.5), while modes 4, 5, and 6 do not split until it reaches a value higher than 0.8. These results will be more clearly displayed later in Fig. 6. Second, we also recognize a strong mode-dependence of the frequency deviation from the original one (e=0). Let us denote the deviations by $\Delta\omega_H \equiv \omega_H - \omega$ and $\Delta\omega_L \equiv \omega - \omega_L$ for the modes shifted in the higher and lower directions, respectively. We can paraphrase the above facts as follows: $\Delta\omega_H > \Delta\omega_L$ for modes 1, 2, 3, and 4, $\Delta\omega_H \simeq \Delta\omega_L$ for modes 5 and 6, and $\Delta\omega_H < \Delta\omega_L$ for modes 7, 8, and 9, when they are compared at a fixed eccentricity. These facts suggest the presence of different mode-splitting mechanisms between the modes near the Γ point, the modes in the middle of the band, and the modes near the band edge. Finally, let us briefly refer to modes 0 and 10. These modes did not split for all the *e* values studied here. However, noteworthy here is that mode 0 slightly increases while mode 10 decreases to a certain extent, according as the eccentricity grows. This fact again suggests the difference in the behavior between the near-Γ-point modes and the near-band-edge modes.

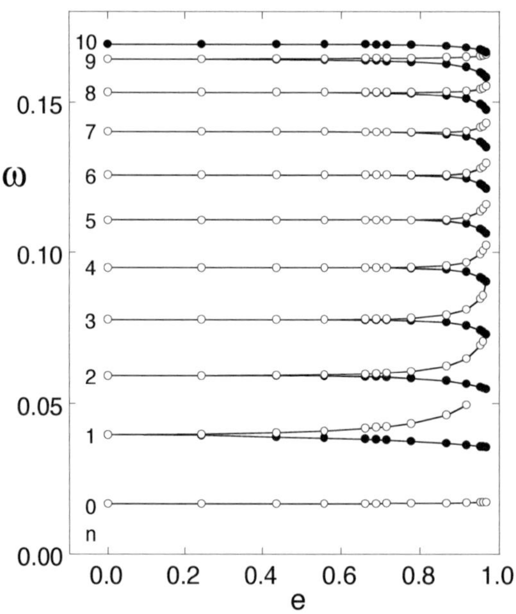

Figure 5. Frequency positions of all modes in the first band as a function of the eccentricity e for the elliptical PA. Here, the upper and lower modes split by the PA modification are denoted by open and closed circles, respectively (except for modes 0 and 10).

Figure 6 shows the mode separation $\Delta\omega \equiv \omega_H - \omega_L$ as a function of the frequency position of the mode, which is evaluated at a variety of the PA eccentricity values. The *e* value for each curve is given in the caption of Fig. 6. As mentioned before, modes 0 and 10 have no split modes. What is intriguing here is that the separation $\Delta\omega$ is not a monotonic function of the mode frequency: it becomes more prominent as they approach the bottom or the top of the band and moreover exhibits a minimum at the middle of the band. When we look at Fig. 6 precisely, the above phenomena are found to occur more pronouncedly for a slightly deformed PA: see, e.g., the results for $e=0.243$, in which the ratio of $\Delta\omega$ between mode 1 and mode 5 reaches as high as 1.7×10^5. On the other hand, in the extremely deformed structures, we find no significant mode dependence of $\Delta\omega$ though these modes have larger frequency splittings. This kind of phenomena has not been observed in the finite-sized optical resonators and even in the similar mode-splitting phenomena [22-24] referred to in Sec. 1 either.

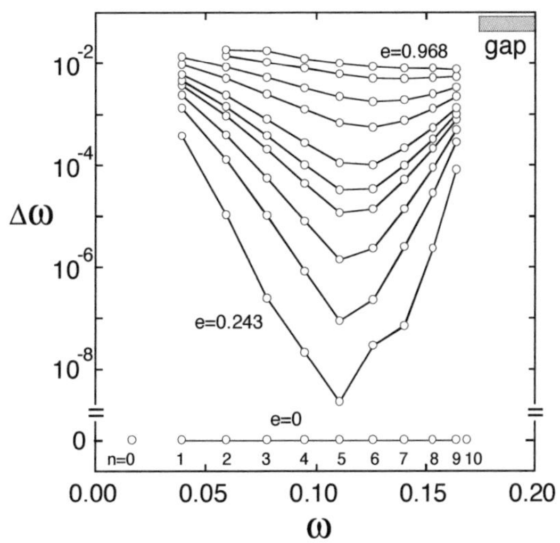

Figure 6. Relations between the split-modes separation $\Delta\omega$ and the mode frequency position for several eccentricity values. Here, the e value ranges from 0, 0.243, 0.436, 0.558, 0.661, 0.714, 0.777, 0.866, 0.916, 0.954, and 0.968, for the curves displayed from bottom to top, respectively.

4.2. Light intensity distributions

The results mentioned in Sec. 4.1 have prompted us to investigate the light field distributions for these modes. Hereafter, we focus on the PA with $e=0.866$ and modes 2, 5, and 9, which have been selected as the optical modes located near the Γ point, in the middle of the band, and near the band edge, respectively. We denote, in what follows, the higher and lower modes split as nH and nL, respectively, for mode n.

Figure 7 shows the light intensity distributions for (a) mode 2L with θ_i=0 , (b) mode 2L with θ_i=90 , (c) mode 2H with θ_i=0 , and (d) mode 2H with θ_i=90 . Here, θ_i is the incident angle of the plane wave of light. The rod array of the PA is also displayed together with the distributions in Fig. 7. Here, the light intensity increases in the order blue, white, yellow, red, and black. Since our calculation is based on the scattering-theoretic method [6], the incident plane wave is included in the distributions as a matter of course. As shown in Fig. 7, these modes have four nodes and loops along the rod chain of the PA and show the weak light confinement because of their shorter lifetimes (see Fig. 4). We find these modes to be excited by the irradiation of light from any directions including 0 *and* 90 shown in Fig. 7, although the maximum intensity of light excited in the PA somewhat differs depending on the directions of incidence. Since its difference reaches only a few times, however, we may conclude that there is no preference in the irradiation direction for their excitation in this case. This fact presents a great contrast to the band-edge modes as shown later (Fig. 8). Although modes 2H and 2L are thus excited by light with any incidence directions, their light distributions depend entirely on the irradiation direction, as shown in Fig. 7. In addition, the oblique incidence with, e.g., θ_i=45 creates light distributions like those obtained by rotating Figs. 7 (a) and 7(c) by 45 . In other words, their wave functions remain uncertain for the unirradiated PAs. This fact is suggestive of the similarity to the electronic mode in an atom. Once the PA is irradiated by a plane wave of light, however, their wave functions are uniquely determined as follows: the incident wave selects their wave functions in such a manner that it can excite the eigen modes based on the symmetry matching between them. The irradiation direction thus works as the quantization axis in the quantum theory. When we look at Fig. 7 in more detail, we find that most light is focused around the downstream side of the PA chain for mode 2L. For mode 2H, on the other hand, we recognize light staying around the upstream side of the PA chain though some light is still around the downstream side. The massive flow of the incident beam generally tends to cause the light distribution to be more highlighted at the downstream side [20], which could correspond to an energetically more stable state. Taking this circumstance into account, we may come to a reasonable conclusion that modes 2L and 2H, respectively—energetically stable and unstable states—concentrate around the downstream and upstream sides. This is true for all incident angles and all other modes near the Γ point. We thus have made clear the difference between the light fields for the modes—located near the bottom of the band—that are split by the modification of the PA structure.

Next, we display the results for the modes near the band edge as a matter of convenience for explanation. Figure 8 shows the light intensity distributions for (a) mode 9L with θ_i=0 , (b) mode 9L with θ_i=90 , (c) mode 9H with θ_i=0 , and (d) mode 9H with θ_i=90 . As can be seen in Figs. 8(a) and 8(d), these modes have 18 nodes and loops along the rod chain of the PA and exhibit somewhat strong light confinement because of their relatively long lifetimes (see Fig. 4). The most striking feature for these modes is found in the pronounced θ_i-dependence of their excitation. In fact, as clearly shown in Fig. 8, mode 9L is excited by the irradiation

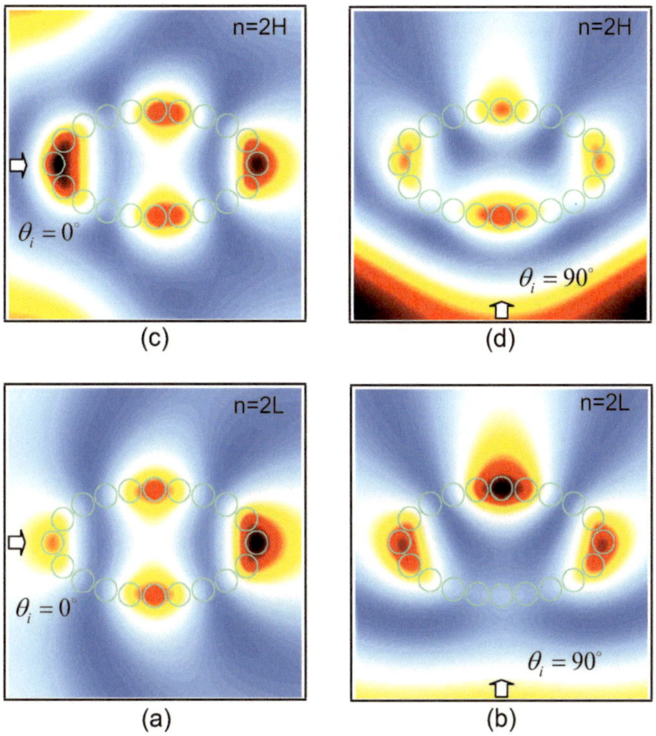

Figure 7. Light intensity distributions for (a) mode 2L with θ_i=0°, (b) mode 2L with θ_i=90°, (c) mode 2H with θ_i=0°, and (d) mode 2H with θ_i=90°. Here, θ_i is the incident angle of the plane wave of light. The rod array of the PA is also displayed together with the distributions. Here, the light intensity increases in the order blue, white, yellow, red, and black.

with the incident angle 0 whereas it is not by the 90 irradiation. In the similar manner, mode 9H is excited by the irradiation with the incident angle 90 whereas it is not by the 0 irradiation. A very high value of several hundreds is reached for the ratio of the maximum intensity of light confined around the rod array between the incident directions causing (e.g., 90 for mode 9H) and not causing (e.g., 0 for mode 9H) its excitation. These results, when viewed from another point, demonstrate that mode 9H is excited efficiently by the irradiation of light that excites mode 9L less efficiently, and vice versa. The same phenomena are confirmed to occur for all other modes near the band edge for the irradiation from any directions and even in the slightly modified PA (e=0.243). From these results, we speculate that these modes are orthogonal to each other since optical mode can be excited only by the light beam with the same symmetricity as the relevant mode. This fact provides a striking contrast to the modes near Γ point (see Fig. 7), which are excited by the irradiation with any incident directions.

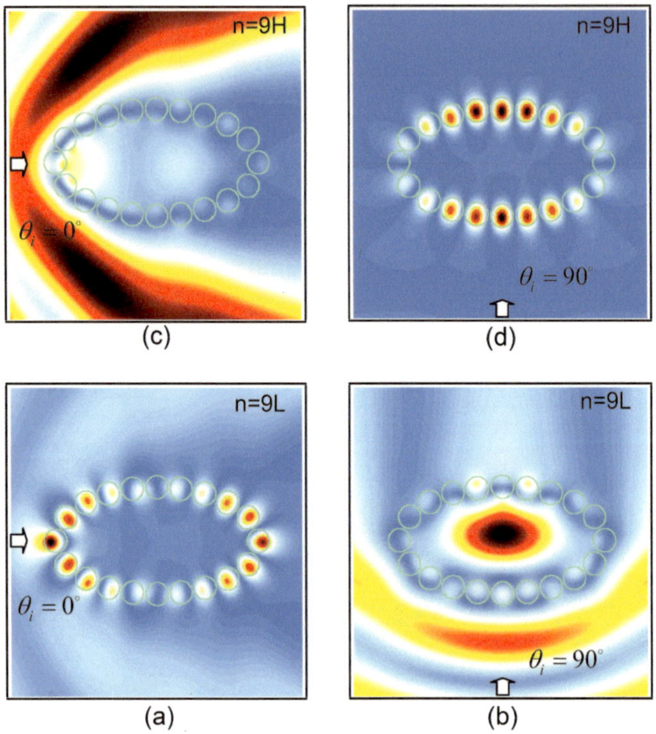

Figure 8. Light intensity distributions for (a) mode 9L with θ_i=0°, (b) mode 9L with θ_i=90°, (c) mode 9H with θ_i=0°, and (d) mode 9H with θ_i=90°.

Finally, we briefly mention the behavior of the modes in the middle of the band. Figure 9 shows the light intensity distributions for (a) mode 5L with θ_i=0 , (b) mode 5L with θ_i=90 , (c) mode 5H with θ_i=0 , and (d) mode 5H with θ_i=90 . These modes have 10 nodes and loops along the rod chain of the PA. In contrast to modes 2L and 2H mentioned before, these modes have no pronounced concentration of the intensity distribution on the upstream or downstream sides of the PA. Moreover, they exhibit no irradiation-direction dependence of the excitation, which has been detected for modes 9L and 9H. These modes are thus known to have the characteristics that are intermediate between the near-Γ-point modes and the near band-edge modes.

4.3. Discussion

Let us discuss the creation of eigen modes and their splitting by the structural modification of the PA resonator. For this purpose, we simplify the discussion by regarding the closed q1D chain as a closed pure-1D system with the position variable x along the circumference. The periodic boundary condition can be applied to this system exactly for its closed structure, and the Bloch theorem for the L-periodicity of the PA. We thus obtain the optical

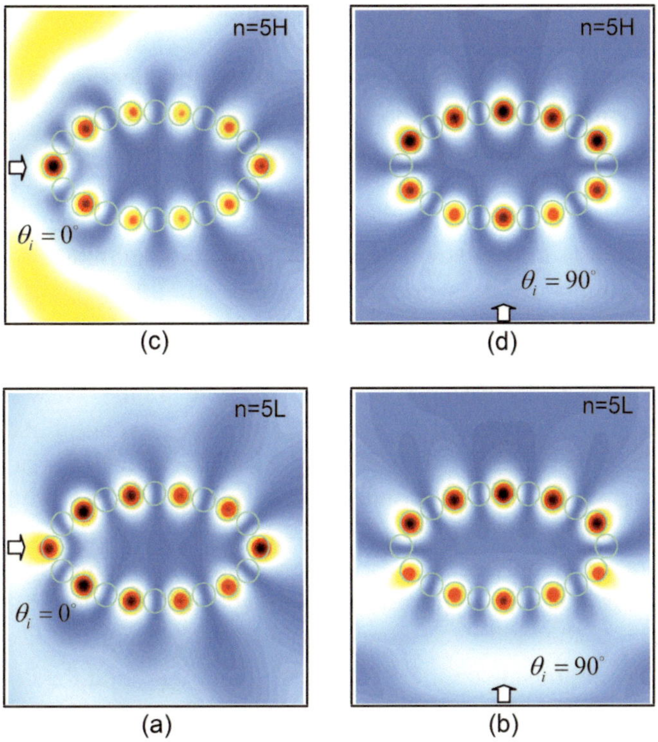

Figure 9. Light intensity distributions for (a) mode 5L with $\theta_i = 0°$, (b) mode 5L with $\theta_i = 90°$, (c) mode 5H with $\theta_i = 0°$, and (d) mode 5H with $\theta_i = 90°$.

modes specified by the wave number $k_n = g(n/20)$ for the wave propagating along the chain with 20 rods, where $g = 2\pi / L$ is the fundamental vector in the reciprocal space. Here, n ranges over –9,…, –1, 0, +1,…, +9, and +10, creating 20 modes. Note that n= –10 is excluded since it is identical to n= +10 in the reciprocal space. The modes with n are thus known to be degenerate with those with $-n$ for n=1, 2,…, 9 and their wave functions are the complex conjugate of each other, which propagate in the opposite directions with the wave numbers $+k_n$ and $-k_n$, respectively. Moreover, we understand that the lowest mode 0 and the band-edge mode 10 are not degenerate: here, we note in passing that the band-edge mode is doubly degenerate for the PA with the odd-numbered rods, e.g., 21 rods.

When we look at Figs. 4-6 together with the above considerations, it is not unusual for modes 0 and 10 to remain single under any perturbations given to the structure because of their nondegeneracy. As for the other modes (n=1, 2,…, 9), it is reasonable to consider that their degeneracy is lifted by the modification of the PA structure. If we apply the group theory to this phenomenon straightforwardly, it may be said that the degeneracy lifting is caused by the reduction of the rotational symmetry in the whole PA structure. Although this is an elementary but important interpretation for these degeneracy-lifting phenomena, we here refer to another perspective. Under the assumption to regard the PA as a very long closed 1D structure—it is actually possible as mentioned before, the waves propagating in the opposite directions ($+k_n$ and $-k_n$) ought to split their frequency only in the presence of some one-way (asymmetric) perturbation along the chain. According to this perspective, these modes should not split by any complicated deformation of the PA structure. This tendency will be magnified in the PA with a larger number of rods, because such a PA—when we focus on its local part—is equivalent to an isolated string of 1D array: no mode-splittings would occur particularly in the extremely large PA (with the infinite number of rods). The PA with a smaller number of rods, on the other hand, will lift the mode degeneracy more easily by the shape modification, since such a PA can no more be regarded as an isolated string of 1D array for the smallness of the whole PA. Actually, we confirmed that the mode splitting $\Delta\omega$ is a rapidly decreasing function of the rod number N: we obtained 0.0097, 0.0060, and 0.0036 for N=10, 20, and 50, respectively, for, e.g., mode 1 at e=0.776. We thus recognize that it is important to take into account the q1D feature of the closed PA structure as well as to consider it from the group-theoretic standpoint. This presents a great contrast to the similar phenomena for the modes in the 2D or 3D structures, for which the group-theoretic considerations would suffice. Next, we would like to refer to the unusual $\Delta\omega - n$ relation in Fig. 4. It is interesting to note that this n-dependence of $\Delta\omega$ is very different from the Stark effect [23] of the Hydrogen atom for which the splitting width varies simply like $\Delta\omega \propto n^2$. For the modes of smaller n, light is loosely bound around the PA because of their shorter lifetime, which ought to render its intensity distributions more sensitive to the PA modification. The modes near the band edge, on the other hand, have the lifetime that is barely retained very long in a fixed (symmetric) PA structure. This implies that their life is vulnerable even to a slight perturbation to the structure and hence its abrupt reduction may cause marked splittings of degenerate modes.

As mentioned on the light field distributions in Sec. 4.2, the structurally deformed PAs have a variety of optical responses. In particular, the band-edge modes (e.g., modes 9H and 9L) exhibit a strong anisotropy of excitation. Moreover, it should be emphasized that this anisotropy is very sensitive to the modification of the structure, i.e., it occurs even under a slight modification of the structure. This implies that optical excitations can be controlled by the mechanical deformation of the structure, which could have a potential to be exploited as high-function devices such as opto-mechanical devices [28]. We therefore believe that the present results will find a number of valuable applications as very high-Q resonators in the state-of-the-art technologies for the optical information systems, which combine the mechanical forces, the electronic phenomena, and the optical processes.

5. Laser oscillations

Because of the scaling rule that holds in our calculation in a similar manner to in the PCs, ω, τ and K_a'' values normalized in the units of $2\pi c/L$, L/c and $2\pi/L$ respectively are determined by f (neither d nor L). We here use the ω, τ and K_a'' values thus normalized.

http://dx.doi.org/10.5772/53435

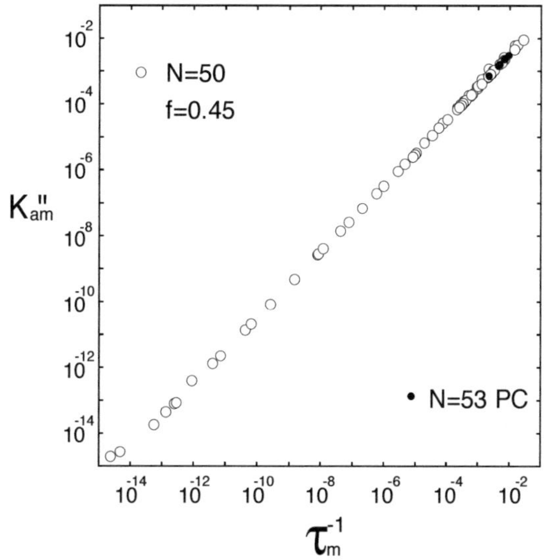

Figure 10. Relation between the threshold amplitude gain K''_{am} and the inverse photon-lifetime τ_m^{-1} for lasing modes. K''_{am} and τ_m are normalized in the units of $2\pi / L$ and L / c, respectively. The closed circles show the results for the band-edge modes near the top of the first band of the 2D PC resonator with 53 rods made of the same material as the rods in the present atoll structure (GaAs). See also Sections 2.2 and 2.3.

We studied the characteristics of a photonic atoll as a laser oscillator by assuming that every rod has the same optical amplitude gain K''_a. The method mentioned in Sec. 2.3 determines the threshold amplitude gain K''_{am} for laser oscillation. In this calculation, we did not take into account absorption and other possible losses in order to isolate the effects inherent to the resonator's geometry. Some modes did not laser-oscillate even when K''_a was very high (they were shown in Fig. 1 by dotted lines, see also Sec. 2.3). Here, we therefore consider only lasing modes (solid lines in Fig. 1). Figure 10 shows the relation between the threshold amplitude gain K''_{am} thus obtained and the inverse lifetime τ_m^{-1} for lasing modes. These points are seen to line up with a slope of 45 on a log-log plot. The inverse proportionality between K''_{am} and τ_m values is reasonable since the increased photon lifetime makes light stay in the resonator for a longer time and drives the laser to oscillate at lower optical gain. Figure 10 may be the first numerical verification of this kind of relation for the photonic-atoll resonator made of closed array of rods, though this kind of relation was also shown for a simple 1D resonator comprising a uniform medium sandwiched between two clear-cut mirrors [4]. The closed circles in this figure also show the results for the band-edge modes near the top of the first band of a 2D PC made of 53 GaAs rods [6]. When we compare the results from the two structures, the threshold gain values obtained for the PC are pessimistically higher than those for the atoll. In other words, laser oscillations with extremely low thresholds can be obtained by using our atoll structures. Noteworthy here is that the two resonators lead to very different results despite the fact that they contain a similar number of rods and that extended modes are responsible for laser oscillations in both resonators. These results again confirm the superiority of the present structure over other PC-based structures.

6. Conclusion

We have theoretically demonstrated that very high *Q* factors and resultant very long photon lifetimes can be achieved by using the closed periodic array of microstructures, which we call a photonic-atoll (PA) resonator. Although other possible losses of light remain to be considered before this structure is put to practical use, the results we obtained suggest that it would be an excellent structure for confining light. In particular, the fact that it does not require a large size to achieve a strong light confinement will prove a great advantage over other ways of light confinement when it is incorporated into optical integrated circuits. Through the investigation for the PAs with a variety of elliptical forms, we found that the photon lifetime is maximized for the symmetric (or circular) form of the resonator. This structure deformation was also shown to give rise to the degeneracy lifting for eigen modes: even a slight deformation created pronounced splitting widths especially for the near-Γ-point modes and the near-band-edge modes whereas it did not for the modes in the middle of the band. Moreover, the band-edge modes split were found to exhibit a striking anisotropy of excitations, while other modes did not show any pronounced anisotropy. These mode splittings should be discussed taking into account the q1D-dimensionality of the structure as well as considering it from the group-theoretic standpoint. We have thus clarified the metamorphoses of the eigen modes split by the modification of the PA structures. Finally, we demonstrated the PA-laser oscillations with very low thresholds, which are much lower than those for the PC band edge lasers. These results would provide much information to understand the relevant resonators more deeply, which we believe will also be possibly exploited as a very high-*Q* resonator in the future optical information processing systems.

Author details

S. Nojima

Department of Nanosystem Science, Graduate School of Nanobioscience, Yokohama City University, Kanazawa, Yokohama, Kanagawa, Japan

References

[1] O. Painter O., Lee RK., Scherer A., Yariv A., O'Brien JD., Dapkus PD., Kim I., Science 1999; 284, 1819.

[2] Chang RK., Campillo AJ., editor. Optical Processes in Microcavities: World Scientific; 1996.

[3] Berman P., editor. Cavity Quantum Electrodynamics: Academic Press; 1994.

[4] Marcuse D., Principles of Quantum Electronics: Academic Press; 1980.

[5] Nojima S., Jpn. J. Appl. Phys. 1998; 37, L565.

[6] Nojima S., J. Appl. Phys. 2005; 98, 043102.

[7] Nojima S., Appl. Phys. Lett. 2001; 79, 1959.

[8] Vučković J., Lončar M., Nabuchi H., Scherer A., Phys. Rev. E 2001; 65, 016608.

[9] Ryu HY., Kim SH., Park HG., Hwang JK., Lee YH., Kim JS., Appl. Phys. Lett. 2002; 80, 3883.

[10] Happ TD., Tartakovskii II., Kulakovskii VD., Reithmaier JP., Kamp M., Forchel A., Phys. Rev. B 2002; 66, 041303.

[11] Akahane Y., Asano T., Song BS., and Noda S., Nature 2003; 425, 944.

[12] Nojima S., Nakahata M., J. Appl. Phys. 2009; 106, 043108.

[13] Nojima S., Yawata M., J. Phys. Soc. Jpn. 2010; 79, 043401.

[14] Takeda MW., Kirihara S., Miyamoto Y., Sakoda K., Honda K., Phys. Rev. Lett. 2004; 92, 093902.

[15] Lin HB., Eversole JD., Campillo AJ., J. Opt. Soc. Am. B 1992; 9, 43.

[16] Gayral B., Gérard JM., Lemaître A., Dupuis C., Manin L., Pelouard JL., Appl. Phys. Lett. 1999; 75, 1908.

[17] Moon HJ., Chough YT., and An K., Phys. Rev. Lett. 2000; 85, 3161.

[18] Armani DK., KippenbergTJ., Spillane SM., Vahala KJ., Nature 2003; 421, 925.

[19] Gmachl C., Capasso F., Narimanov EE., Nöckel JU., Stone AD., Faist J., Sivco DL., Cho AY., Nature1998; 280, 1556.

[20] Nojima S., J. Phys. Soc. Jpn. 2007; 76, 023401.

[21] Nojima S., Phys. Rev. B 2002; 65, 073103.

[22] Nojima S., J. Phys. Soc. Jpn. 2005; 74, 577.

[23] Landau LD., Lifshitz EM., Quantum Mechanics: Non-relativistic Theory: Butterworth-Heinemann; 1981.

[24] Post EJ.,Rev. Mod. Phys. 1967; 39, 475.

[25] Abramowitz M., Stegun IA., Handbook of Mathematical Functions: Dover; 1972.

[26] Villeneuve P R and Piché M, Phys. Rev. B 1992; 46, 4969.

[27] Nojima S., Usuki M., Yawata M., Nakahata M., Phys. Rev. A 2012; 85, 063818.

[28] Eichenfield M., Chan J., Camacho RM., Vahala KJ., Painter O., Nature 2009; 462, 78.

Chapter 8

Dynamic Characteristics of Linear and Nonlinear Wideband Photonic Crystal Filters

I. V. Guryev, J. R. Cabrera Esteves, I. A. Sukhoivanov,
N. S. Gurieva, J. A. Andrade Lucio,
O. Ibarra-Manzano and E. Vargas Rodriguez

Additional information is available at the end of the chapter

http://dx.doi.org/10.5772/54118

1. Introduction

In the chapter, we give results of investigation of dynamics of linear and nonlinear photonic crystals (PhC).

It is well-known fact that modern semiconductor electronic data processing systems are experiencing fundamental problems with further improvement of the microprocessors productivity. One of the alternative ways is to use hybrid or all-optical circuits on the basis of PhCs.

The heart of such all-optical circuit is nonlinear PhC which may provide the basis for logic, memory cells, switching, local routing, power limiters, isolators, etc. Therefore, it is of crucial importance to understand the processes taking place in such components and optimize their characteristics. One of the most important points of view to the PhCs is their interaction with short and ultra-short pulses which may limit the productivity of an optical circuit. Recently, there have been proposed a great number of PhC components bases on different operating principles. However, being resonant-transmitting structures, PhCs themselves reduce the possibility to work with ultra-short pulses.

In the papers of the authors, it have been proposed to use wideband PhC filters instead of high-Q ones [9], [6]. Lower resonant properties as compared to high-Q filters, allow to reduce distortion of the signal passing through such filters. In this chapter we present the investigation results and analysis of the temporal response of different kinds of wideband PhC filters. Namely, we consider filters made of linear optical materials which can be used for local multiplexing and routing and the ones made of nonlinear optical materials which properties strictly depend on the radiation intensity.

We explain the computation process of such characteristics of the PhC filters as transmission spectra, eye-diagrams and the band structure.

The chapter is organized as follows:

In the first section of the chapter, we briefly explain theoretical background under the computation of dynamic characteristics of micro-devices. Then, in the second section, we give the results of investigation of linear wideband PhC filters. We concentrate attention on transmission spectra and an eye-diagram of such filters. Finally, we demonstrate application of the PhC filters to the wavelength division multiplexing and analyze their limitations. The third section of the chapter is dedicated to nonlinear PhC filters and their characterization. We first present one of the methods of the band structure computation of nonlinar PhCs. After this, we investigate such important application of the nonlinear PhC filters as all-optical flip-flop which may become the basis of optical data processing systems.

Although we do not provide here the detailed description of the physical processes below the presented characteristics, the reader can find them in the book "Photonic Crystals: Physics and practical modeling" [8]

2. Computing the temporal response of the PhC filter

The term PhC is usually used to define infinite periodic structure. However, such structures do not have many practical applications since they only possess artificial reflecting and refracting properties and cannot control effectively the radiation flow. To implement effective radiation flow control, we have to create at least one defect of the periodic structure to be able to localize the radiation. However, in real devices, we should be able to provide light guiding, localization and dynamic routing.

Therefore, speaking of PhC devices we are usually assume their complex structure which cannot be represented by strictly periodic variation of the refractive index. In this situation, the only way to find the field distribution inside the PhC device is to apply numerical methods. Due to recent advance in computing technologies, there have been appeared a wide variety of numerical methods giving time-dependent field distribution in complex nonlinear media. Most of them are highly time- and resource-consumable. However, the most easy to implement and, yet, quite effective is the finite difference time-domain (FDTD) method which allows computing field distribution in nonlinear complex media such as PhC devices.

In general, there have to be considered complete system of Maxwell's equations which, in linear case represents six (or even twelve [1]) equations. One simplification can be made though. Namely, most of the models are based on 2D PhC of different configuration since they possess wide photonic band gap and, on the other hand, provide enough flexibility to design wide variety of the components.

The system of Maxwell's equations can be reduced to 2D case considering certain polarization. Namely, in case of TM polarization (as referred to in [7]), we have the following system of equations [11]:

http://dx.doi.org/10.5772/54118

$$\begin{aligned}
\frac{\partial}{\partial t}H_x &= -\frac{1}{\mu_0}\frac{\partial}{\partial y}E_z, \\
\frac{\partial}{\partial t}H_y &= \frac{1}{\mu_0}\frac{\partial}{\partial x}E_z, \\
\frac{\partial}{\partial t}E_z &= \frac{1}{\varepsilon_0}\left(\frac{\partial}{\partial x}H_y - \frac{\partial}{\partial y}H_x - J_z\right),
\end{aligned} \tag{1}$$

and in case of TE polarization:

$$\begin{aligned}
\frac{\partial}{\partial t}E_x &= \frac{1}{\varepsilon_0}\left(\frac{\partial}{\partial y}H_z - J_x\right), \\
\frac{\partial}{\partial t}E_y &= \frac{1}{\varepsilon_0}\left(-\frac{\partial}{\partial x}H_z - J_y,\right) \\
-\frac{\partial}{\partial t}H_z &= \frac{1}{\mu_0}\left(\frac{\partial}{\partial x}E_y - \frac{\partial}{\partial y}E_x\right),
\end{aligned} \tag{2}$$

where $\vec{J}$ is an electric current density which, properly defined, determines nonlinearity of the material.

Particularly, in case of non-saturable Kerr nonlinearity polarization current density is given in following form [4]:

$$\vec{J} = \frac{\partial \vec{P}}{\partial t} = \frac{\partial}{\partial t}\varepsilon_0\chi^{(1)}\vec{E} + \frac{\partial}{\partial t}\varepsilon_0\chi^{(3)}|\vec{E}|^2\vec{E} \tag{3}$$

where $\chi^{(1)}$ and $\chi^{(3)}$ are the terms of linear and nonlinear susceptibility.

However, materials usually possess non-saturable Kerr properties only within low radiation intensity range and, therefore, we consider Kerr-saturable nonlinear materials and nonlinear susceptibility terms. Assuming slowly varying amplitude of the field $\left(\frac{\partial}{\partial t}|\vec{E}|^2 \approx 0\right)$, we can present nonlinear polarization term, by the analogy with [2], in following form:

$$\vec{P} = \varepsilon_0\chi_0^{(3)}\frac{|\vec{E}|^2}{1 + |\vec{E}|^2/I_{sat}}\vec{E}, \tag{4}$$

and corresponding polarization current takes form:

$$\vec{J} = \frac{\partial}{\partial t}\varepsilon_0\chi^{(1)}\vec{E} + \frac{\partial}{\partial t}\varepsilon_0\chi_0^{(3)}\frac{|\vec{E}|^2}{1 + |\vec{E}|^2/I_{sat}}\vec{E} \tag{5}$$

where nonlinearity term is now presented in form of saturable function.

Applying the FDTD technique expanded with auxiliary differential equation for the nonlinear medium [4] with polarization current given in form of (5) and assuming perfectly-matched layer [1] at the boundary of computation region, we can compute time-dependent electromagnetic field distribution in nonlinear saturable media.

3. Passive wideband PhC filters

Modern trends in data processing and transmission systems require new compact and high-speed solutions for all-optical circuits. Particularly, this concerns precise spectral filtering which can be implemented on the basis of PhCs. Recently, two wide categories of the PhC filters have been investigated, namely, high-Q and wideband ones. The first kind of filters possesses incredible spectral characteristics and suppose to be used in telecommunication for dense WDM demultiplexing. However, such filters have several disadvantages which make them hardly implemented in the nearest future. Particularly, recently designed high-Q filters require technology precision which is only possible in laboratory conditions. Moreover, due to their resonant nature, such filters cannot be used in the systems utilizing ultra-short pulses.

On the other hand, wideband filters which Q-factor is much lower than the one of the high-Q filter, possess comparatively low resonant properties which makes them suitable for ultra-short pulses application. Moreover, their characteristics are not affected too much by slight variation of the geometric parameters.

3.1. PhC filters spectrum

When designing wideband PhC filters, first thing we need to know is their spectral properties. Various numerical methods can be applied to compute such characteristics.

Particularly, when using the FDTD method, there are two different ways to find transmission or reflection spectrum of the PhC device. The first one is based on analysis of the response to continuous wave (CW) radiation. The second method uses Fourier analysis of the pulsed signal.

Analysis of the pulsed response is fast and accurate way to compute the spectrum. Basically, it is computed as a Fourier transform of the temporal response of the structure taken in certain spatial point.

However, this technique is only suitable when dealing with transversally-confined radiation (i.e. in case of the PhC waveguides as shown in Figure 1a). When it is necessary to find the spectrum in case of scattered radiation distribution (as presented if Figure 1b), the spectrum should be computed for each spatial point of interest.

3.1.1. Analysis of the CW response of the structure

CW signal usually possesses very narrow spectra and, therefore, computing the structure response to the CW we find its transmission of reflection at a specific wavelength. To compute the whole spectrum of the structure the response should be obtained at several wavelengths according to required spectrum.

To provide high accuracy of the method, several criteria should be satisfied:

http://dx.doi.org/10.5772/54118

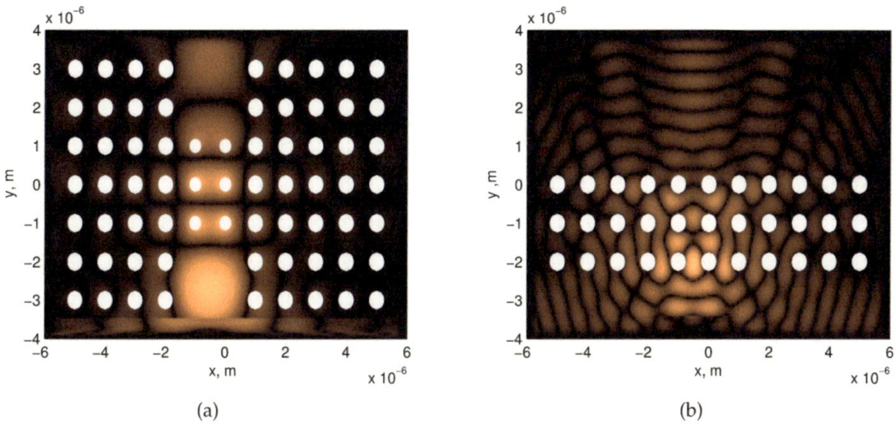

Figure 1. Confined or concentrated (a) and scattered (b) field distribution

- Every moment of time the radiation should be integrated along all the area of interest.
- Computation time should be long enough to achieve constant radiation intensity at the output (i.e. to pass all the transition processes).
- Spectral points should be selected close enough to avoid discontinuities of the final spectrum.

In case of PhC devices, the main application of such technique is computing the transmission spectra of bulk PhCs where the radiation is scattered.

3.1.2. Analysis of the pulse response of the structure

Unlike the CW radiation, pulsed one possesses wide spectrum which can be easily found from its Fourier transform.

To find the transmission spectrum of the PhC filter, it is first necessary to find temporal response of the filter to the launched Gaussian pulse (or any other wide-spectrum pulse). The temporal response is taken at a single spatial point. After a certain time, the Fourier transform of the temporal response can be taken. However, the spectrum obtained in such a way depends on the spectrum of the pulse launched to the system. Therefore, to find the final spectrum of the structure it is necessary to divide it by the spectrum of the initial pulse.

In general, to implement the method, certain steps should be made:

1. Set up the structure (i.e. define the refractive index distribution)
2. Set up the launch field before the structure
3. Run the simulation for a time much longer than the pulse lasts. The longer the simulation lasts, the higher the resolution of the spectrum will be.
4. Save the temporal distribution of one the field components after the structure
5. Find the Fourier transform of this temporal distribution.
6. Normalize the frequency.

3.2. Building an eye-diagram

In the electronic devices design and testing, it is usually used an eye-diagram to determine the quality of the transition characteristics. In fact, an eye diagram is represented by the series of the device work cycles drawn one over another. During this cycles the device is randomly turned on and off. Resulting characteristic resembles a human eye. If an "eye" is "closed" this points to poor quality of the device. In its "open" state, an eye's dimensions define parameters of the device such as bit error rate (BER).

In case of active and passive PhC wideband filters, an eye diagram can be used as well to define the quality of the device. Since a PhC possesses resonant transmission (i.e. the radiation is propagated from one element to another) after the device working cycle a fraction of an optical radiation is still remaining in the PhC elements. If this fraction is large enough, it will interfere with the next pulse resulting in radiation accumulation from pulse to pulse. After several pulses, the remaining radiation level can be large enough to affect the functioning of the nonlinear device or produce an error bit.

To detect such effects and also to find the pulse shape variation at the output of the PhC filter, an eye-diagram of the device can be built and analyzed.

Let us consider the process of building an eye-diagram of a simple nonlinear PhC wideband filter working at the edge of the photonic band gap. The filter is confined with linear PhC waveguide. We will now investigate its response to the sequence of Gaussian signals of different periods. In the first case, the period is large enough to release the radiation completely. The second pulses series possesses higher frequency.

The response of the filter in both cases is given in top of the Figure 2. To build an eye-diagram, it is important to know the period of the pulses (which in most cases is not obvious from the response). Since we know the repetition rate of the input pulses, we will use it. Now we only need to skip the transition time of the filter and split the response characteristic into equal pieces. Here we present the Matlab program which builds an eye-diagram from the response to the random pulses series. The response should be saved into a separate file in form of sequence "Time   E(Time)".

```
%The program is intended to represent computed
%temporal response of an optical structure to
% a series of pulses, as an eye-diagram. The
%response should be given in form of amplitute-vs-
%time

%Number of pulses in the response
num_pulses=30;
%Number of time-points within one pulse
%It should be determined from a computation method
period=1280;
%Initial point of the series (non-zero due to
%finite value of the speed of light)
t0=period;
```

```
%Loading the data with stored temporal response
data=load('response.dat');

%Creating a figure
figure;
subplot(2,1,1);
%Plotting the response
plot(data(:,1), data(:,2));
ylabel('|E|^2');

ax=subplot(2,1,2);

hold on;
%The data for the X-axis (time within the period)
time=data(1:period+1,1);
%Plotting every period in the same figure
for i=0:num_pulses-1
   plot(time, data(t0+(i*period:(i+1)*period),2), 'o')
end
set(ax, 'XGrid', 'on');
set(ax, 'YGrid', 'on');
xlabel('Time, s');
ylabel('|E|^2');
```

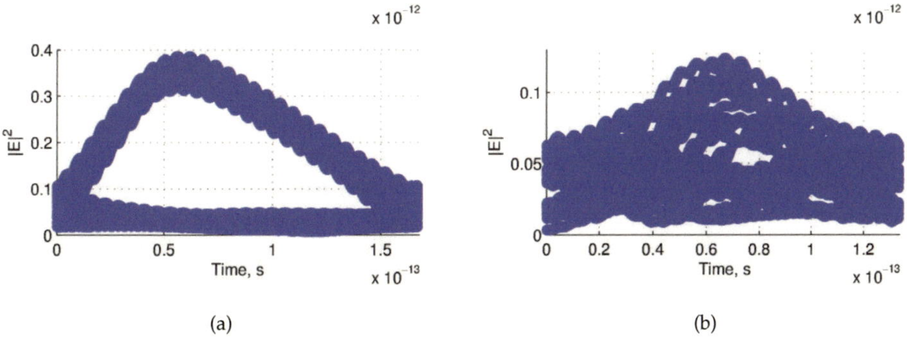

Figure 2. Examples of generated eye-diagrams. a) With weak pulse perturbation and b) with strong pulse perturbation

In the bottom parts of the Figure 2 we give two cases of an eye-diagram computed for a single PhC filter at different pulse duration. In the first case, an "eye" is clearly opened which tells us that the radiation is not accumulated in the filter. On the other hand, when pulse repetition rate is too high, the radiation is accumulated within the PhC which is reflected in the diagram (an "eye" is closed in Figure 2b). Investigating the shape of the pulse, we can also make a conclusion about how much does filter distort the pulse shape. For instance, even in the Figure 2a, the output pulse shape is obviously non-Gaussian due to distortions introduced by the filter.

3.3. PhC wavelength division demultiplexer

One of the most basic and, on the other hand, important applications of a passive PhC is a wavelength division demultiplexer. In multi-wavelength systems it provides spatial separation of the wavelength-mixed signal.

One of possible structures providing two-channel demultiplexing is presented in the Figure 3a. A signal containing two wavelengths enters through the bottom part of the device, travels to the coupler where it is separated by the wideband filters.

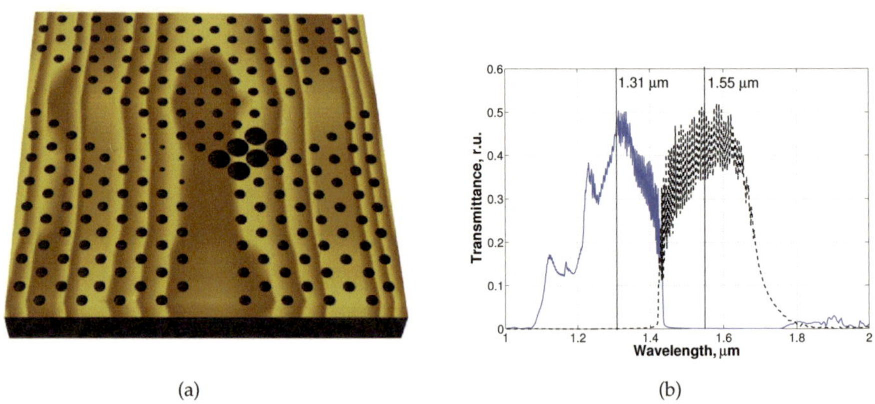

Figure 3. Structure of the PhC demultiplexer (a) and the spectra of the output channels (b)

Resulting spectra found by analyzing temporal response of the filters is presented in Figure 3b.

However, knowing spectral properties is not enough to characterize the demultiplexer completely. Since each PhC device possesses resonant properties, it is necessary to investigate distortions introduced to the signal when passing this device. This can be done by computing an eye-diagram of each wavelength channel (see Figure 4). Here we presented the diagrams computed for the pulses sequence with period $T = 160\ fs$ and pulse width of about $\tau = 80\ fs$

Presented eye-diagrams demonstrate that the demultiplexer can be used to process the ultra-short pulses. However, in case of $\lambda = 1.55\ \mu m$ the filter introduces more distortion into a pulse shape (i.e. pulse shape is not Gaussian at the output of the filter). This fact does not affect too much if a single device is used. However, when implementing an integrated optical circuit including series of linear and nonlinear filters, such distortions should be minimized to prevent data losses in the circuit.

http://dx.doi.org/10.5772/54118

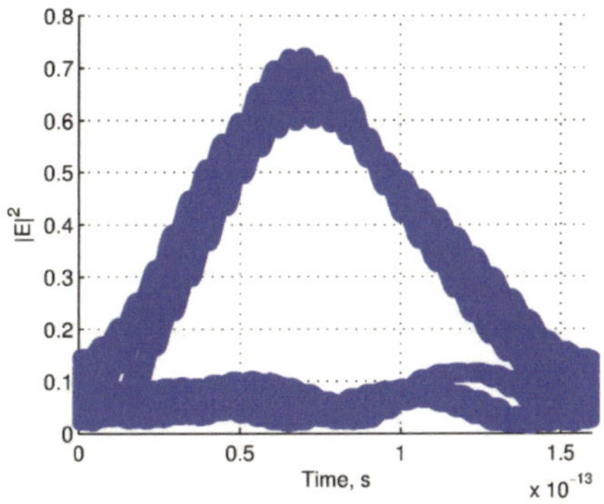

(a)

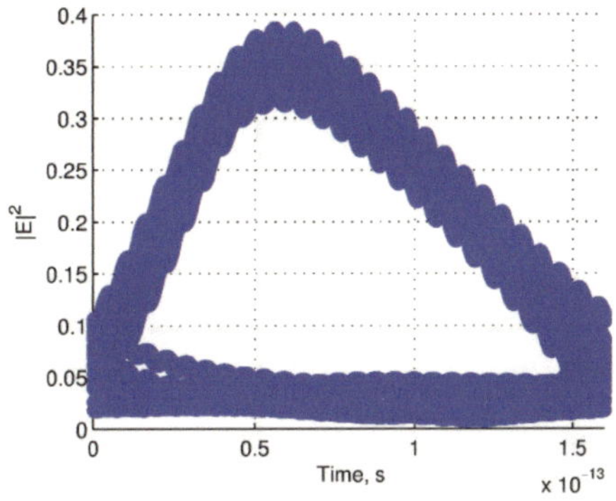

(b)

Figure 4. Eye diagrams of the PhC demultiplexer at 1.31 μm (a) and 1.55 μm (b)

4. Temporal characteristics of active PhC filters

4.1. Nonlinear PhC band structure

The band structure of the PhC can be computed by means of different methods. Among them, the most fast is the plane wave expansion (PWE) method. However, PWE has drawbacks which do not allow it to be applied to active PhCs. Namely, it is impossible to take into consideration the chromatic dispersion, absorption and gain as well as nonlinear material properties when refractive index depends on the radiation intensity.

One of possible ways to overcome the problem is to apply the FDTD technique. In contrast to the PWE method, the FDTD allows to take into account the refractive index variation during the computation process [11] and, therefore, to compute the light propagation in the nonlinear materials.

Here, we consider basic principles underlying the band structure computation by means of FDTD technique which are briefly discussed, for example, in [12].

In general, PBG computation using FDTD should be carried out as follows:

1. Determine the computation area.
2. Set up periodic boundary conditions.
3. Define the radiation excitation function. The radiation spectrum should be wide enough to cover whole investigated frequency range.
4. Carry out the spectral analysis of the time-dependent response of the structure on the probe pulse by searching all of local maxima and plotting them over frequency axis.
5. Repeat steps from 2 to 4 at different values of the phase shift in periodic boundary conditions corresponding to all selected points within the PhC Brillouin zone the band structure is computed for.

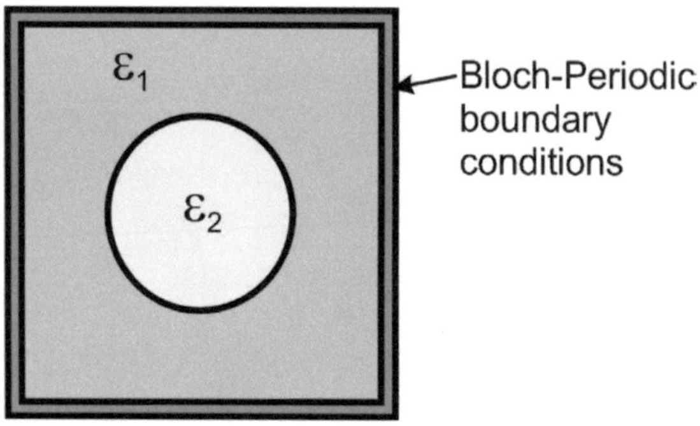

Figure 5. Computation domain for 2D PhC with square lattice

Let us now consider in details each step.

To perform any FDTD simulation, it is first necessary to determine the computation domain. However, since the PhC is considered as an infinite structure and computation over an infinite structure takes infinite time, the response of such a structure is impossible to find. Solution in this case is considering a single unit cell since it carries the information about whole structure. The computation domain in case of 2D PhC with square lattice is presented in figure 5.

The periodicity of the structure is achieved by setting up periodic boundary conditions at the edges of the computation domain.

Besides the translation emulation the periodic boundary conditions should provide simulation of electro-magnetic field propagation with certain wave vectors. Such kind of periodic boundary conditions are referred to as Bloch periodic boundary conditions [10]. The expressions of Bloch's periodic boundary conditions for electric and magnetic field components take following form:

$$\begin{aligned} \vec{E}(x+a, y+b, z+c) &= \vec{E}(x,y,z) \cdot e^{-\vec{i}\cdot k_x \cdot a - \vec{j}\cdot k_y \cdot b - \vec{k}\cdot k_z \cdot c}, \\ \vec{H}(x+a, y+b, z+c) &= \vec{H}(x,y,z) \cdot e^{\vec{i}\cdot k_x \cdot a + \vec{j}\cdot k_y \cdot b + \vec{k}\cdot k_z \cdot c}. \end{aligned} \tag{6}$$

where a, b, c are linear dimensions of the unit cell along X, Y and Z axes respectively; k_x, k_y, k_z are wave vector components.

When applying simple periodic boundary conditions, the electric or magnetic field intensity is taken from one boundary of the computation region and is added to the corresponding field component at the opposite boundary. However, in contrast to simple periodic conditions, the Bloch ones include phase shift achieved by multiplying the field intensity by the exponential function which argument contains radiation wave vector.

Therefore, setting up the Bloch periodic boundary conditions provides the possibility to investigate propagation of radiation possessing different wave vectors to compute the band structure.

The next important moment is an input signal parameters.

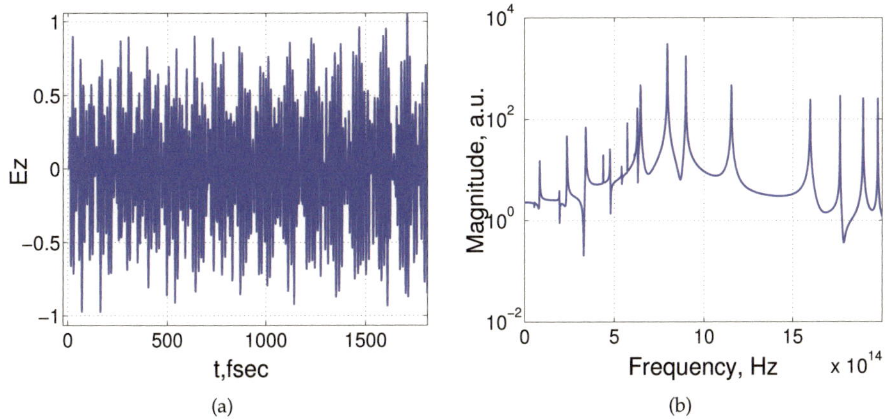

Figure 6. Temporal response (a) and its spectrum obtained by FFT (b)

The radiation can be introduced to the structure in various ways. However, we will consider excitation from a single point of the computation region. Since we are going to search for the resonant frequencies of the PhC within wide spectrum range, an input pulse should possess wide spectrum. The simplest signals used in this case are delta-pulse and Gaussian pulse. The delta-pulse in introduced in a single moment of time while the modulated Gaussian signal should be excited continuously during all simulation. It is obvious that using delta-pulse is the most simple case which we will use in our example:

$$\delta\left(t - t_0, x - x_0, y - y_0, z - z_0\right) = 1. \tag{7}$$

It is widely known that the spectrum of the delta-pulse is infinitely wide so it gives structure responses at any frequency. After the delta-pulse is introduced, the excitation is turned off, however, due to periodic boundary conditions, radiation exists infinitely long time in the structure without absorption.

After the pulse response of the structure is obtained, it should be properly analyzed. This analysis gives eigen-states of the PhC.

The spectral analysis of the time dependent pulse response can be carried out by Fourier transform. The accuracy of the method achieves its maximum when computation time is infinite. However, since we have finite computer resources, the computation time is taken long enough just to prevent spurious solutions.

Fast Fourier transform (FFT) [3] is usually used within this technique since the response function is discrete one and in this case the FFT performance is much faster then general Fourier transform. As a result of the FFT, we have the discrete spectrum as well. The example of the FFT of the structure response to the delta-pulse excitation is shown in figure 6

The eigen-states of the structure are searched for as local maxima at the response spectrum. Detailed analysis should be made to avoid spurious solutions. Such spurious solutions are usually appear as inessential peaks at the spectrum. Therefore, the local maximum does not always correspond to the eigen-state. The maxima corresponding to spurious solutions are just a little bit higher than neighbor spectrum points while valid solutions values give peaks with magnitudes several times larger than the one of neighbor points.

Here, we present the Matlab program for computation of the band structure of 2D PhC with square lattice. For simplicity, we consider a PhC made of linear material. To obtain the band structure of a nonlinear PhC, an auxiliary differential equation technique should be added in the FDTD part.

```
%The program is intended to compute
%the band structure of 2D PhC by means of
%the FDTD method.

%Cleaning up previous workspace
clear all;

%Setting up parameters of the PhC
%PhC period in each direction
maxX=1e-6;
maxY=1e-6;
%Radius of the cilinder
r=maxX*0.3;
%Refractive index of the cilinder
eps1=9;
%Background refractive index
eps2=1;
%Permeability (is always 1 for non-magnetic materials)
mu=1;

%Speed of light
c=3e8;

%Setting up FDTD parameters
%Computation time
maxT=2^13;

%Number of spatial points in the grid in each direction
accuracyX=16;
accuracyY=16;
%Number of k-points between high-symmetry points
accuracyK=5;

%Defining statial and temporal steps
Dx=maxX/accuracyX;
Dy=maxY/accuracyY;
Dt=Dx/2/c;
```

```
%Defining the unit cell permittivity distribution
%Central coordinates
x0=maxX/2;
y0=maxY/2;

%For smooth permittivity profile, defining the width of
%transition zone
dd=sqrt(Dx^2+Dy^2);
%Correcting the radius according to transition zone
r=r-dd/2;
%Generating the permittivity profile
eps=ones(accuracyX, accuracyY)*eps2;
for i=1:accuracyX
   for j=1:accuracyY
      if sqrt((i*Dx-x0)^2+(j*Dy-y0)^2)<r
         eps(i,j)=eps1;
      elseif sqrt((i*Dx-x0)^2+(j*Dy-y0)^2)-r<dd
         if(eps1>eps2)
            eps(i,j)=eps1-abs(eps1-eps2)/dd*(sqrt((i*Dx-x0)^2+...
                  (j*Dy-y0)^2)-r);
         else
            eps(i,j)=eps1+abs(eps1-eps2)/dd*(sqrt((i*Dx-x0)^2+...
                  (j*Dy-y0)^2)-r);
         end

      end

   end
end

%For faster FDTD computation we find the coefficients
%in the FD equations
eps_x=c*Dt/Dx./eps;
eps_y=c*Dt/Dy./eps;

%Defining k-path
kx(1:accuracyK+1)=0:pi/maxX/accuracyK:pi/maxX;
ky(1:accuracyK+1)=zeros(1,accuracyK+1);

kx(accuracyK+2:accuracyK+accuracyK+1)=pi/maxX;
ky(accuracyK+2:accuracyK+accuracyK+1)=...
            pi/maxY/accuracyK:pi/maxY/accuracyK:pi/maxY;

kx(accuracyK+2+accuracyK:accuracyK+accuracyK+1+accuracyK)=...
          pi/maxX-pi/maxX/accuracyK:-pi/maxX/accuracyK:0;
ky(accuracyK+2+accuracyK:accuracyK+accuracyK+1+accuracyK)=...
          pi/maxY-pi/maxY/accuracyK:-pi/maxY/accuracyK:0;
```

```
%Crating the figure
figure;
ax1=axes;
hold on;

%Counter of the wave vector points
curr_vector=0;

%% The cycle for all the points in k-path
for phase=1:length(kx)
  curr_vector=curr_vector+1;
  %Computing phase shift for a specific wave vector
  rotatex=(exp(-1i*(kx(phase)*maxX)));
  rotatey=(exp(-1i*(ky(phase)*maxY)));

  %Cleaning the computation region
  Ez=zeros(accuracyX,accuracyY);
  Hx=zeros(accuracyX,accuracyY);
  Hy=zeros(accuracyX,accuracyY);

  %Ecxitation is defined as Delta-function in a single point
  Ez(round(accuracyX/3),round(accuracyY/4))=100;

  %% Cycle for time
  for t=0:Dt:maxT*Dt

    %% Computing H-field

    %Defining periodic boundary conditions for H-field

    for x=1:accuracyX
      Hx(x,1)=Hx(x,1)-c*Dt/mu/Dy*(Ez(x,1)-rotatey*...
                              Ez(x,accuracyY));
    end
    for y=2:accuracyY
      Hx(1,y)=Hx(1,y)-c*Dt/mu/Dy*(Ez(1,y)-Ez(1,y-1));
    end

    for x=2:accuracyX
      Hy(x,1)=Hy(x,1)+c*Dt/mu/Dx*(Ez(x,1)-Ez(x-1,1));
    end
    for y=1:accuracyY
      Hy(1,y)=Hy(1,y)+c*Dt/mu/Dx*(Ez(1,y)-rotatex*...
                              Ez(accuracyX,y));
    end
```

```
    %% Computing the H–field distribution

    for y=2:accuracyY
       for x=2:accuracyX
          Hx(x,y)=Hx(x,y)–c*Dt/mu/Dy*(Ez(x,y)–Ez(x,y–1));
          Hy(x,y)=Hy(x,y)+c*Dt/mu/Dx*(Ez(x,y)–Ez(x–1,y));
       end
    end

    %% Computing E–field

    for y=1:accuracyY–1
       for x=1:accuracyX–1
          Ez(x,y)=Ez(x,y)+eps_x(x,y)*(Hy(x+1,y)–Hy(x,y))–...
                         eps_y(x,y)*(Hx(x,y+1)–Hx(x,y));
       end
    end

    %% Defining periodic boundary conditions for E

    for x=1:accuracyX–1
       Ez(x,accuracyY)=Ez(x,accuracyY)+...
     eps_x(x,accuracyY)*(Hy(x+1,accuracyY)–Hy(x,accuracyY))–...
     eps_y(x,accuracyY)*(Hx(x,1)/rotatey–Hx(x,accuracyY));
    end

    for y=1:accuracyY–1
       Ez(accuracyX,y)=Ez(accuracyX,y)+...
     eps_x(accuracyX,y)*(Hy(1,y)/rotatex–Hy(accuracyX,y))–...
     eps_y(accuracyX,y)*(Hx(accuracyX,y+1)–Hx(accuracyX,y));
    end
    Ez(accuracyX,accuracyY)=Ez(accuracyX,accuracyY)+...
       eps_x(accuracyX,accuracyY)*...
       (Hy(1,accuracyY)/rotatex–Hy(accuracyX,accuracyY))–...
       eps_y(accuracyX,accuracyY)*...
       (Hx(accuracyX,1)/rotatey–Hx(accuracyX,accuracyY));

    Eres(round(t/Dt)+1)=Ez(round(accuracyX/3),round(accuracyY/7));
    Time(round(t/Dt)+1)=t;

end

%% Analyzing the temporal response
%Computing the Fourier transform of the response
fourier=abs(fft(Eres));

%Normalizing frequency
```

```
    f=1/Dt*(0:length(fourier)-1)/length(Eres);

    %eigen-frequencies counter
    wcount=1;

    %Analyzing the first point of the spectrum
    if(fourier(1)/(max(fourier(2:4)))>1.1)
       weigen(curr_vector, wcount)=f(1);
       wcount=wcount+1;
    end

    %Analyzing the rest of the spectrum
    for u=3:length(fourier)-3
       if(fourier(u)/(max(fourier(u+1:u+2)))>1.01)&&...
                 (fourier(u)/max(fourier(u-2:u-1))>1.01)
             weigen(curr_vector, wcount)=f(u);
             wcount=wcount+1;
       end
    end

    %Plotting 5 solutions maximum
    if(wcount-1>=5)
     plot(curr_vector,abs(weigen(curr_vector,1:5))*maxX/c,'ob');
    else
     plot(curr_vector,abs(weigen(curr_vector,1:wcount-1))*maxX/c,'ob');

    end
    %Decoraring the plot
    set(ax1,'xtick',[1 accuracyK+1 2*accuracyK+1 3*accuracyK+1]);
    set(ax1,'xticklabel',['G';'X';'M';'G']);
    ylabel('Frequency \omegaa/2\pic','FontSize',14);
    xlabel('Wavevector','FontSize',14);
    set(ax1,'XGrid','on');

    drawnow;

end
```

The results computed by the presented code are given in figure 7 . The parameters in the program are selected to eliminate the spurious solutions. However, if an input power is changed, one should change the accuracy in the spectrum analysis part.

4.2. Bistable nonlinear PhCs

Nonlinear PhC filters with properly selected parameters, possess bistability and can be used as logical gates in all-optical data processing systems. In this section, we give an example of such filters on the basis of 2D PhC [5].

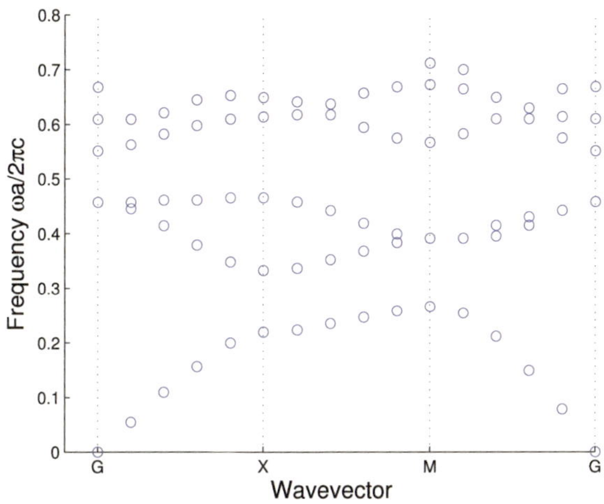

Figure 7. Band structure of 2D PhC computed by the FDTD method

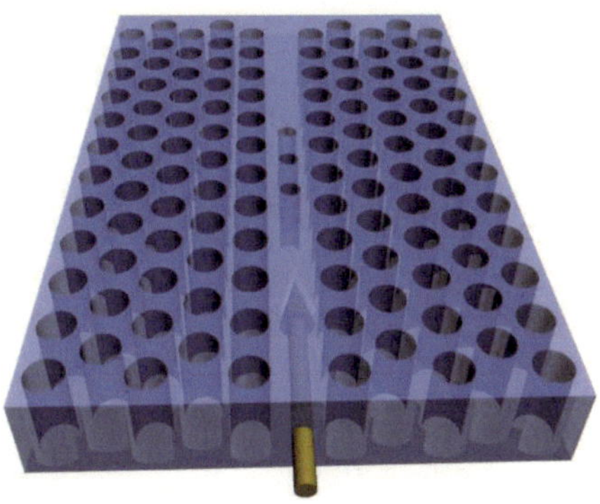
(a)

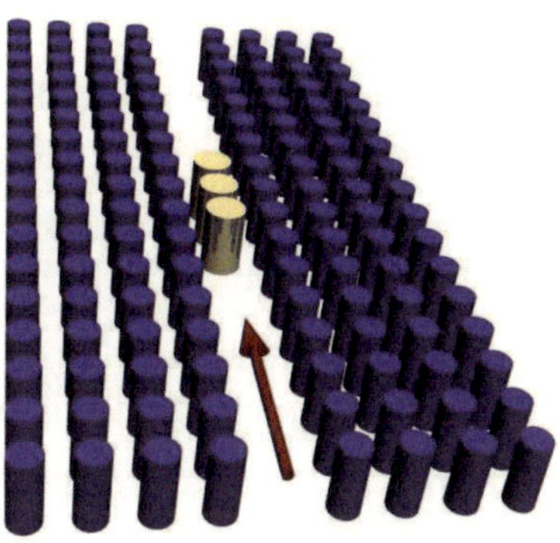
(b)

Figure 8. The structure of the investigated nonlinear PhC confined by the PhC waveguide on the basis of hexagonal (a) and square (b) lattice

The schematic of investigated nonlinear PhC filter confined by the PhC waveguide is shown in the figure 8.

In both cases, the PhC filter is presented by three PhC elements with parameters similar to the confined PhC. The photonic bandgap (PBG) of the filters are shifted as respect to the background PhC. As it has been demonstrated for the linear PhCs, such filters almost do not disturb an ultra-short pulses shape.

Initially, the operating wavelength $\lambda = 1.05\ \mu m$ is selected to fall at the PBG of both background PhC and filter. However, due to optical nonlinearity the spectral characteristic of the filter appears to be shifted when increasing the radiation intensity.

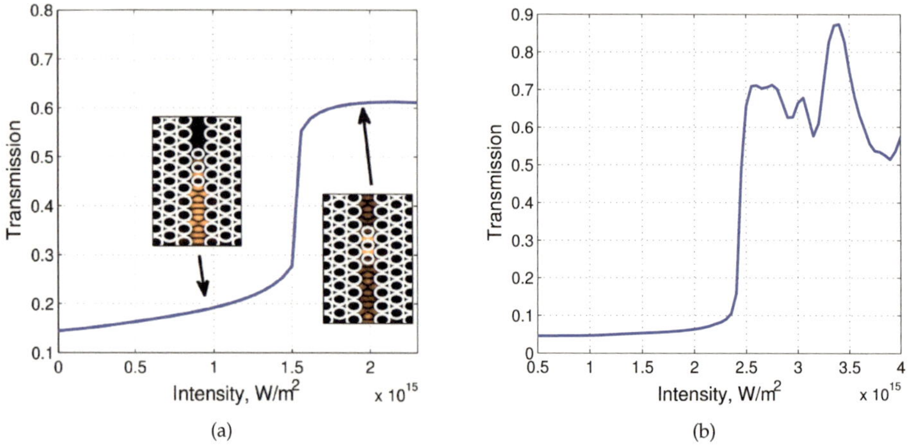

Figure 9. Transmission of the filter as a function of the input intensity in case of hexagonal PhC (a) and square PhC (b)

The reaction of the structure to increasing radiation intensity is presented in figure 9. The figure represents filter transmission as a function of the intensity of the CW monochromatic radiation. The insertions demonstrate the field behavior inside the waveguide with nonlinear PhC filter in case of low and high radiation intensity.

The results in the figure 9 allow to conclude that starting at certain value of the intensity, the radiation wavelength appears outside the PBG which increases transmission of such a filter dramatically. The further growth of the nonlinearity no longer increases the transmission as is seen from the figure 9. Therefore, in our investigation we have selected operating intensity slightly below the switching-on intensity.

Since the PhCs and, particularly, PhC-based filters possess resonant radiation transmission (the radiation is propagating from one element to another), the nonlinear spectrum shift require certain time which does not rely on response time of the nonlinear material. In order to investigate such a phenomena, we carried out the study of temporal response of such a filter to Gaussian pulses of different durations. Each of the pulses in serie possesses the same magnitude but different duration. The photorefractive properties of the materials remain the same for all pulses. This allows investigating only the contribution of the resonant processes inside the PhC into the bistability.

After the temporal response is obtained, it is represented in form of the dependence of output intensity on the input one as presented in figure 10. The intensity growth corresponds to the lowest branch of the hysteresis loop and lowering of the intensity stands for highest branch. Both in case of hexagonal and square lattices such characteristics look almost the same and, therefore, we provide here only the ones for the hexagonal PhC.

In case of linear optical materials, the branches are coincide since no processes affect the properties of the PhC. However, in case of nonlinear materials the light trapped inside the filter due to resonances holds the refractive index of the nonlinear material and, consequently, the transmission of the filter, high, thus, providing the difference in propagation of the leading and trailing edges of the pulse.

http://dx.doi.org/10.5772/54118

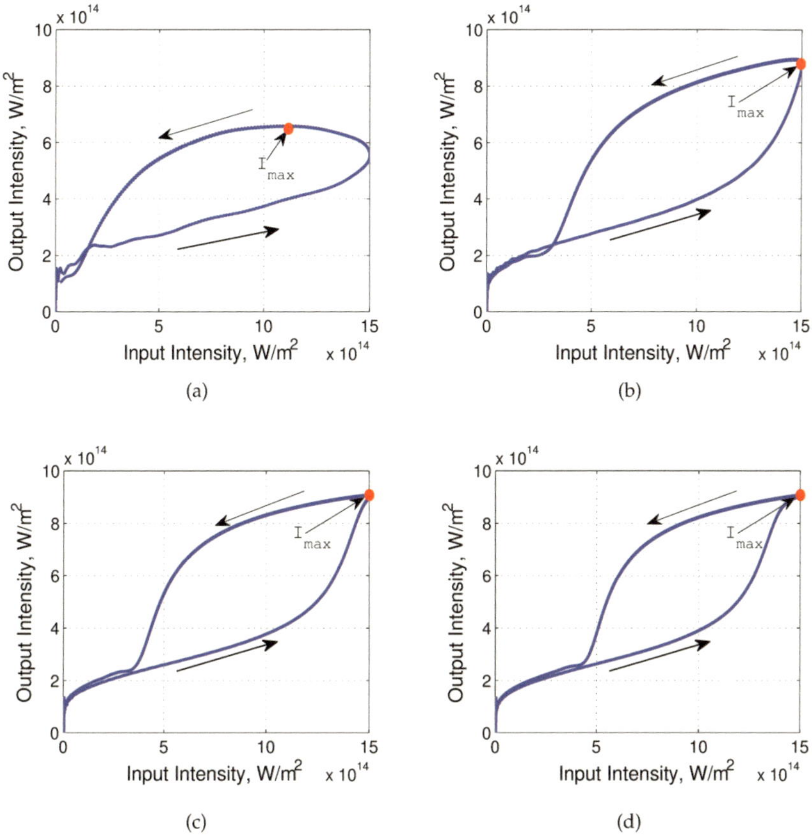

Figure 10. Hysteresis loops at different durations of the Gaussian pulse: a) $\tau = 50\ fs$, b) $\tau = 200\ fs$, c) $\tau = 400\ fs$, d) $\tau = 800\ fs$

Due to finite saturation time of the resonances in the PhC, the minimum allowed pulse duration exists for a specific PhC filter. Normally, during the front edge of Gaussian pulse, the intensity inside the filter grows which causes refractive index changes and, consequently, the changes in the filter characteristic. However, if the pulse duration is lower than the time required to excite the eigen-state in the filter, the significant nonlinear effects such as transmission growth appear after the input pulse maximum (see figure 10(a)). On the other hand, when the pulse duration is large, it is enough to excite the filter and, therefore, the maximum intensity of the input and output pulses are coincide.

Thus, the study of the temporal responses carried out in the work demonstrates the possibility of all-optical switching of the filter by the pulses which increase of reduce the intensity temporarily and, consequently, change the filter state.

After this we have studied its nonlinear switching dynamics. For this reason the continuous wave pump signal is launched into the waveguide. Then, with certain delays, the Gaussian control signals are launched which turn on and turn off the transmission of the filter. The power of the pump signal corresponds to the maximum magnitude of the hysteresis loop.

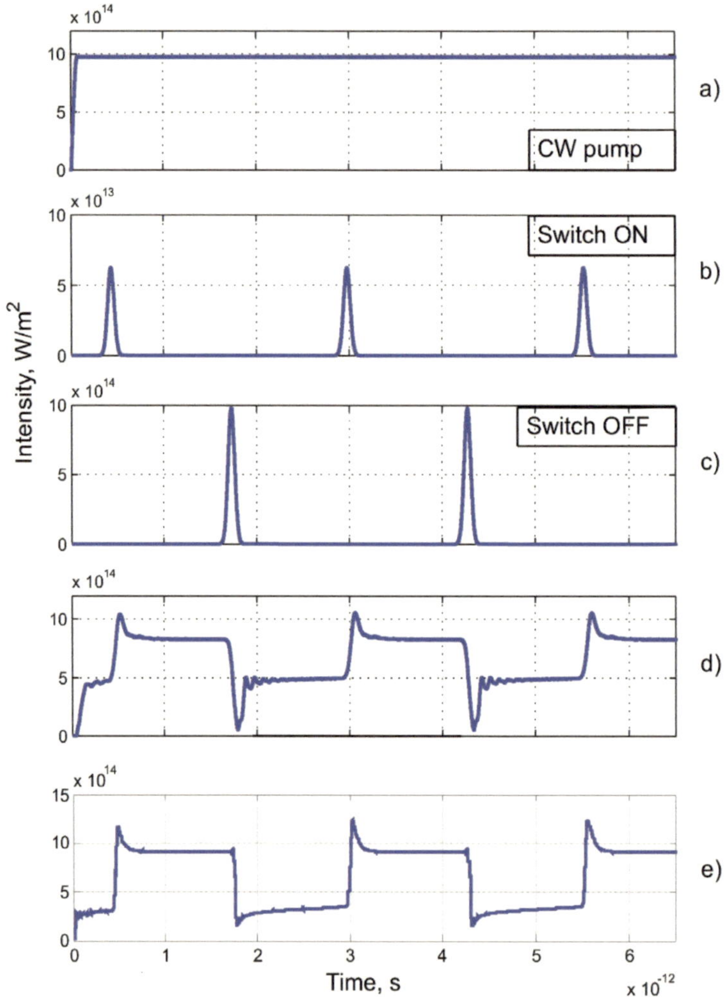

Figure 11. The CW pump (a), ON (b), OFF (c) and resulting signal in hexagonal (d) and square (e) PhC

Switching ON occurs when the Gaussian signal is launched with the same phase as the pump one. If the signal possesses opposite phase, switching OFF occurs.

The temporal response of the investigated filter is demonstrated in the figure 11. The topmost figure shows the intensity of the pump signal. The figures 11(b) and 11(c) demonstrate turn on and turn off pulses sequences. In two lowest figures, the resulting output signals are shown in case of hexagonal and square PhC lattices.

The pump signal intensity is slightly below the nonlinearity threshold. Therefore, switching ON requires low intensity Gaussian pulse. On the other hand, when switching OFF, the signal intensity should be reduced down to $2 \cdot 10^{14}\ W/m^2$ as follows from the figure 10(d). Therefore, the switching OFF signal maximum intensity is taken the same as that of the pump signal.

http://dx.doi.org/10.5772/54118

Comparing two nonlinear wideband filters we can mention their efficiency as bistable devices. However, at the same conditions transition time is larger in case of hexagonal PhC. On the other hand, in square PhC the lower radiation level is not as stable as the higher one and the output intensity in this case grows raising the probability of bit error.

Nevertheless, both these filters can be used as a basic logic in all-optical data processing circuits and the choice will be determined only by the technological factors.

4.3. Conclusions

In the chapter, we have demonstrated several applications of temporal characteristics as well as their computation method for both linear and nonlinear PhC wideband filters. Such micro-devices as wideband filters, wavelength division multiplexers and bistable elements may become a basis for future all-optical integrated circuits.

Presented Matlab codes for building an eye-diagram and for computing the band structure of 2D PhC by means of the FDTD methods will be useful for master and PhD students working on design and optimization of PhC-based micro-devices.

Author details

I. V. Guryev*, J. R. Cabrera Esteves,
I. A. Sukhoivanov, N. S. Gurieva, J. A. Andrade Lucio,
O. Ibarra-Manzano and E. Vargas Rodriguez

* Address all correspondence to: guryev@ieee.org

University of Guanajuato, Campus Irapuato-Salamanca, Division of Engineering, Mexico

References

[1] Berenger, J. [1994]. A perfectly matched layer for the absorption of electromagnetic waves, *Journal of Computational Physics* 114: 185–200.

[2] Bian, S., Martinelli, M. & Horowicz, R. J. [1999]. Z-scan formula for saturable kerr media, *Opt. Commun.* 172: 347–353.

[3] Brenner, N. & Rader, C. [1976]. A new principle for fast fourier transformation, *IEEE Trans. Acoust. Speech Signal Process* 24: 264–266.

[4] Greene, J. H. & Taflove, A. [2006]. General vector auxiliary differential equation finite-difference time-domain method for nonlinear optics, *Opt. Express* 14(18): 8305–8310.

[5] Guryev, I., Sukhoivanov, I., Lucio, J. A. & Rodriguez, E. V. [2012]. All-optical flip-flop in wideband phc-based flter with kerr saturable nonlinearity, *Applied Physics B* 106(3): 645–651.

[6] Guryev, I. V., Sukhoivanov, I. A., Lucio, J. A. A., Manzano, O. G. I. & Rodriguez, E. V. [2012]. Comprehensive analysys of the phcbased filters for optical micro-devices engineering, *Optical Engineering* 51(7): 074002.

[7] Joannopoulos, J., Meade, R. & Winn, J. [1995]. *Photonic Crystals: Molding the Flow of Light*, Princeton University Press, Princeton.

[8] Sukhoivanov, I. & Guryev, I. [2009]. *Photonic crystals – physics and practical modeling*, Optical sciences, 1 edn, Springer, Berlin.

[9] Sukhoivanov, I., Guryev, I., Andrade-Lucio, J., Alvarado-Méndez, E., Trejo-Durán, M. & Torres-Cisneros, M. [2008]. Photonic density of states maps for design of photonic crystal devices, *Microelectronics Journal* 39: 685–689.

[10] Sukumar, N. & Pask, J. E. [2009]. Classical and enriched finite element formulations for bloch-periodic boundary conditions, *International Journal for Numerical Methods in Engineering* 77(8): 1121–1138.

[11] Taflove, A. & Hagness, S. [2005]. *Computational Electrodynamics: The Finite-Difference Time-Domain Method*, 3 edn, Artech House, Norwood, MA.

[12] Yu, H. & Yang, D. [1996]. Finite difference analysis of 2-d photonic crystals, *IEEE Transaction On Microwave Theory And Techniques* 44(12): 2688–2695.

Section 2

Experiments and Applications

Chapter 9

Photonic Crystal Coupled to N-V Center in Diamond

Luca Marseglia

Additional information is available at the end of the chapter

http://dx.doi.org/10.5772/54138

1. Introduction

In this work we aim to exploit one of the most studied defect color centers in diamond , the negatively charged nitrogen vacancy (NV^-) color center, a three level system which emits a single photon at a wavelength of 637*nm* providing a possible deterministic single photon emitter very useful for quantum computing applications. Moreover the possibility of placing a NV^-in a photonic crystal cavity will enhance the coupling between photons and NV^-center. This could also allow us to address the ground state of the NV^-center, whose spin, could be used as qubit. It is also remarkable to notice that for quantum computing purposes it is very useful to increase the light collection from the NV^-centers, and in order to do that we performed a study of another structure, the solid immersion lens, which consists of an hemisphere whose center is at the position of an emitter, in this case the NV^-center, increasing the collection of the light from it. In order to create these structures we used a method called focused ion beam which allowed us to etch directly into the diamond many different kinds of structures. In order to allow an interaction between these structures and the NV^-centers we need to have a method to locate the NV^-center precisely under the etched structures. We developed a new technique ([1]) where we show how to mark a single NV^-center and how to etch a desired structure over it on demand. This technique gave very good results allowing us to etch a solid immersion lens onto a NV^-previously located and characterized, increasing the light collection from the $NV^-$of a factor of $8\times$.

2. Introduction to nitrogen vacancy center in diamond

Diamond has emerged in recent years as a promising platform for quantum communication and spin qubit operations as shown by [2], as well as for "quantum imaging" based on single spin magnetic resonance or nanoscopy. Impressive demonstrations in all these areas have mostly been based on the negatively-charged nitrogen vacancy center, NV^-, which consists of a substitutional nitrogen atom adjacent to a carbon vacancy. Due to its useful optical and magnetic spin selection properties, the NV^-center has been used by [3] to demonstrate a stable single photon source and single spin manipulations ([4]) at room temperature. A single-photon source based on NV^-in nano-diamond is already commercially available, and a ground state spin coherence time of 15*ms* has been observed in ultra-pure diamond at

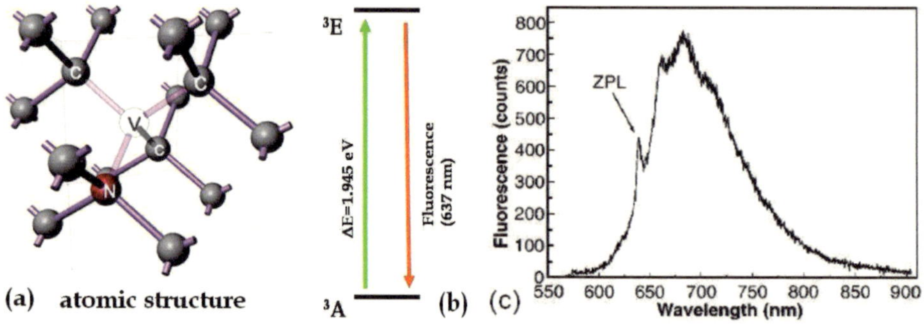

Figure 1. a)Atomic structure of NV$^-$center in diamond(N=nitrogen V=Vacancy, C=Carbon) b) Energy level scheme of NV$^-$center. c) Fluorescence spectrum of a single NV$^-$defect center. The wavelength of the zero phonon line (ZPL) is $637nm$ ($1.945eV$). Excitation was at $514nm$ [6]

room temperature. At present, one of the biggest issues preventing diamond from taking the lead among competing technologies is the difficulty in fabricating photonic devices to couple and guide light. For the realization of large-scale quantum information processing protocols (e.g. via photonic module approaches) or for quantum repeater systems, it will be necessary to connect NV$^-$centers through "flying" qubits such as photons. To achieve this, micro-cavities and waveguides are needed to enable the transfer of quantum information between the electron spin of the NV$^-$center and a photon. In this work I will show some applications of diamond useful for quantum computing. Synthetic diamonds can be doped in order to create implanted NV$^-$center which interacts with light, as described further. From its discovery, it has not been very clear if the NV$^-$were a proper two level system. Recently it has been shown that it has properties more typical of a three level system with a metastable level. In its ground state it has spin $s = 1$ and different emission rates for transitions to the ground states, so NV$^-$center can be also exploited in order to achieve spin readout.

3. Interaction of N-V center with light

The NV$^-$center in diamond occurs naturally or is produced after radiation damage and annealing in vacuum. As described earlier is made by substitutional nitrogen atom adjacent to a vacancy in carbon lattice in the diamond as depicted in Fig.1a. The NV$^-$center has attracted a lot of interest because it can be optically addressed as a single quantum system as discussed by [5]. The NV$^-$center behaves as a two level system with a transition from the excited state to the ground state providing a single photon of $637nm$, as shown in Fig.1b. This is a very useful characteristic for quantum information purposes because it can be used as single photon source. Let us remember that a characteristic of the NV$^-$center is a zero-phonon line (ZPL), in the spectrum at room temperature, at $637nm$ as shown in Fig.1c, the zero-phonon line constitutes the line shape of individual light absorbing and emitting molecules embedded into the crystal lattice. The state of NV$^-$center ground state spin strongly modulates the rate of spontaneous emission from the $^3E \leftrightarrow ^3 A$ sub-levels providing a mechanism for spin read out as discussed by [4]. We have recently shown theoretically ([7]) that spin readout with a small number of photons could be achieved by placing the NV$^-$centre in a subwavelength scale micro-cavity with a moderate Q-factor($Q \sim 3000$). So one of our aims is to optimize the output coupling of photons from diamond color

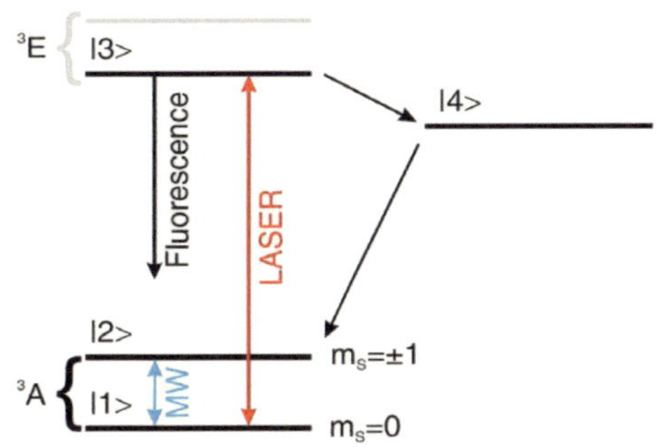

Figure 2. Energy level scheme of the nitrogen vacancy defect center in diamond. The greyed out lines correspond to the $m_s = \pm 1$ sublevels [10]

centers into waveguides and free space to increase the efficiency of single photon sources and to enable faster single spin read-out. In order to do that we want to study resonant structures. These structures confine the light close to the emitter allowing cavity-QED effects to be exploited to direct an emitted photon into a particular spatial mode and will allow us to enhance the ZPL. An improvement of the photon emission rate and photon indistinguishability for NV^-can be achieved due to the (coherent) interaction with the highly localized photon field of the cavity. In principle a high-Q micro-cavity can be realized directly in diamond but the first experimental demonstrations with micro-disk resonators and photonic crystal cavities, made for example by [8], suffered from large scattering losses due to the poly-crystalline nature of the diamond material used. The fabrication of high-Q cavities in single crystal diamond is very challenging because vertical optical confinement within diamond requires either a $3D$ etching process or a method for fabricating thin single crystal diamond films. We want analyze photonic crystal structures in diamond and fabrication methods to achieve efficient spin read-out in low-Q cavities. Electronic spin resonance (ESR) experiments performed by [9] has shown that the electronic ground state of NV^-center (^3A) is paramagnetic. Indeed the electronic ground state of the NV^-center is a spin triplet that exhibits a $2.87GHz$ zero-field splitting defining the z axis of the electron spin. An application of a small magnetic field splits the magnetic sublevel $m_s = \pm 1$ energy level structure of the NV^-center, as we can see in Fig.2. Electron spin relaxation times (T_1) of defect centers in diamond range from millisecond at room temperature to seconds at low temperature. Several experiments have shown the manipulation of the ground state spin of a NV^-center using optically detected magnetic resonance (ODMR) techniques, the main problem in using ODMR is that detection step involves observing fluorescence cycles from the NV^-center which has a probability of destroying the spin. Another characteristic of NV^-center useful for quantum information storage is the capability of transferring its electronic spin state to nuclear spins. Experiments performed by [5] have shown the possibility of manipulating nuclear spins of NV^-. Nuclear spins are of fundamental importance for storage and processing of quantum information, their excellent coherence properties make them a superior qubit candidate even at room temperature.

4. Beyond the two level system model

In order to study the dynamics of the NV^-center, remembering that $m_e << m_C$ where m_e is the value of the mass of the electrons and m_C is the value of the mass of carbon atom,

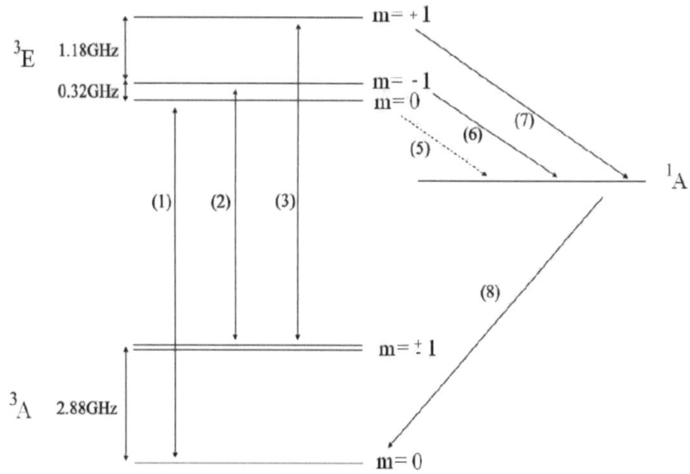

Figure 3. Experimentally measured energy level diagram of the NV center in diamond showing the experimentally determined ground and excited state splitting [17, 18]. The defect has zero phonon line at 637 nm, with width of order MHz at low temperatures [19](image taken from [7]).

we make the so called Born-Oppenheimer approximation, in which we consider the nuclei fixed in a crystal geometry and the coordinates of the electrons are considered with respect to them. When a defect is present it breaks down the crystal symmetry and, regarding the NV^-center, we have a contribution of one electron from each carbon atom, the nitrogen contributes two electrons, and an extra electron comes from the environment as described by [11], possibly given by substitutional nitrogen, so ending with a total number of six electrons. The excited level of the $NV^-$is not yet very well understood explicitly but there are many theoretical descriptions using group theory and partially confirmed by experimental results. The joint use of group theory and numerical calculations has led to predictions of the ordering of the levels of the ground state and excited states of the NV^-. Taking account the coulomb interaction, spin-orbit effect and spin-spin coupling it explains the splitting of the levels initially degenerate giving rise to different transition between them. Without entering in to detailed group theory calculations we can summarize by stating that the hamiltonian of the system is composed of three elements, the coulomb interaction H_C, the spin orbit interaction H_{SO} and the spin-spin interaction H_{SS}.

$$H = H_C + H_{SO} + H_{SS} \tag{1}$$

The dynamics of the system is resolved by solving the hamiltonian in free space with coulomb interaction as potential and the spin-orbit and spin-spin interaction were eventually added as perturbations. We show the detailed energy level structure of the NV^-center in Fig.3. The overall effect can be summarized as follows, the coulomb interaction, splits the degeneracy between singlet and triplets of the ground state and the first excited state, the spin orbit interaction splits the states which have $M_{s=0}$ and $M_{s=\pm 1}$ and finally the spin-spin splits the A levels. The end result is that the optical transitions between ground and excited states $(1-3)$ occur at different energies. In the absence of external fields the ground state is a spin triplet split by 2.88 GHz due to spin−spin interactions. [12] showed that the excited state is a triplet split by spin-spin interactions, but with the further addition of spin-orbit coupling . Recent experimental evidence performed by [13], has uncovered this excited state structure (Fig.3). The net effect of spin-spin and spin-orbit interactions is to create a detuning $\approx$ 1.4 GHz (6μeV) between the transitions $^3A_{(m=0)} \rightarrow ^3E_{(m=0)}$ (transitions 1)and $^3A_{(m=+1)} \rightarrow ^3E_{(m=+1)}$ (transition 3). A similar detuning of $\approx$ 2.5 GHz (10μeV) exists for the $^3A_{(m=-1)} \rightarrow ^3E_{(m=-1)}$ (transition 2), the rates for these three transition is $k_1 = k_2 = k_3 = 77MHz$ which gives a spontaneous emission (SE) lifetime $\tau \approx 13ns$. The energy level structure is not simply a ground and excited triplet state, there also exists an intermediate singlet state ^1A arising from Coulomb interactions. There is a probability of the transition $^3E \rightarrow ^1 A$, with different rates depending on the spin. For the $^3E_{m=\pm 1}$ states (transitions 6,7) both theoretical predictions and experimental results suggest that the decay rate is around $k_6 = k_7 = 30MHz$ giving a spontaneous emission (SE) lifetime $\tau \approx 30ns$. For the $^3E_{m=0}$ state (transition 5) theoretically the rate of decay to the singlet should be zero, however, experimental observations made by [14] have shown the rate to be $\approx 10^{-4} \times 1/\tau$. Since the ^1A singlet state decays preferentially to the $^3E_{m=0}$ state (transition 8), then it is clear from the rates above that broadband excitation leads to spin polarization in the spin zero ground state. Since transition 8 is non-radiative then there will be a dark period in the fluorescence when ^1A becomes populated, and as the decay rate from $^3E_{m=\pm 1}$, $k_8 = 3.3MHz$, to the singlet state is much larger than from $^3E_{m=0}$, the change in intensity measures the spin state. Clearly using fluorescence intensity to detect the spin state has a probability to flip the spin, therefore it would seem necessary for a scheme to suppress this. However, spin-flip transitions are essential to initialize the system. Thus a compromise is required between the perfectly cyclic spin preserving transitions required for readout and the spin flip transitions needed for reset.

5. Photonic crystals

To take advantage of atom-photon coupling using NV^-, as required by many quantum protocols, cavity structures are required. Again, concentrating on monolithic diamond solutions, photonic crystal cavities are the most natural structures to explore. A photonic crystal structure modulates the propagation of light in a way that is analogous to the way a semiconductor crystal modulates the motion of electrons. In both cases a periodic structure gives rise to 'band-gap' behavior, with a photon (electron) being allowed or not allowed to propagate depending on its wave vector. In photonic crystals the periodicity is comprised of regions of higher and lower dielectric constants. The basic physical phenomenon is based on diffraction, the period needs to be of the order of a half-wavelength of the light to be confined. For visible light the wavelength goes from $200nm$ (blue) to $650nm$ (red), leading to a real challenge in order to make the fabrication of optical photonic crystals because of the small dimensions. Breaking the periodicity in a controlled way creates nanocavities that confine light to extremely small volumes in which the lightmatter interaction is dominated by cavity quantum electrodynamic. We have previously described the characteristics of the NV^-center, a three level system which is promising as an efficient room temperature source of single photons at a wavelength of $637nm$. We pointed out that the NV^-center looks very promising for performing quantum spin readout, which is also useful for quantum computing purposes. Zero-phonon emission, at $637nm$, accounts for only a small fraction ($\sim$ 4%) of NV^-fluorescence, with the majority of emitted photons falling in the very broad ($\sim 200nm$) phonon-assisted sideband. By coupling the NV^-center to a photonic crystal cavity, spontaneous emission in the phonon sideband can be suppressed and emission in the zero-phonon line can be enhanced ([15]) so the photonic crystals offers a controllable electromagnetic environment, ideal for the compact integration and isolation of the fragile quantum system. The challenges of engineering the parameters of the photonic crystal in diamond at this scale are not trivial, as described further where we will show how to tune a cavity to increase the efficiency of light collection from an emitter placed in it. Indeed, a single photon emitted by a NV^- could then interact with another NV^- allowing entanglement between both qubits represented by the spin of the NV^- centers ase described by [16]. High-Q resonators of different kinds have been fabricated in non-diamond materials and coupled to NV^- emission from nano-diamonds. Since we are concerned here with developing monolithic photonics, it is necessary to fabricate cavities in the diamond itself. It should be noted that photonic crystal cavities have been fabricated in diamond films and an un-coupled Q-factor as high as 585 at $637nm$ has been measured by [8]. The polycrystal nature of the material used in those demonstrations makes it unsuitable for our purposes due to enhanced scattering and background fluorescence. We aim to fabricate photonic crystal cavities in ultra-high-purity type IIa single-crystal diamond (Element Six) grown by chemical vapor deposition. This material has extremely low levels of nitrogen (less than $1ppb$), and very few native NV^- centers, making it the ideal material for creating NV^- centers in a controlled fashion by implantation and annealing. In order to have strong coupling we need to have a cavity with high Q factor and small modal volume, but a cavity with a more moderate Q would still be useful. In particular, a scheme for reading out the ground state spin of an NV^- center has been described by our group ([7]), that requires a Q (before coupling) of only ~ 3000. This scheme exploits the zero-field splitting in the NV^- center ground state and uses narrow band resonant excitation to achieve high-fidelity read-out of the ground state spin with just a few excitation cycles.

6. Two-dimensional hexagonal photonic crystal structure

Our aim is to fabricate a structure which will behave as resonant cavity for the single photon emission of the NV^- center. The best choice to pursue this goal would have been a $3D$ photonic crystal structure with a NV^- placed in its center, but unfortunately the fabrication of this kind of structure is challenging. So we decide to follow a different path using a quasi$-3D$ structure. In fact combining the photonic crystal feature and the total internal reflection (TIR), we obtain a structure which confines the light in the three directions XYZ. Indeed the light is confined by distributed Bragg reflection in the plane of periodicity (XY) and by total internal reflection in the perpendicular plane (Z), so we aim to fabricate a photonic crystal in a thin membrane. Recently [17]we have described in detail the study of a photonic cristal cavity, here we report just the results obtained for the $L3$ structure which consist of a mebrane of $185nm$ with two-dimensional hexagonal photonic crystal structure which has a bandgap centered to wavelength of $637nm$, namely a PC structure resonant with the NV^- center emission, as shown in is shown in Fig.4a with its calculated resonance frequency shown in Fig.4b.

7. Focussed ion beam milling

The system we used to etch structures into the diamond is a well-studied and developed technology known as focussed ion beam milling (FIB)[18]. It was initially developed during the late $1970s$ and the early $1980s$, and the first commercial instruments were introduced

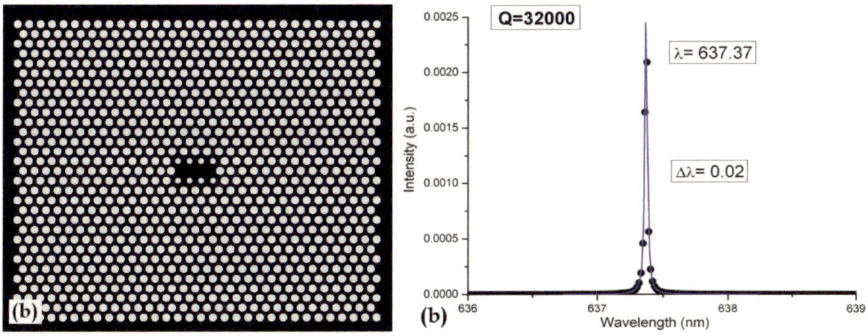

Figure 4. a)$L3$ photonic crystal cavity structure modified with smaller radius and nearest holes shifted. b) Fourier Transform of the E_x plot with Lorentzian Fit leading to an estimation of $Q = 32000$

more than a decade ago [19]. The technology enables localised milling and deposition of conductors and insulators. A schematic diagram of a FIB column is shown in Fig. 5. The structure of the column is similar to that of a scanning electron microscope, with the difference that the FIB machine uses Ga^+ ions instead of electrons. Inside the column a vacuum of about $1 \times 10^{-7} mbar$ is maintained. In the FIB system a highly focussed ion beam (diameter $\sim 10nm$) is aimed at a target area on the sample, the ion beam is generated from a liquid-metal ion source ($LMIS$) by the application of a strong electric field.

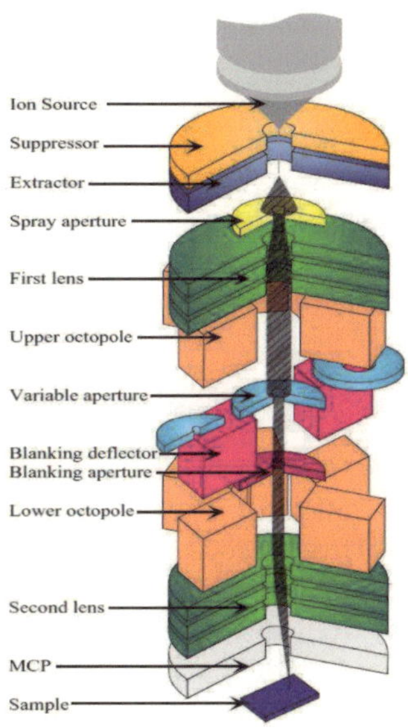

Figure 5. : Schematic diagram of a FIB column (image taken from [20]).

This electric field causes the emission of positively charged ions from a liquid gallium cone, which is formed on the tip of a tungsten needle. After being focused with the first collimating lenses, the ion beam is tuned with a variable aperture, generating a beam current which is typically in a range from $1pA$ to $1nA$, leading to the possibility of performing very fine high resolution imaging or beam milling. When the focused gallium ion beam is raster scanned

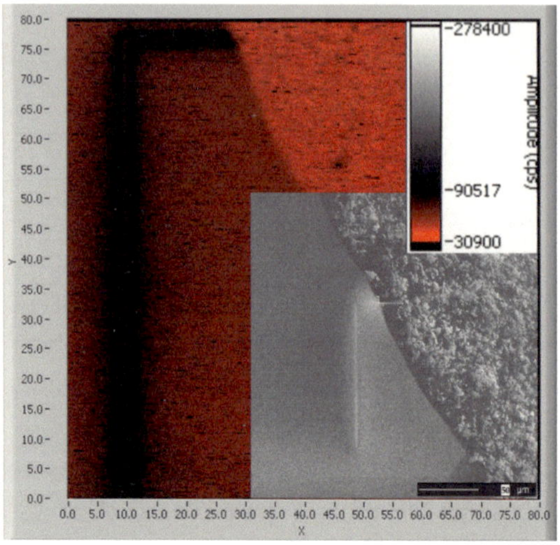

Figure 6. Confocal microscope image (pump $532nm$ filter $637 \pm 67nm$) of the silver coated zone with a line etched on and next to it. (inset) secondary electron image.

over a substrate it causes electron emission which is collected on a biased high gain detector, called, a microchannel plate (MCP). The detector bias is a positive or a negative voltage, respectively, for collecting secondary electrons or secondary ions. Detection can then be used to build up an image of the sample as the beam is scanned across it. The imaging taken in this way has resolution limited only by the focal spot size ($\sim 10nm$). Moreover if the current of the beam is high, another effect also occurs; physical sputtering or milling of the sample material. By scanning the beam over the substrate, arbitrary shapes can be etched. The FIB offers the ability to perform nanopatterning, allowing design and prototyping of new micro or nanostructures. The ion beam itself could also be used to perform spatially confined Ga-doping.

8. Fabrication of structures near N-V centers, preliminary studies

In [1] we have introduced the new procedure we used to create structures on demand coupled to NV^-centres , this new technique but before relies on some promising preliminary results we are going to show here. These results indicate that etching with FIB next to the NV^-centres ($\sim 1\mu m$) does not affect their emission. In order to be sure that the FIB has not damaged the implanted NV^-centres like for example in [21], we first cover the surface of the diamond with silver in the zone where the NV^-centres were located. The silver cover zone acted as a shield for the gallium implantation and allowed us to check the difference in light emission from the NV^-centres exposed to the Ion beam. Then we etched, with FIB, a deep long line next to the silver cover zone as shown in Fig. 6. After that we removed the silver coating, and analysed the possible damage to the NV^-array. Fig. 7 shows the behaviour of count rate from single NV^-centres taken at two different points, one in the region exposed to gallium and the other in the region covered by the silver after cleaning. Once we were assured that there is no damage to the array, we etched another line closer to the NV^-array which was used as a reference in future procedures, and we repeated the procedure, confirming again that the FIB etching does not affect the NV^-centre emission

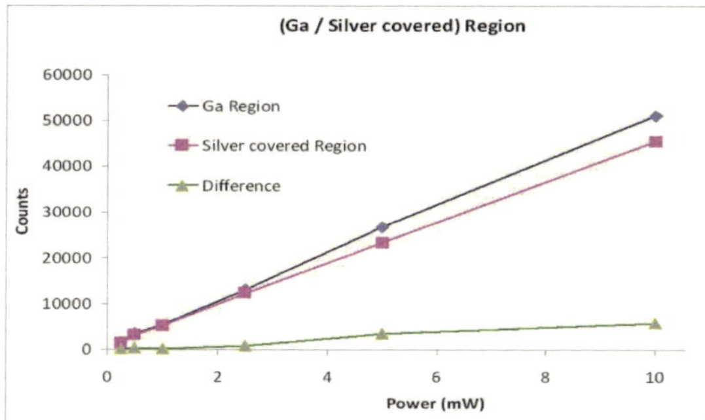

Figure 7. Count rate as function of laser power, measured after cleaning (section 10) the surface

because there were no changes in the count rates from the NV^-. This procedures proves that etching next to the NV^-centres (within $5\mu m$) does not create any damage to them.

8.1. Platinum deposition

Another important feature of the FIB machine is represented by the possibility of performing metal deposition onto the sample. The way it works is similar to chemical vapour deposition(CVD) and the occurring reactions are comparable to, for example, laser induced CVD. The main difference is the better resolution but lower deposition rate of FIB. The metals that can be deposited on commercially available machines are platinum(Pt) and tungsten(W). We mainly used platinum deposition in our etching sessions. By coating in Platinum we produce a conducting surface on the diamond which decreases the build up of charge due to deposition of Ga^+. This then reduces the beam shift due to accumulation of the gallium ion charge during the milling process. The deposition process works as follows: a platinum-bearing organometallic precursor gas is sprayed on the surface by a fine needle (nozzle), where it adsorbs. In a second step, the ion beam decomposes the adsorbed precursor gases. Then the volatile reaction products desorb from the surface and are removed through the vacuum system leaving mostly platinum and small amount of organic solid at the surface. Auger electron observations have shown that this Platinum (27%) is highly contaminated with Carbon (65%) originated during the metal organic precursor decomposition and Gallium (8%) coming form the ion beam. The deposition result is illustrated in Fig. 8 where we can see the platinum coated zone and the mouth of the injector needle. We used the platinum deposition in order to perform precise etching of desired pattern as we will see in the next section.

9. Fabricating photonic crystals using focus ion beam etching

Having simulated photonic crystal structure cavities we began fabrication via focused ion beam etching (FIB) [22, 23] our aim being to create a suspended membrane with the "Noda" cavity described previously. Other groups have performed fabrication with FIB in diamond of different kind of photonic crystal structures, some group performed also a measure of a Q factor of $Q = 535$ [24]. Here we try different approach in order to fabricate the "Noda" Cavity. In the first fabrication step, the diamond crystal is undercut by turning side-on and etching to obtain a $200nm$ thick slab attached to the bulk (a suspended slab). In this stage

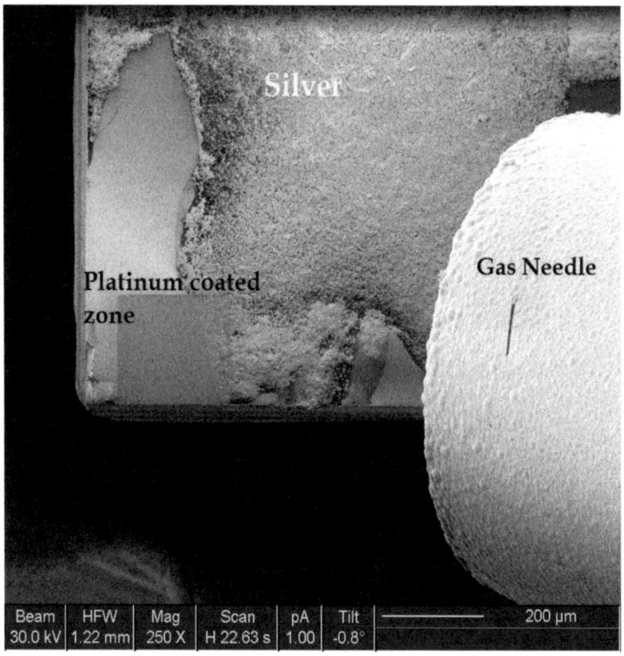

Figure 8. Secondary electron image of the sample with a square of Platinum coated zone made with gas deposition technique, the gas needle is also visible on the right hand of the picture.

the use of platinum deposition, as described before, is crucial in order to obtain precise structures. We needed to etch the membrane first, because if we had made the photonic crystal structure first, at the stage in which we etch the membrane some sputtering could have filled the holes. In order to etch the membrane we mounted the sample on a stage, and then tilted it to 90°. After we covered the NV^-centre array zone with silver, in order to protect the implanted NV^-swe etched a thin membrane of $200nm$ according to the results of the simulation shown previously. In Fig.9(a) we can see a top view of the membrane and in 9(b) we can see an image of the same membrane tilted by 45°. After we made the membrane we repositioned the sample horizontally and finally we etched the hexagonal air hole array with cavity formed from three filled holes. Fig. 10 shows two views, tilted 45° at different magnifications, of the resulting structure. Both were secondary electron images taken with *FIB* after the etching. In Fig. 10(a), we can see the photonic crystal cavity etched in the membrane. This is more evident in Fig. 10(b) where we have a scan over a larger area which shows the size of the cavities compared to the suspended membrane. In the top view, shown in Fig. 11(a), we can observe the cavity and notice some imperfections in it due to the *FIB* technique which creates deposits of etched material during the scanning. In Fig. 11(b) we can see an image taken with a confocal microscope, in which blue colour means low intensity and red colour means high intensity. Fig. 11(b) is remarkable because we can clearly see reduced fluorescence in the unetched zone, forming the cavity. Because there is no (or less) etch damage in these regions. This is an encouraging because it means that if there were an NV^-centre in the cavity we might be able to see it. We performed some measurement of the spectrum of the light emitted from the cavity region. Unfortunately we were not able to see any enhancement of the signal as we might expect from a cavity resonance, but just a broad emission as shown in Fig. 12. At this stage we decided to take a step back and to

http://dx.doi.org/10.5772/54138

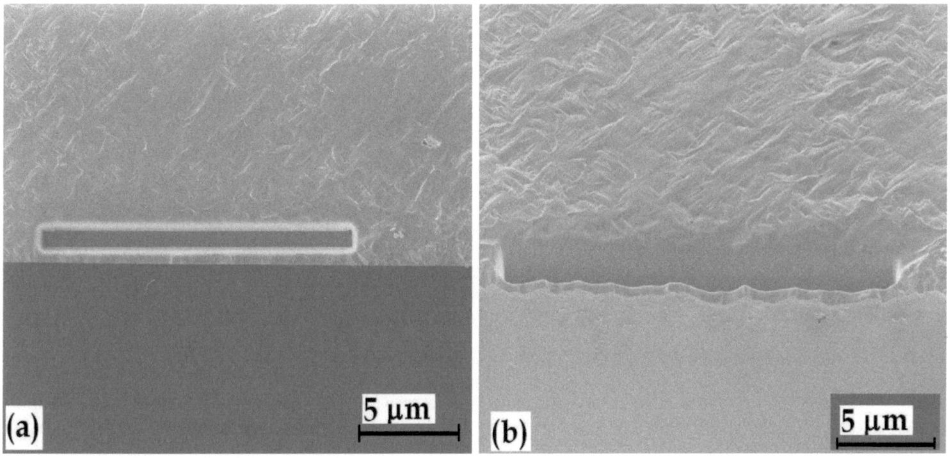

Figure 9. secondary electron image of a etched membrane in the diamond sample. a) a top view of the etched membrane b) 45° Tilted view of the etched membrane

Figure 10. Secondary electron image of a etched photonic crystal structure in the diamond sample: a) Tilted view of the *L*3 cavity taken with *FIB* at different tilt and magnitude. b) larger image of the membrane and the cavities.

perform a preliminary study about the real possibility of coupling a single NV^-centre to a larger structure etched in the diamond with *FIB*. This motivated our studies of the solid immersion lens as will be discussed in the next section.

10. Fabricating solid immersion lenses using focused ion beam etching

In this chapter we have studied how to fabricate the photonic crystal structure in order to create a cavity to enhance the coupling of NV^-centres in diamond. As already discussed, NV^-photon collection efficiency is severely reduced by losses due to the high refractive index of diamond. This is a problem regardless of the application, or the particular defect centre of interest. A possible solution we presented [25], is represented by the fabrication of hemispherical integrated solid immersion lenses (*SILs*) etched directly into the diamond surface. In order to avoid any scattering and absorption of the light emitted at high angles we need to have a *SIL* surrounded just by air. Hence we etch a *SIL* surrounded by a ring trench.

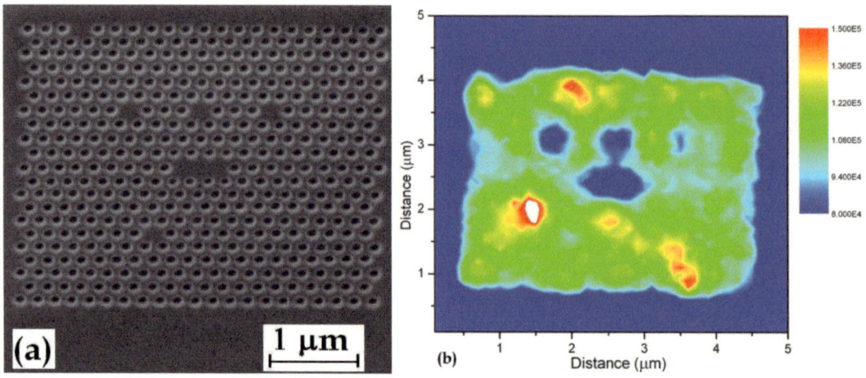

Figure 11. Photonic crystal structure in the diamond sample: a) Secondary electron emission image of the top view of the $L3$ cavity (inset:zoom of the cavity). b) Fluorescence image taken with the confocal microscope (colour red: high intensity, colour blue:low intensity).

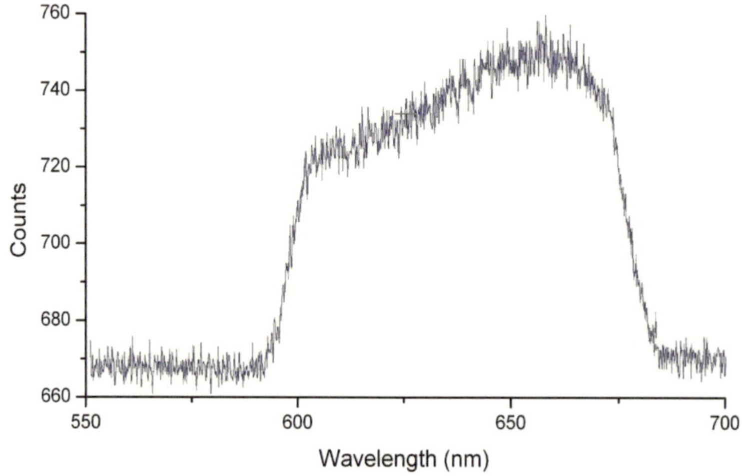

Figure 12. Image of the emission spectrum taken at the centre of the photonic crystal cavity, taken with a pass-band filter(semrock $FF01-675/67-25$) tilted a little bit in order to avoid raman scattering from the diamond. No narrow peaks are seen as would be expected from cavity resonance effects.

In order to do this, we have split the whole procedure into two different steps, etching a ring trench the depth of the SIL and then etching the SIL inside the ring. We etched a ring of a depth of $4\mu m$ with a current of $6.6nA$. This part of the procedure is a routine etch with FIB. The challenge is represented by etching a hemispherical SIL in the centre of the ring. In order to do that we milled a series of concentric rings of increasing depth and diameter at the current of $1nA$. We etched the material in a way different from the usual one: instead of etching from the top to bottom of the structure we milled from the centre moving the ion beam in a spiral way (trepanning), and then we repeated this for many concentric rings. The spiral technique allowed us to avoid a possible shift due to the etching process, giving us a SIL placed in the ring etched previously. Fig. 13(a) shows images of the $SILs$ taken using secondary electron emission in the FIB system. We can see a a SIL in the centre of the image, just below the square, and a SIL inside a ring in the lower right/hand corner of the image. Fig. 13b shows a 45o tilted view of a SIL. Recently [1] we have shown how to fabricate a SIL

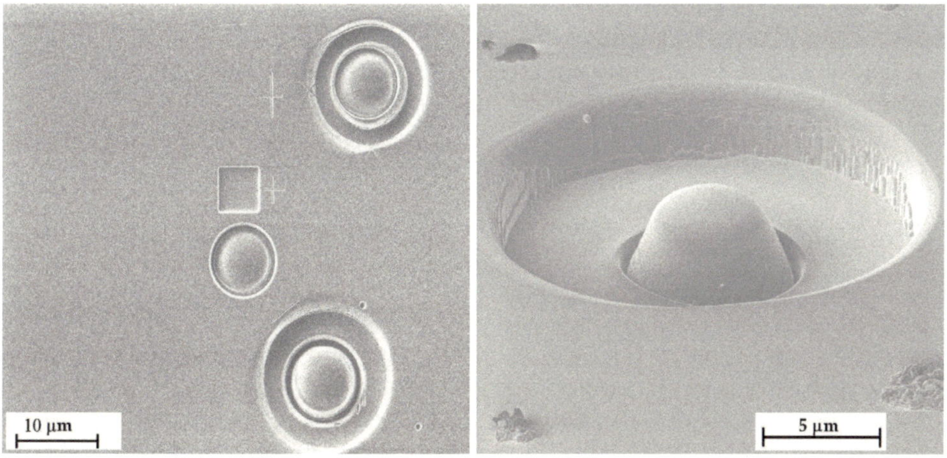

Figure 13. secondary electron emission images showing: a) top view of the SIL etched in the diamond b) 45° tilted view of the SIL etched in the diamond

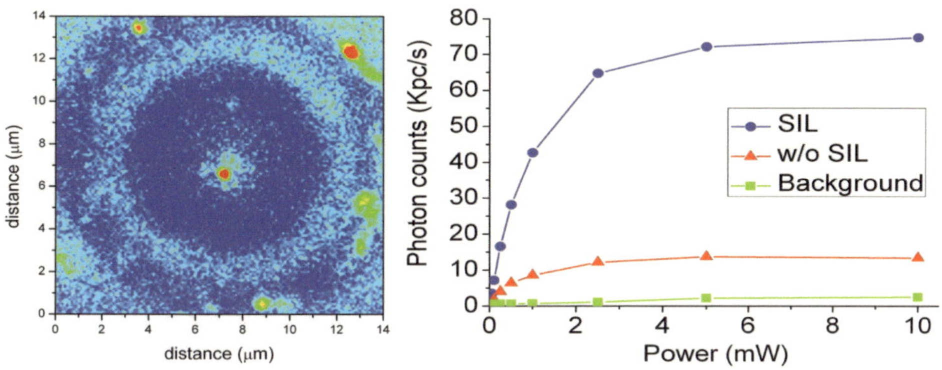

Figure 14. a)Photoluminescence image of the SIL with the enhanced emission from the NV^-center taken at temperature $T = 4.2K$, b)Comparison of the different photoluminescence count rates as function of the laser intensity (image taken from [1]).

located over a previously identified NV^-centre, without entering in detail, we just report the increasing of the collection of the light of $8\times$, as shown in Fig.14b, from a Single NV^-centre coupled to the SIL fabricated on demand on it as shown in Fig.14a

10.1. Chemical cleaning of the diamond

After each etching procedure it was essential to thoroughly clean the diamond surface. This dramatically reduced the background fluorescence intensity by removing all sputter deposits of gallium, platinum and other organic contaminants. We firstly rinsed the sample in a hydro-sonic bath three different times, each with a different component: the first time with acetone in order to remove the main dust, the second time with isopropanol to remove the acetone and finally with distilled water in order to remove the isopropanol. After we cleaned the most visible impurities, we put the diamond in a bath of $100ml$ of Sulphuric Acid (H_2SO_4) and then heated it till it reached the temperature of $200°C$. Once the right temperature was achieved we added 5mg of potassium nitrate (KNO_3) leaving it for 20 minutes. The resulting reaction, is shown in Eq. 2. It is very aggressive and fast and creates a mixture of sulphuric acid and nitric acid at high temperature which removes all remaining contaminants leaving a very clean sample.

$$H_2SO_4 + NO_3^- \rightarrow HSO_4^- + HNO_3 \tag{2}$$

11. Conclusion

In this work we discussed about the feasibility of NV^-centers as single photon emitters and how to use its spin as qubit for quantum computing applications, remarking the many advantages that the use of NV^-center in diamond would produce. One of the key challenges in order to perform a real implementation of a quantum computer concerns the possibility of handling the qubit. The spin of the ground state of the NV^-center shows some characteristics we have described earlier in this work, which looks very promising for this purposes. One of the crucial step in order to perform spin readout and non demolition measurement of the spin of the NV^-center is represented by increasing the coupling between the light and the solid state system. We showed a way to increase the coupling between the $NV^-$and the light by placing the NV^-center in a photonic crystal cavity. We characterized the photonic crystal cavity tuning it to be resonant with the NV^-center emission, having had encouraging results in the simulation and fabrication of the cavity, in that we reached a reasonable high value of quality factor and small modal volume. Another very important aspect in order to build a quantum computer is represented by the possibility of handling a single photon source. In order to use NV^-as single photon emitter, one of the challenges is represented by the light collection. We discussed in this work a solution we developed in order to increase the light collection from the NV^-center by etching a solid immersion lens around it, proposing a technique ([1]) in order to locate NV^-centers with accuracy of $10nm$, and fabricate structure around them. In future works we will use the technique we have recently developed, in order to create a photonic crystal around a single NV^-center. Another important path we want to follow consist in exploiting another useful color center in the diamond, the chromium center. This center acts as a single photon emitter as well but with a narrower spectrum. It has a resonant wavelength of $755nm$ which is in the wavelength range for the Si photon counting detector, allowing us to detect them with high efficiency. We are very interested in using the technique we have developed in order to etch photonic crystal around single chromium center and coupling with it, this would allow us to increase the light collection from the chromium center permitting it to be used as an ultra bright single photon emitter. As it emits at $755nm$ it is compatible with integrated photonic circuits being developed in our group by [26]. Similarly the nickel-nitrogen complex (NE8) center in diamonds, studied by [27, 28], has narrow emission bandwidth of $1.2nm$ at room temperature with emission wavelength around $800nm$, again suitable for Si detectors and quantum photonic circuits. In addition, in this spectral region little background light from the diamond bulk material is detected, which made it an interesting possible candidate for single photon source. Once we are able to locate an $NE8$ (or other suitable narrowband) center we will extend the registration procedure developed to allow fabrication of photonic crystal structure around individual defects, ending in the measurement of Q-factors and Purcell enhanced emission. In order to handle and guide the light emitted from the source a detailed study of parameters of a photonic crystal waveguide in the diamond will be required as demonstarted by [29]. We will simulate the behavior of the electromagnetic field inside the cavity and how it will couple with the waveguide. A good response will lead us to fabricate and then measure the effective coupling. We will also explore different etching techniques such as Reactive-Ion Etching (RIE) which will allow us to create membranes and very thin structures in diamond with a high precision. This will be useful in order to create different structures around registered NV^{-}centers, for instance with photonic crystal nanobeam cavities studied by [30]. This kind of structures as the remarkable advantage to be very easy to fabricate offering a huge quality factor and a very small modal volume.

Author details

Luca Marseglia

Institut für Quantenoptik, Universität Ulm, Ulm, Germany

References

[1] L Marseglia, J P Hadden, A C Stanley-Clarke, J P Harrison, B Patton, Y-L D Ho, B Naydenov, F Jelezko, J Meijer, P R Dolan, J M Smith, J G Rarity, and J L O'Brien. Nano-fabricated solid immersion lenses registered to single emitters in diamond. *Appl. Phys. Lett.*, 98:133107, 2011.

[2] T Gabel, M Dohman, I Popa, C Wittmann, P Neumann, F Jelezko, J R Rabeau, N Stavrias, A D Greentree, S Prawer, J Meijer, J Twamley, P R Hemmer, and J Wrachtrup. Room-temperature coherent coupling of single spins in diamond. *Nature Physics*, 2:408 – 413, 2006.

[3] C Kurtsiefer, S Mayer, P Zarda, and H Weinfurter. Stable solid-state source of single photons. *Phys. Rev. Lett.*, 85:290, 2000.

[4] R Hanson, F M Mendoza, R J Epstein, and D D Awschalom. Polarization and readout of coupled single spins in diamond. *Phys. Rev. Lett.*, 97:087601, 2006.

[5] E van Oortt, N B Manson, and M Glasbeekt. Optically detected spin coherence of the diamond n-v centre in its triplet ground state. *J. Phys. C: Solid State Phys.*, 21:4385–4391, 1988.

[6] A Gruber, A Dräbenstedt, C Tietz, L Fleury, J Wrachtrup, and C von Borczyskowski. Scanning confocal optical microscopy and magnetic resonance on single defect centers. *Science*, 276, 1997.

[7] A Young, C Y Hu, L Marseglia, J P Harrison, J L O'Brien, and J G Rarity. Cavity enhanced spin measurement of the ground state spin of an nv center in diamond. *New Journal of Physics*, 11:013007, 2009.

[8] C F Wang, R Hanson, D D Awschalom, E L Hu, T Feygelson, J Yang, and J E Butler. Fabrication and charcterization of two-dimensional photonic crystal microcavities in nanocrystalline diamond. *Appl.Phys.Lett.*, 91:201112, 2007.

[9] F Jelezko, T Gaebel, I Popa, A Gruber, and J Wrachtrup. Observation of coherent oscillation of a single nuclear spin and realization of a two-qubit conditional quantum gate. *Phys. Rev. Lett.*, 93:7, 2004.

[10] F Jelezko, T Gaebel, I Popa, A Gruber, and J Wrachtrup. Observation of coherent oscillations in a single electron spin. *Phys. Rev. Lett.*, 92:7, 2004.

[11] A Gali, M Fyta, and E Kaxiras. Ab initio supercell calculations on nitrogen-vacancy center in diamond: Electronic structure and hyperfine tensors. *Physical Review B*, 77:155206, 2008.

[12] N B Manson, J P Harrison, and M J Sellars. Nitrogen-vacancy center in diamond: Model of the electronic structure and associated dynamics. *Phys. Rev. B*, 74:104303, 2006.

[13] P Tamarat, N B Manson, J P Harrison, R L McMurtrie, A Nizovtsev, C Santori, R G Beausoleil, P Neumann, T Gaebel, F Jelezko, P Hemmer, and J Wrachtrup. Spin-flip and spin-conserving optical transitions of the nitrogen-vacancy centre in diamond. *New Journal of Physics*, 10:045004, 2008.

[14] F Jelezko and J Wrachtrup. Read-out of single spins by optical spectroscopy. *J. Phys.: Condens. Matter*, 16:104, 2004.

[15] C H Su, A D Greentree, and L C L Hollenberg. Towards a picosecond transform-limited nitrogen-vacancy based single photon source. *Optics Express*, 16:6240, 2008.

[16] P Neumann, N Mizuochi, F Rempp, P Hemmer, H Watanabe, S Yamasaki, V Jacques, T Gaebel, F Jelezko, and J Wrachtrup. Multipartite entanglement among single spins in diamond. *Science*, 320:1326, 2008.

[17] L Marseglia. Photonic crystals - innovative systems, lasers and waveguides. *InTech*, ISBN 978-953-51-0416-2, 2012.

[18] A A Tseng. Recent developments in micromilling using focused ion beam technology. *J. Micromech. Microeng.*, 14:15–34, 2004.

[19] J Melngailis. Critical review: focused ion beam technology and applications. *J. Vac. Sci. Technol. B*, 5:469, 1987.

[20] S Reyntjens and R Puers. A review of focused ion beam applications in microsystem technology. *J. Micromech. Microeng.*, 11:287?300, 2001.

[21] S Rubanov, P Munroe, S Prawer, and D Jamieson. Surface damage in silicon after 30kev ga fib fabrication. *Microscopy and Microanalysis*, 9:884, 2003.

[22] D P Adams, M J Vasile, T M Mayer, and V C Hodges. Focused ion beam milling of diamond: Effects of h2o on yield, surface morphology and microstructure. *J. Vac. Sci. Technol. B*, 21, 2003.

[23] J Taniguchi, N Ohno, S Takeda, I Miyamoto, and M Komuro. it focused-ion-beam-assisted etching of diamond in xef2. *J. Vac. Sci. Technol. B*, 16, 1998.

[24] C F Wang, Y S Choi, J C Lee, E L Hu, J Yang, and J E Butler. Observation of whispering gallery modes in nanocrystalline diamond microdisks. *Appl. Phys. Lett.*, 90:081110, 2007.

[25] J P Hadden, J P Harrison, A C Stanley-Clarke, L Marseglia, Y-L D Ho, B R Patton, J L O'Brien, and J G Rarity. Strongly enhanced photon collection from diamond defect centres under micro-fabricated integrated solid immersion lenses. *Appl. Phys. Lett.*, 67:241901, 2010.

[26] A Politi, M J Cryan, J G Rarity, S Yu, and J L O'Brien. Silica-on-silicon waveguide quantum circuits. *Science*, 320:646–649, 2008.

[27] J R Rabeau, Y L Chin, S Prawer, F Jelezko, T Gaebel, , and J Wrachtrup. Fabrication of single nickel-nitrogen defects in diamond by chemical vapor deposition. *Appl. Phys. Lett.*, 86:131926, 2005.

[28] T Gaebel, I Popa, A Gruber, M Domhan, F Jelezko, and J Wrachtrup. Stable single-photon source in the near infrared. *New Journal of Physics*, 6:98, 2004.

[29] B-S Song, S Noda, and T Asano. Photonic devices based on in-plane hetero photonic crystals. *Science*, 300:1537, 2007.

[30] P B Deotare, M W McCutcheon, I W Frank, M Khan, and M Loncar. Coupled photonic crystal nanobeam cavities. *Appl. Phys. Lett.*, 95:031102, 2009.

Chapter 10

Silicon Nitride Photonic Crystal Free-Standing Membranes: A Flexible Platform for Visible Spectral Range Devices

T. Stomeo, A. Qualtieri, F. Pisanello, L. Martiradonna,
P.P. Pompa, M. Grande, D'Orazio and M. De Vittorio

Additional information is available at the end of the chapter

http://dx.doi.org/10.5772/54846

1. Introduction

Two-dimensional (2D) photonic crystal (PhC) technology is well established at telecommunication bands, and materials such as Silicon (Si), Gallium Arsenide (GaAs) or Indium Phospide (InP) represent a common solution for applications at these wavelengths [1-5]. However, the interest of scientific community on structures operating in other spectral regions, such as the visible one, is growing up for both linear and non- linear applications [6-8]. Indeed, 2D-PhC resonant cavities in the visible spectral range are considered a promising tool to boost photonic devices performance in several fields, such as biosensing, integrated optics, quantum communications, solar energy, etc. As a consequence of this wide area of interest, a photonic platform able to answer to the needs of all these fields would be attractive for scientific and technical communities.

Trying to develop such technological platform, the first problem one should face is the material choice. In principle it should be transparent in the whole visible spectral range with a relatively high refractive index (n), economical, compatible with silicon based technologies, robust, biocompatible and suited for easy functionalization with several biological species.

In past years several materials have been proposed with this purpose; among them, of remarkable interest are Gallium Nitride (GaN) [9], Gallium Phospide (GaP) [10], polymers [11] and Silicon Dioxide (SiO_2) [12, 13]. Another appealing material is Silicon Nitride (Si_3N_4), which answers to most of the above-mentioned requirements. Indeed, stoichiometric silicon nitride is transparent in the visible spectral range with a refractive index $n \sim 1.9$ (@ $\lambda = 600$nm), it can be grown on Si with low-cost and widely diffused growth facilities such as Plasma Enhanced

Chemical Vapor Deposition (PECVD) [14], it is compatible with Si based electronics (it is used as insulator in MOSFET gates [15,16]), it is biocompatible and can be functionalized with several kinds of proteins [17–19].

Several Si_3N_4 2D-PhC cavities have been already proposed in past years [6,20,21], showing a maximum experimental quality factor (Q) of ~ 5000 in the case of a double heterostructure nanocavity [21]. Moreover, recent advances in the development of nanobeam cavities have led to extremely high quality factors also in the visible range, with a maximum Q of ~ 55000 [22–24].

This chapter is devoted to the use of Si_3N_4 PhC resonators as a flexible platform to realize photonic devices based on the engineering of nanoemitters spontaneous emission in the visible spectral range. First of all, the nanocavity design based on the closed band-gap principle will be presented and discussed. The chapter will then be focused on the nanotechnological procedures developed in recent years to realize high quality Si_3N_4 PhC cavities and the coupling of these structures with organic and inorganic nanoemitters. The versatility of the examined approaches will be also reviewed, showing how it is possible to couple several types of quantum light emitters to the two photonic states allowed in a closed band-gap single point defect nanocavity [25,26]. At the end of the chapter, a case of study on PhC-based biosensors [27] will be used to make the reader conscious of the possibility to realize advanced photonic devices in the visible spectral range exploiting the Si_3N_4 PhC technology. In the conclusions, we will discuss how improvements in modeling and processing of PhC structures in Silicon Nitride, which are highly compatible with both biological materials and inorganic quantum emitters, can further boost device performance, envisioning a broader application of two-dimensional PhC nanocavities in the visible spectral range.

2. Microcavity design: The closing band-gap and the modal selective tuning

2.1. The closing band-gap for low refractive index materials

A system composed by a quantum light emitter coupled to a resonant optical mode can be modeled as two interacting oscillators. The strength of this interaction can lead to two different coupling regimes known as strong and weak coupling. In weak coupling regime, the free-space spontaneous emission rate (Γ_0) is modified by the so-called Purcell effect: the coupled system emits with a rate $\Gamma_C = F\Gamma_0$, where $F = 3/(4\pi^2)Q/V\ (\lambda/n)^3$ is called Purcell Factor (Q and V are the quality factor and the modal volume of the photonic mode, respectively). When the system is instead in the strong coupling regime, the confined excitons and photons coherently exchange energy with a coupling strength, g, inversely proportional to V, i.e. $g \propto 1/\sqrt{V}$. Thus the properties of the photonic mode and, in particular, the electromagnetic field confinement in both time and spatial domains strongly affect the dynamic of the coupled system.

At visible wavelengths, these phenomena have been observed by means of several optically confined systems [28,29], but 2D-PhCs represent the most promising structures, since they give the best control on the optical properties of the resonators. To date, PhC cavities for visible spectral range are based on various geometries [6,11,21] and on higher-order modes of the widely studied H1 defect [30] (sketched in Fig. 1(a), inset). This resonator consists of a missed hole in a triangular PhC lattice, and it allows two orthogonally polarized resonant modes in the photonic band gap (hereafter referred to as x- and z-pole modes, on the base of the orientation of the wave vector).

The cross polarization of x- and z-pole modes and the absence of higher-order states represent non-negligible advances for applications in quantum optics [31,34]. Moreover, the H1 cavity presents the lowest V among PhC point defects, thus enhancing quantum electrodynamic (QED) phenomena in both strongly and weakly coupled systems. However, obtaining small V at visible wavelengths is a challenging goal, because of the low refractive index of transparent materials in this spectral range, which reduces the effectiveness in localizing the optical modes. Nevertheless, the aforementioned advantages, together with the increasing interest toward the realization of efficient emitting devices in the visible spectral range, foster theoretical and experimental studies to find alternative routes to improve light confinement in low-index H1 systems.

In the following, we consider a resonator consisting of a point defect H1 in a triangular lattice of air holes (period *a*, radius *r*) realized in a silicon nitride slab having refractive index n=1.93 and a thickness t. Plane-Wave Expansion (PWE) and 3D Finite Difference Time Domain (FDTD) algorithms [32] were used to investigate the electromagnetic response of such structure. All the calculations were restricted to modes with non-negligible components of the electric field along x and z and a non-negligible component of the magnetic field along y (hereafter referred to as TE- like modes).

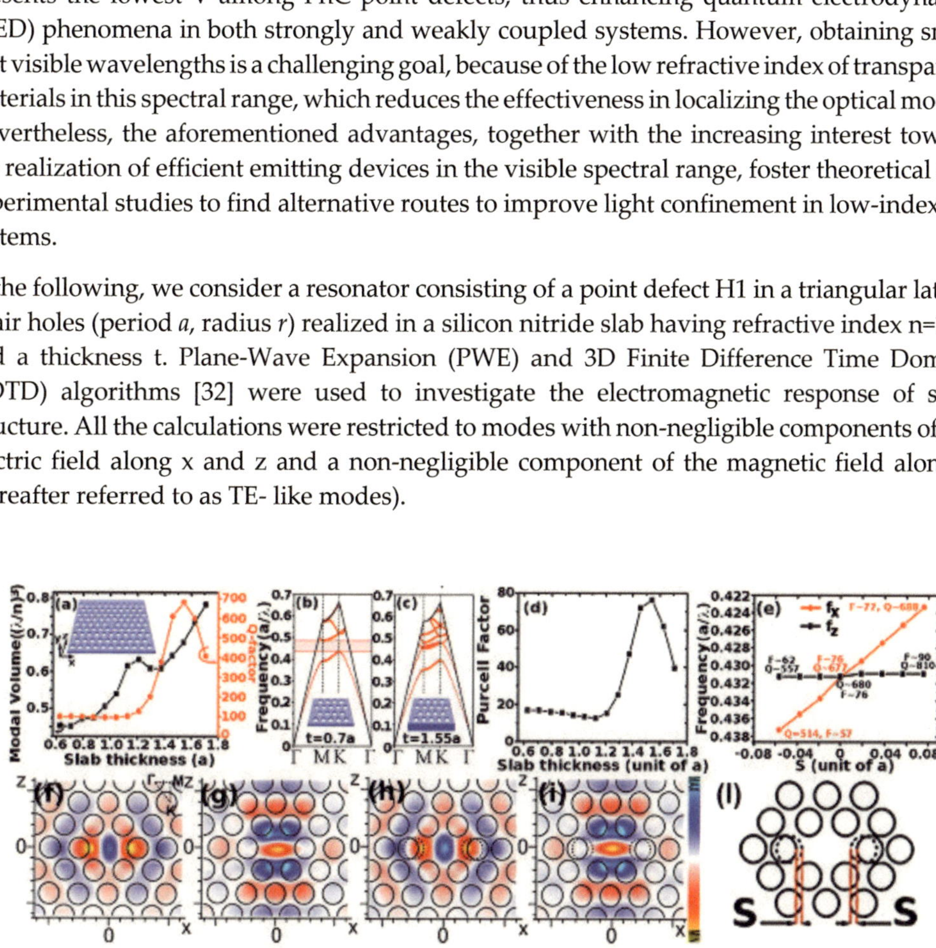

Figure 1. (a) Dependence of modal volume and Q factor on the thickness of the slab *t*. Inset, photonic crystal H1 cavity. (b), (c) Photonic band structure of the structure for $t = 0.7a$ and $t = 1.55a$, respectively. (d) Dependence of the Purcell factor on t (for S = 0). (e) Modification of the resonant frequencies and of the Q factor of the degenerated modes when two cavity neighboring holes are moved, as shown in the inset. The holes are moved closer to (farther from) the center for S < 0 (S > 0). (f) E_z for the x-pole mode in an unmodified H1 cavity. (g) E_x for the z-pole mode in an unmodified H1 cavity. (h) E_z for the x-pole mode in a H1 cavity with S=0.2*a*. (i) E_x for the z-pole mode in a H1 cavity with S = 0.2*a*. (l) Definition of S.

One way to realize ultrasmall-volume PhC cavities while keeping high Q- factors in the visible range and preserving the dipole-like shape of the modes is the so-called closing band-gap technique [33], involving PhC slab thickness (t) optimization. As shown in Fig. 1(a), the x-pole mode Q-factor has a maximum for $t = 1.55a$, while it is almost constant for $t < 1.2a$. The Photonic Band-Gap (PBG) existing for $t = 0.7a$ disappears when t is increased to $1.55a$ (see Figs. 1(b) and 1(c)). In agreement with the closing band-gap principle [33], this effect is assigned to a new nature of the electromagnetic confinement in the xz plane: it is not still due to the PBG, but it has to be assigned to the momentum space mismatch between the cavity mode and the second guided mode in the PhC slab. The increased thickness of the slab leads to slight variations of the x- and z-pole modal profiles along y, thus leading to a wider modal volume, as shown in Fig. 1(a). However, these variations of V are negligible with respect to the increase in Q, since the modal extension in the xz plane is preserved. Indeed the Purcell factor [Fig. 1(d)] follows the Q-factor behavior: for $t = 1.55a$, F is maximized to F ~ 78 with V ~ 0.68 $(\lambda/n)^3$ and Q ~ 700. A similar trend has been found for the z-pole mode.

2.2. Modal selective tuning

The x- and z-pole modes engineering would foster many applications based on H1 nanocavities operating at visible wavelengths. For instance, the degeneracy of x- and z-pole modes may be useful for entangled photon generation [34]. Other applications, such as single-photon sources or PhC-based optical read out of lab-on-chip devices [27], require well-defined and linearly polarized non-degenerate resonances. Several solutions have been reported in past years to break the energy degeneracy of the optical modes or to recover it [30,35-37]. A promising strategy to obtain a control on x- and z-pole modes is displayed in Fig. 1(e): by acting on two cavity neighboring holes, the resonant frequency of the x-pole mode (f_x) can be significantly modified while keeping constant the z-pole mode one. This finding can be ascribed to the selective modification of the wavevector $\mathbf{k} = (k_x, k_y, k_z)$ along a specific axis. Indeed x- and z-pole modes have the strongest component of k oriented along the x and z axes, respectively. If two holes are moved one toward each other along the x axis (S<0, see Fig. 1(l) for definition), k_x is modified without affecting k_z. As a consequence, f_x increases while f_z does not change. In the same way f_x decreases for $S > 0$, while keeping f_z constant. Figures 1(f-i) display x- and z-pole modal profiles for $S = 0$ and $S = 0.2a$: the electric field component along x (E_x) of the z-pole mode profile remains unchanged when the holes are moved far from the center [Figs.1(g) and 1(i)]. The shift instead results in the elongation of the x-pole modal function along x [Figs. 1(f) and 1(h)], thus modifying its resonant frequency.

It is important to notice that such alterations of field distributions modify the modal Q factors [Fig. 1(e)]: when $S < 0$, abrupt changes are introduced near the electric field maximum of the z-pole mode function, resulting in an increase in radiation losses and in a smaller Q factor (Q ~ 557 for $S = -0.057a$) [38]. In contrast, if $S > 0$ these abrupt variations are avoided, the radiative energy in the light-cone minimized, and the Q-factor of the z-pole mode enhanced together with almost preserved V and f_z. The optimized Q-factor turns out to be Q ~ 810 for $S = 0.075a$, and the Purcell factor is assessed as F ~ 90.

These findings are confirmed by the analysis in the z-pole momentum space, obtained by using a 2D Fourier Transform (2DFT), reported in Figs 2(a) and (b) for $S = -0.1a$ and $S = 0.2a$, respectively.

The white circle of Figs 2 (a) and (b) delimits the leaky region, defined by the light cone [38-40]: the stronger the components within this area, the higher the radiation losses along y. For S = −0.1*a* (Figs. 2(a) and (e)) a sharp peak is present at the center of the leaky region, affecting the value of Q; instead if S = 0.2*a* (Figs. 2(b) and (f)), the 2DFT is almost constant inside the light cone.

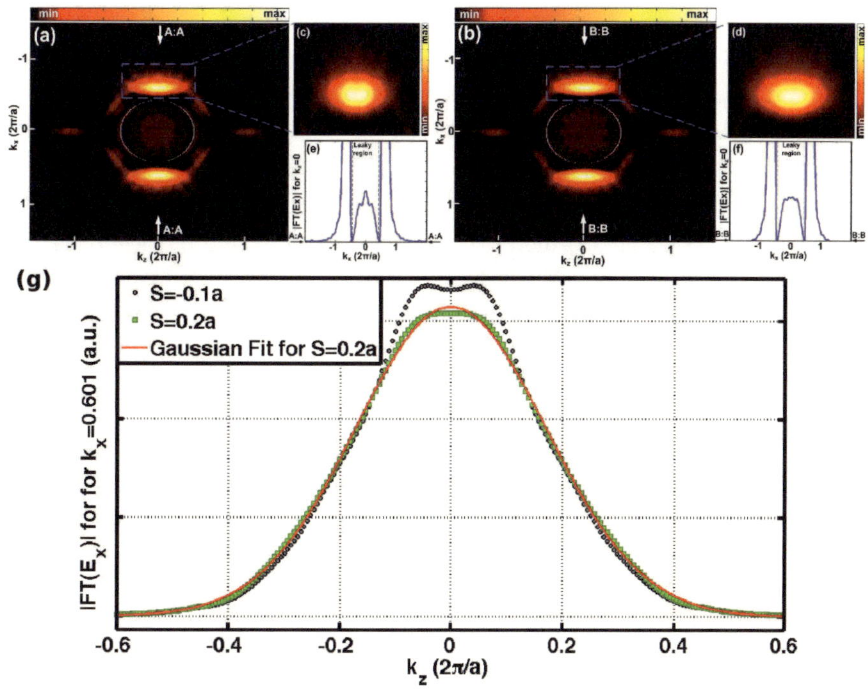

Figure 2. E_x field-distribution in momentum space for the z-pole mode. The white circle represents the light-cone. (a) |2DFT(E_x)| for S = −0.1*a*. (b) |2DFT(E_x)| for a S = 0.2*a*. (c) Zoom to a specific area of (a). (d) Zoom to a specific area of (b). (e) Momentum function extracted from (a) for k_z = 0. (f) Momentum function extracted from (b) for k_z = 0. (g) Cross section of the lobe for S = −0.1*a* shown in Fig. 2(c) (black circles) and of the lobe S = 0.2*a* displayed in Fig. 2(d) (green squares). Red continuous line represents a gaussian fit of the case S = 0.2*a*

As demonstrated in [38], the Q-factor depends also on the shape of the two peaks outside the light cone. A Gaussian shape typically leads to higher Q-factor, giving a direct measure of the energy not coupled with the radiation mode. Figs. 2(c) and (d) show a zoom of Figs. 2(a) and (b), respectively: the behavior for S = −0.1*a* is far from a 2D Gaussian function. For S = 0.2*a* (Fig. 2(d)), it is instead clear that by moving two holes far from the center, a 2D Gaussian function for these peaks is obtained, as also confirmed by the 1D gaussian fitting of the cross section of these lobes reported in Fig. 2(g). Therefore, by increasing the z-pole Q-factor, a positive S does not substantially affect the position, modal volume and resonant frequency of the z-pole electric field main lobe, leading to a straightforward increase of the Purcell factor of microresonators.

This therefore verifies that momentum space engineering, a strategy exploited to improve the confinement of defect states localized within the PBG [11,38,42], can also be efficient for cavity resonances without PBG.

3. Fabrication of Si_3N_4 PhC nanocavities

The 2D-PhC nanocavities were fabricated into a 400-nm-thick Si_3N_4 layer deposited on a silicon substrate by means of Plasma Enhanced Chemical Vapor Deposition (PECVD) technique. Si_3N_4 refractive index was measured through spectrophotometric methods (performed with a Varian Cary 5000 spectrophotometer), giving a value of $n_{SiN} = 1.93@\lambda = 600$ nm.

The PhC geometry was defined using a Raith150 e-beam lithography tool (equipped with a Gemini Column) operating at 30 kV. A thickness of 400 nm of ZEP520-A resist was chosen to ensure sufficient durability as a mask for pattern transfer into the underlying Si_3N_4 and, at the same time, to ensure a good resolution of the e-beam writing. The key issue in the fabrication process is to achieve PhC devices with well-controlled patterns size.

A preliminary dose-test was performed to define the optimum layout since the actual size of the pattern is influenced by the electron dose. A proximity error correction (PEC) was also applied to accomplish this target and the final dose was determined through Scanning Electron Microscope (SEM) inspections at 10 kV. Moreover, in order to achieve smoother and circular holes and faster exposure, the EBL system was used in "circular mode". In this mode, every circular hole is exposed by the deflection of the beam along concentric circles.

The patterns defined in the ZEP were then transferred into the Si_3N_4 layer using inductive coupled plasma reactive ion etching (ICP-RIE) in fluorine chemistry until the silicon substrate surface was reached. The membrane structure was released by wet etching of the underlying Si substrate in a TetraMethylAmmonium Hydroxide (TMAH) solution. Each sample consisted of an array of H1 nanocavities, whose dimensions were scaled according to the lattice period a (in the range $a = 257$ nm – 277 nm) thus allowing spectral shifting of the resonant wavelength. Fig. 3(a) shows a Scanning Electron Microscope (SEM) image of the realized resonators.

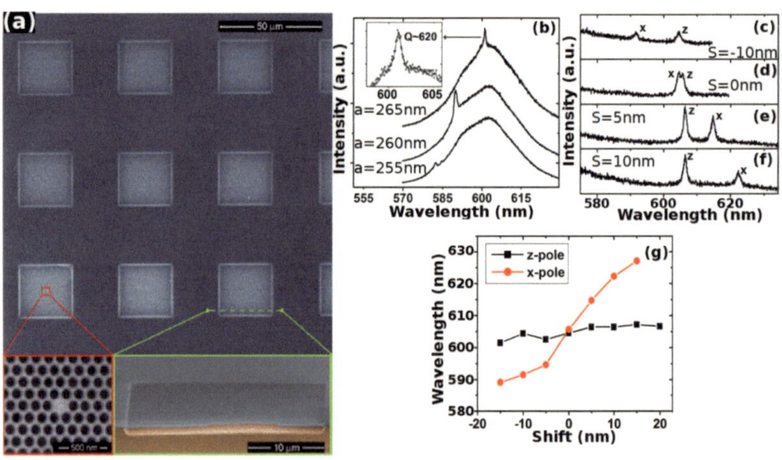

Figure 3. a) Top and bird's eyes view of the realized nanocavities acquired by SEM. (b) Resonances obtained by drop casting colloidal nanocrystals on the structure for three different values of a and $r = 0.308a$. The inset shows the Lorentzian fitting of the resonant peak for $a = 265$ nm. (c)-(f) Resonance of x- and z-pole modes for different values of the hole shift and $a = 270$ nm. (g) Resonant frequencies of x- and z-pole modes as a function of the hole shift.

http://dx.doi.org/10.5772/54846

4. Coupling of H1 nanocavity with organics and inorganics emitters

The theoretical findings described in section 2 have been experimentally demonstrated by coupling to the nanocavities both cyanine 3 (Cy3) fluorophore and core/shell CdSe/CdS colloidal dot-in-rod (DR) nanocrystals. Room temperature microphotoluminescence (μPL) characterization was therefore performed to investigate the properties of the coupled system.

4.1. Nanoemitters deposited on top of the cavity

A micromolar solution (10^{-6}mol/l) of DRs in toluene was prepared by using the synthesis procedure described in [43] by L. Carbone and co-workers and drop-casted on the realized microcavities. Figure 3(b) displays three resonances for three different values of *a*. The resonant peaks are well fitted by a Lorentzian function [Fig. 3(b), inset] and result in a maximum Q ~ 620 for an unmodified H1 cavity (*a* ~ 265 nm).

To explore the mode shifting over a wide spectral range, an organic fluorophore (Cy3) with broad emission spectrum was immobilized on the device. The μPL spectra for different values of S reported in Figures 3(c-f) show that the z-pole mode is almost unaffected by holes shifting, while x-pole resonant wavelength can be broadly tuned by means of S. Polarization-resolved measurements were carried out to identify the two modes, and their resonant wavelengths (λ_x and λ_z) as a function of S are displayed in Fig. 3(g). In agreement with the theoretical results of Fig. 3(e), the x-pole mode is tunable over a range $\Delta\lambda_x$ ~ 40 nm. Small discrepancies between experimental results and theoretical calculations have been observed in terms of slight variations of λ_z and weak nonlinearity of λ_x; since these variations do not show a clear dependence on S, they could be reasonably attributed to unavoidable fabrication imperfections. The theoretical findings about the influence of the holes position on the z-pole Q-factor have been confirmed by the experiments. For S = 15 nm Q ~ 750 has been measured, while for S = −20 nm the z-pole Q-factor falls down to a value of ~ 200.

4.2. Colloidal nanocrystals localized in the maximum of the electric field distribution

It is well known that in 2D-PhC slabs the in-plane confinement (xz) is due to the photonic band gap produced by the PhC periodicity, while in the out-of-plane direction (y) the confinement is due to the total internal reflection. As already mentioned, in the xz plane the electromagnetic field is localized in the center of the cavity (see figures 1(f-i)); FDTD simulations show also that along y the main lobe of the confined radiation is in the center of the slab (see Fig. 4).

The coupling reported in section 4.1 is thus not optimized, as the nanocrystals and the organic molecules are deposited on top of the cavities.

A viable strategy to approach the maximum allowed Purcell factor is to localize the nanoemitters in the center of the slab. This has been done with colloidal dot-in-rod nanocrystals using the same fabrication process described in section 3 and splitting the growth procedure of the Si_3N_4 slab in two steps. Figure 5(a) shows a sketch of the fabrication procedure. First of all, a 200 nm thick Si_3N_4 slab was grown on a Si substrate. A thin layer of colloidal DRs, with a molar concentration of ~ 10^{-6}mol/l was then spin-coated on it with a rotating speed of 500 rpm, thus obtaining a thickness lower than 10 nm as assessed by SEM inspection. After solvent evaporation, a second 200 nm thick layer of Si_3N_4 was grown on top of the sample. We verified the uniformity of the deposited layer by exploiting both the morphological characterization and the photoluminescence maps collected by a confocal microscope. A SEM cross-section of the resulting sandwiched structure is shown in Fig. 5(b). The nanocavities were then realized through electron beam lithography and dry and wet etching processes by following exactly the fabrication procedure reported in paragraph 3 (Fig. 5(c)).

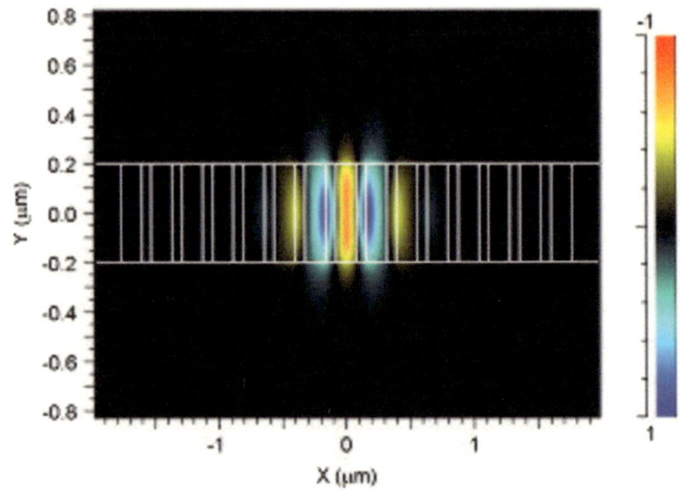

Figure 4. Cross view of the electric field x-component with the superposition of the slab cross-section at z = 0.

Also in this case, the optical measurements of the nanocavities were carried out by the OLYMPUS FluoView 1000 confocal laser scanning microscope, with a spatial resolution of 200 nm. A CW laser diode emitting at wavelength λ_{ex} = 405 nm was used as excitation source. In Fig. 5(d) are reported the photoluminescence spectra collected from the 2D-PhC H1 nanocavities with different lattice constants *a*. Superimposed to the broad emission spectrum (FWHM ~ 30 nm) of NCs uncoupled to the cavity, sharp peaks with a quality factor of about 600 are clearly detected, assessing the modulating effects of the PhC nanocavity on the emission of NCs coupled to the optical mode localized in the defect. Moreover, the normalized frequency a/λ of the experimental results was found to be about $a/\lambda \sim 0.46$ against the expected value of a/λ of ~ 0.431. As already suggested in case of the modal selective tuning, this slight difference can be mainly attributed to the effects of fabrication imperfections, inducing unavoidable uncontrolled variations in the optical properties of the PhC nanocavities [45].

The efficient coupling between the semiconductor nanocrystals layer and the dielectric cavity is due to the fact that the nanocrystals layer can be precisely positioned in the maximum of the confined electric field in the vertical direction. Indeed, in this case the Purcell effect results optimized [46] and the spontaneous emission rate strongly increased, leading to the possibility to measure a better Q-factor [47]. At the same time it is noteworthy to point out how the introduction of a guest material, embedded in two Si_3N_4 layers, does not affect the optical properties of the nanocavity as shown by the good match of the calculated and measured Q-factors (equal to 680 and 600, respectively).

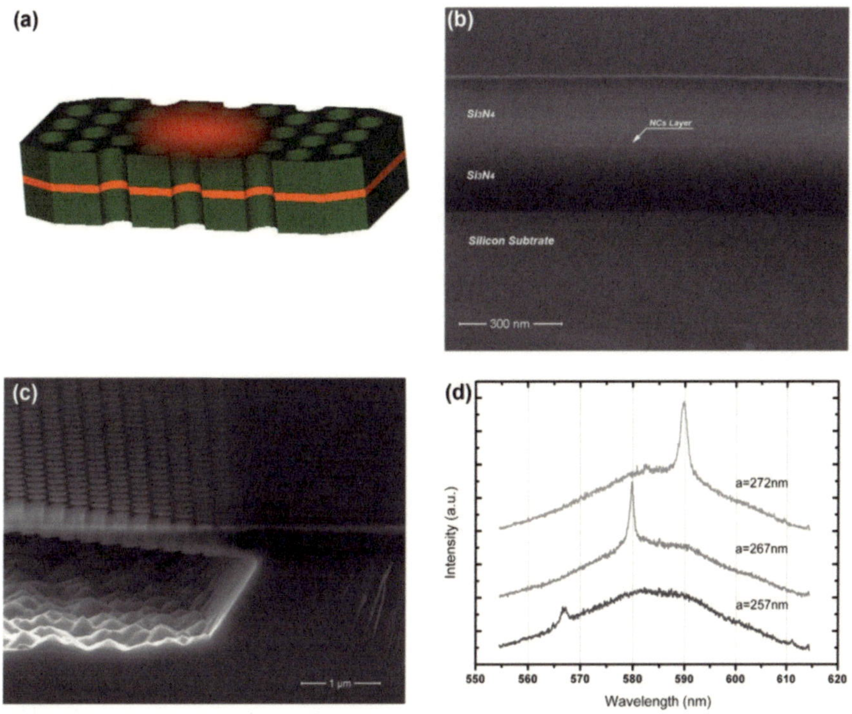

Figure 5. a) Sketch of the fabrication process, (b) cross-section SEM image of the un-patterned sample; (c) bird's eye view of the fabricated Si_3N_4 2D-PhC H1 nanocavity membrane and (d) photoluminescence spectra collected from the 2D-PhC H1 nanocavities with different lattice constants *a*.

5. A biosensor based on Si_3N_4 PhC nanocavities

Recently the light molding properties of PhC have been profitably exploited to boost the performance of optical sensors and transducers for biochemical analyses [48-50]. This paragraph proposes the idea of exploiting the sharp resonances of PhC nanocavities to assign unique spectral features to fluorophore-labeled bioanalytes, thus allowing their identification through wavelength-resolved light detection. Spectral tagging of organic dyes through photonic crystal nanocavities is experimentally proved to bring important benefits to cutting edge devices for biodiagnostics, such as DNA and protein biochips, in terms of improved sensitivity, efficiency and multiplexing capability.

5.1. Working principle

PhC nanocavities can be embedded in a two-dimensional array, to realize an improved optical detection system of a miniaturized assay for genomic and proteomic analyses, (DNA or protein microarray). Fig. 6(a) is a three-dimensional sketch of the biochip architecture including different nanocavities, each having a different resonant wavelength. Moreover, a one-to-one correspondence is also preserved between a cavity and a group of specific bio-molecules (probes) immobilized on the surface (as shown in the expanded view of Fig. 6(a)). The as-realized chip can be exposed to a biological solution containing unknown target species, or analytes; conjugation between the analytes and their complementary probes takes place on the device surface [51]. Since the target analytes are typically labeled with fluorescent markers, the binding events can be revealed through optical inspection of the biochip readout area, thus allowing a complete compositional analysis of the assay [52].

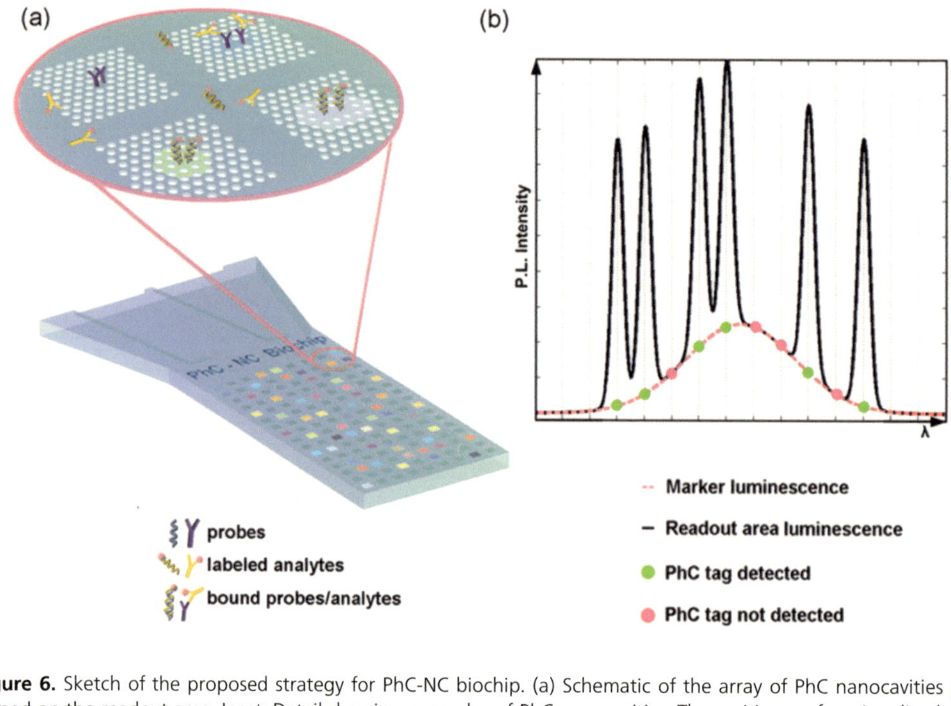

Figure 6. Sketch of the proposed strategy for PhC-NC biochip. (a) Schematic of the array of PhC nanocavities patterned on the readout area. Inset: Detail showing examples of PhC nanocavities. The cavities are functionalized with different probes molecules, that specifically interact with complementary target analytes labeled with fluorescent markers. The signal is collected from this area and spectrally discriminated in order to identify the different spectral tags univocally associated to each nanocavity and thus to each bioprobe. (b) Example of a possible luminescence detected from the whole readout area (black line) as compared to the unmodified broad marker luminescence (red dotted line). The presence of each peak in the spectrum reveals the presence of the corresponding analyte in the investigated assay.

The recourse to a microarray configuration already allows the simultaneous analysis of a certain number of analytes thanks to spatial discrimination [52,53]. Here we upgrade the allowed degree of parallelization by assigning a peculiar spectral signature, given by the resonating behavior of each cavity, to each bioprobe immobilized on the surface. This gives the possibility to distinguish the spectral response of each target analyte bound to the corresponding probe, albeit a single common fluorophore is used for the labeling of the whole unknown solution. Fig. 6(b) exemplifies a possible spectral scan of the signal collected from the whole readout area. Different peaks can be observed on the emission spectrum of the fluorescent marker, each revealing the presence of a specific target analyte in the investigated assay. Besides the spatial discrimination implemented in microarray configurations, in this case the spectral distinction contributes substantially to the parallelization of the device. We also expect a beneficial effect given by Purcell effect, which increases the radiative emission rate of emitting materials interacting with quantum confined systems [54, 55]. Hence, a significant increase in the luminescence intensity of the markers coupled to the PhC cavities is envisioned, leading to a significant improvement of the signal-to-noise ratio and of the overall sensitivity of the biochip detection.

5.2. Experimental results

PhC nanocavities resonating in the visible spectral range were fabricated in Si_3N_4 membranes on a Si substrate, exploiting the modified single defect H1 nanocavity described in section 2 [33,43,47,57]. Several chips were fabricated, each containing an array of optimized H1 resonators with variable lattice constant *a*, thus tuning the corresponding resonant wavelengths. We tested the proposed architecture both with single-stranded DNA (ss-DNA) and antibody probes immobilized on the Si_3N_4 surface of two different devices. Complementary DNA targets or specific secondary antibodies, labeled with cyanine 3 (Cy3) and rodhamine (TRITC) fluorophores, respectively, were then allowed to recognize the immobilized probes, thus obtaining a uniform fluorescent monolayer of the biomolecular species.

The effects of fluorescence enhancement and peak sharpening in resonant conditions are clearly observed in the emission spectra reported in Fig. 7(a) for PhC nanocavities treated with TRITC-labeled proteins and in Fig. 7(b) for DNA-functionalized nanocavities (five uppermost lines, compared to the lowest spectrum corresponding to the emission of Cy3- DNA strands without photonic resonators). In both cases it is evident that the change of the lattice period *a* of the photonic crystal resonator leads to the modification of the spectral response coming from target analytes conjugated by the same broad emitting organic dye: a specific spectral feature is thus attributed to the target analytes captured on different cavities. The best measured Q-factor obtained in the PhC-nanocavities DNA-chip prototype is ~ 725, corresponding to a full-width at half maximum of ~ 0.9 nm. Taking into account the spectral resolution limits, a conservative estimate suggests the possibility to distinguish up to 150 different resonant peaks within the 150nm bandwidth of the Cy3 emission spectrum. This means that up to 150 parallel analyses can be simultaneously performed with one single spectral scan of the readout area of the biochip, thus drastically decreasing the time required for a complete compositional identification.

By confocal microscopy it is also possible to visualize the effects of emission enhancement in resonant conditions, as reported in the photoluminescence maps reported in Fig. 7(c). In this array of five different nanocavities, functionalized with ss-DNA and hybridized with Cy3-

labeled complementary DNA sequences we have performed a spectral scanning of the acquisition wavelength with a resolution of 2 nm. When the detection wavelength matches one of the five resonating wavelengths of the nanocavities, marked from λ_1 to λ_5 in Fig. 7(b), it is possible to distinguish a bright spot in the center of each nanocavity.

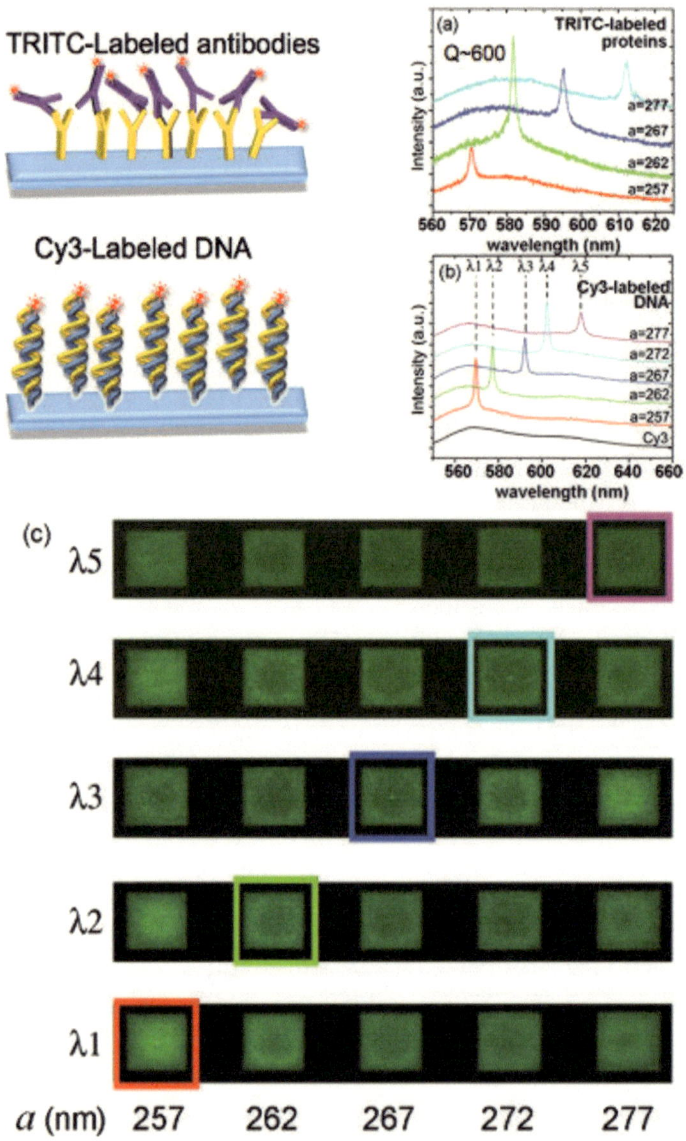

Figure 7. a) PL spectra collected from TRITC-labeled proteins captured onto the PhC nanocavities; (b) PL spectra collected from Cy3–labeled DNA (colored lines) onto the PhC nanocavities as compared to Cy3 emission spectrum collected on a PhC pattern (black line); (c) PL maps of an array of five Cye-labeled DNA-PhC nanocavities, collected at different wavelengths [also indicated in (b)]. For each spectrum and PL map, the reported *a* value indicates the lattice period of the measured PhC nanocavity.

http://dx.doi.org/10.5772/54846

In order to quantify the enhancement effect of each photonic crystal pattern, in Fig. 8 it is reported a three-dimensional intensity profile collected on a Cy3-labeled DNA functionalized nanocavity in resonant conditions. The central bright spot corresponding to the H1 defect cavity reveals a signal improvement as high as 160 as compared to the luminescence coming from unpatterned Si_3N_4 surface. A major role of the Purcell effect [54,55] can be envisioned, by virtue of the strong optical quantum confinement performed by the H1-shifted nanocavities.

Noteworthy, the photonic crystal pattern itself causes an improvement of fluorescence emission as compared to the surrounding unpatterned Si_3N_4 layer, although the immobilization and hybridization processes have been homogeneously performed on the whole sample surface. In this case, an enhancement of ~ 20 times is achieved. This behavior may be ascribed to the combination of two effects. First, the free-standing membrane layer makes available a larger surface area to the probes immobilization (about a factor of 4 more than the unpatterned layer), resulting in a higher number of immobilized Cy3-labeled analytes in the PhC regions. Second, in 2D-PhC patterns an efficient transfer channel between externally radiated light and energy trapped in the membrane is represented by the so-called leaky modes [48,57,58]. The coupling of such modes with the absorption or emission bands of neighboring emitters may lead to a significant increase of their luminescence. Although the photonic crystal pattern has not been specifically optimized to maximize such effect, the role of leaky modes localized on the PhC pattern for the further increase of the luminescence experimentally observed is not negligible.

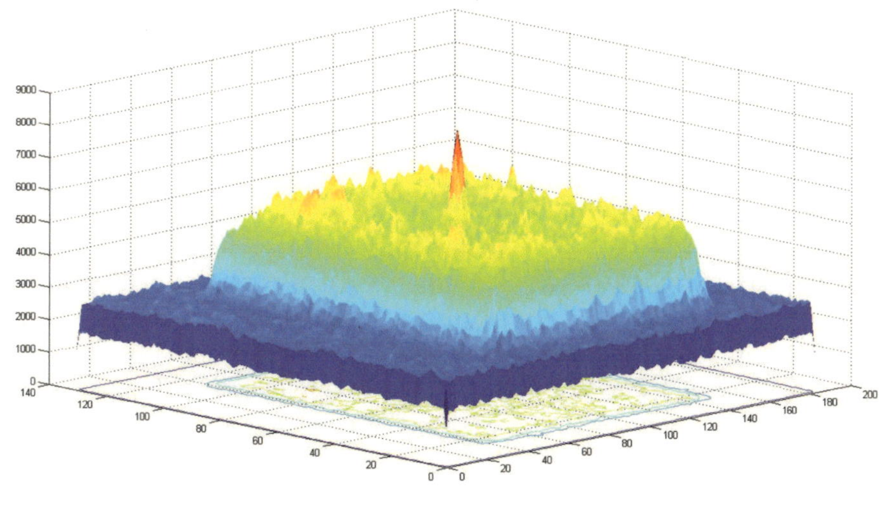

Figure 8. Three-dimensional intensity profile of photoluminescence collected from Cy3- labeled DNA captured by a functionalized nanocavity. Emission outside the PhC pattern has been normalized to unit. A 20-fold luminescence enhancement due to the PhC pattern, as compared to the unpatterned Si_3N_4 surface, has been measured. The cavity confinement further enhances Cy3 emission up to 160-fold.

The insertion of PhC cavities in classical biochip architectures leads, therefore, to a huge increase of the emission intensity of fluorescent markers, thus providing higher sensitivity, and allowing detection of very small amounts of target biomolecules in the investigated solution. In addition, the nanocavities attribute peculiar spectral features to the target analytes captured by their surface, so that the presence of specific species in the solution can be inferred by a simple spectral analysis of the optical response of the read-out region. This enables parallel detection of multiple elements, thus accelerating the analysis time.

6. Conclusions and perspectives

This chapter details the use of Si_3N_4 2D-PhC nanocavities as flexible platform to realize photonic devices based on the engineering of spontaneous emission of nanoemitters in the visible spectral range. The versatility of the approach is demonstrated by coupling several types of emitters to the two photonic states allowed in a closed band gap single point defect nanocavity. In particular, DNA strands and antibodies marked with Cy3 and TRITC organic dyes have been immobilized on top of the nanocavities, while colloidal quantum dots emitting in the visible spectral range have been dropcasted on the devices and also positioned in the resonators at the maximum of the localized photonic mode. The optical measurements, carried out by μPL confocal microscopy, revealed maximum quality factors close to the theoretical estimations for all the emitters. Improvements in modeling and processing of PhC structures in Silicon Nitride, which is highly compatible with both biological materials and inorganic quantum emiters, let us envision a broader application of two-dimensional PhC nanocavities also in the visible spectral range. In particular, the coupling of a single colloidal dot-in-rod nanocrystal with a photonic crystal cavity would be an important milestone to reach in next years, and would allow further improvements of single photon rate and stability.

Author details

T. Stomeo[1], A. Qualtieri[1], F. Pisanello[1,2], L. Martiradonna[1], P.P. Pompa[1], M. Grande[3], D'Orazio[3] and M. De Vittorio[1,4,5]

1 Center for Bio-Molecular Nanotechnology UniLe, Istituto Italiano di Tecnologia, Arnesano (Lecce), Italy

2 Center for Neuroscience and Cognitive Systems UNITN, Italian Institute of Technology, Rovereto (TN), Italy

3 Dipartimento di Elettrotecnica ed Elettronica, Politecnico di Bari, Bari, Italy

4 National Nanotechnology Laboratory (NNL), Istituto di Nanoscienze-CNR, Via Arnesano, Lecce, Italy

5 Dipartimento Ingegneria dell'Innovazione, Università del Salento, Lecce, Italy

References

[1] Tanabe T., Notomi M., Mitsugi S., Shinya A., Kuramochi E. All-optical switches on silicon chip realized using photonic crystal nanocavities. Applied Physics Letters 2005; 87(15) 151112.

[2] Van Laere F., Stomeo T., Cambournac C., Ayre M., Brenot R., Benisty H., Roelkens G., Krauss T.F., Van Thourhout D., Baets R. Nanophotonic Polarization Diversity Demultiplexer Chip. J. Lightwave Technology 2009; 27 417-425.

[3] Stomeo T., Van Laere F., Ayre M., Camburnac C., Benisty H., Van Thourhout D., Baets R., Krauss T.F. Integration of grating couplers with a compact photonic crystal demultiplexer on an InP membrane. Optics Letters 2008; 33(8) 884-886.

[4] Stomeo T., Grande M., Rainò G., Passaseo A., D'Orazio A., Cingolani R., Locatelli A., Modotto D., De Angelis C., De Vittorio M. Optical filter based on two coupled PhC GaAs-membranes. Optics Letters 2010; 35 (10) 411-413.

[5] D'Orazio A., De Sario M., Marrocco V., Petruzzelli V., Prudenzano F. Photonic crystal drop filter exploiting resonant cavity configuration. IEEE Transactions on Nanotechnology 2008; 7 (1) 10-13.

[6] Makarova M., Vuckovic J., Sanda H., Nishi Y. Silicon based photonic crystal nanocavity light emitters. Applied Physics Letters 2006; 89 221101.

[7] Shambat G., Rivoire K., Lu J., Hatami F., Vuckovic J. Tunable-wavelength second harmonic generation from GaP photonic crystal cavities coupled to fiber tapers. Optics Express 2010; 18 12176-12184;

[8] Corcoran B., Monat C., Grillet C., Moss D. J., Eggleton B. J., White T. P., O'Faolain L., Krauss T. F. Green light emission in silicon through slow-light enhanced third-harmonic generation in photonic-crystal waveguides. Nature Photonics 2009; 3 206-210.

[9] Choi Y.-S., Hennessy K., Sharma R., Haberer E., Gao Y., DenBaars S. P., Nakamura S., Hu E. L., Meier C. GaN blue photonic crystal membrane nanocavities. Applied Physics Letters 2005; 87 243101.

[10] Rivoire K., Faraon A., Vuckovic J. Gallium phosphide photonic crystal nanocavities in the visible. Applied Physics Letters 2008; 93 063103.

[11] Martiradonna L., Carbone L., Tandaechanurat A., Kitamura M., Iwamoto S., Manna L., De Vittorio M., Cingolani R., Arakawa Y. Two-dimensional photonic crystal resist membrane nanocavity embedding colloidal dot-in-a-rod nanocrystals. Nano Letters 2008; 8 (1) 260-264.

[12] Gong Y., Vuc◎ković J. Photonic crystal cavities in silicon dioxide. Applied Physics Letters 2010; 96 031107.

[13] Grande M., O'Faolain L., White T. P., Spurny M., D'Orazio A., Krauss T. F. Optical filter with very large stopband (approximate to 300 nm) based on a photonic-crystal vertical-directional coupler. Optics Letters 2009; 34 (21) 3292-3294.

[14] Yoon D.H., Yoon S.G., Kim Y.T. Refractive index and etched structure of silicon nitride waveguides fabricated by PECVD. Thin Solid Films 2007; 515 (12) 5004-5007.

[15] Mui D. S. L., Liaw H., Demirel A. L., Strite S. and Morkoç H. Electrical characteristics of Si sub 3 N sub 4 /Si/GaAs metal-insulator-semiconductor capacitor. Applied Physics Letters 1991; 59 2847-2849.

[16] Chen A., Young M., Li W., Ma T.P., Woodall J.M. Metal-insulator-semiconductor structure on low-temperature grown GaAs. Applied Physics Letters 2006; 89 233514.

[17] Gao H., Luginbühl R., Sigrist H. Bioengineering of Silicon-Nitride. Sensor and Actuator B, Chemical 1997; 38 (1-3) 38-41.

[18] Diao J., Ren D., Engstrom J. and Lee K. A surface modification strategy on silicon nitride for developing biosensors. Analytical Biochemistry 2005; 343 (2) 322–328.

[19] Dauphas S., Ababou-Girard S., Girard A., Le Bihan F., Mohammed-Brahim T., Vié V., Corlu A., Guguen-Guillouzo C., Lavastre O. and Geneste F. Stepwise functionalization of SiNx surfaces for covalent immobilization of antibodies. Thin Solid Films 2009; 517 6016-6022.

[20] Barth M., Kouba J., Stingl J., Löchel B., Benson O. Modification of visible spontaneous emission with silicon nitride photonic crystal nanocavities. Optics Express 2007; 15 (25) 17231-17240.

[21] Barth M., Nüsse N., Stingl J., Löchel B. and Benson O. Emission properties of high-Q silicon nitride photonic crystal heterostructure cavities. Applied Physics Letters 2008; 93 021112.

[22] Eichenfield M., Camacho R., Chan J., Vahala K. and Painter O. A picogram-and nanometer-scale photonic-crystal optomechanical cavity. Nature 2009; 459 (7246) 550–555.

[23] Gong Y., Makarova M., Yerci S., Li R., Stevens M., Baek B., Nam S., Dal Negro L. and Vuckovic J. Observation of transparency of erbium-doped silicon nitride in photonic crystal nanobeam cavities. Optics Express 2010; 18 13863–13873.

[24] Khan M., Babinec T., McCutcheon M., Deotare P. and Lonc◎ar M. Fabrication and characterization of high-quality-factor silicon nitride nanobeam cavities. Optics letters 2011; 36 (3) 421–423.

[25] Pisanello F., Martiradonna L., Pompa P. P., Stomeo T., Qualtieri A., Vecchio G., Sabella S., De Vittorio M. Parallel and high sensitive photonic crystal cavity assisted readout for DNA-chips. Microelectronic Engineering 2010; 87 (5-8) 747-749.

[26] Qualtieri A., Pisanello F., Grande M., Stomeo T., Martiradonna L., Epifani G., Fiore A., Passaseo A., De Vittorio M. Emission control of colloidal nanocrystals embedded in

Si_3N_4 photonic crystal H1 nanocavities. Microelectronic Engineering 2010; 87 (5-8) 1435–1438.

[27] Martiradonna L., Pisanello F., Stomeo T., Qualtieri A., Vecchio G., Sabella S., Cingolani R., De Vittorio M., Pompa P. P. Spectral tagging by integrated photonic crystal resonators for highly sensitive and parallel detection in biochips. Applied Physics Letters 2010; 96 113702.

[28] Qualtieri A., Morello G., Spinicelli P., Todaro M. T., Stomeo T., Martiradonna L., Giorgi M., Quélin X., Buil S., Bramati A., Hermier J.-P., Cingolani R., De Vittorio M. Nonclassical emission from single colloidal nanocrystals in a microcavity: a route towards room temperature single photon sources. New Journal of Physics 2009; 11 033025.

[29] Le Thomas N., Woggon U., Schöps O., Artemyev M., Kazes M. and Banin, U. Cavity quantum electrodynamic (QED) with Semiconductor Nanocrystals. Nano Letters 2006; 6 (3) 557–561.

[30] Kitamura M., Iwamoto S. and Arakawa Y. Enhanced light emission from an organic photonic crystal with a nanocavity. Applied Physics Letters 2005; 87 151119.

[31] Bennett C., Brassard G.et al., Proceedings of IEEE International Conference on Computers, Systems and Signal Processing, vol. 175 (Bangalore, India, 1984).

[32] Yee K. Numerical solution of inital boundary value problems involving maxwell's equations in isotropic media. IEEE Transactions on Antennas and Propagation 1966; 14 (3) 302–307.

[33] Tandaechanurat A., Iwamoto S., Nomura M., Kumagai N. and Arakawa Y. Increase of Q-factor in photonic crystal H1-defect nanocavities after closing of photonic bandgap with optimal slab thickness. Optics Express 2008; 16 448–455.

[34] Larqué M., Karle T., Robert-Philip I. and Beveratos A. Optimizing H1 cavities for the generation of entangled photon pairs. New Journal of Physics 2009; 11 (3) 033022.

[35] Painter O., Srinivasan K., O'Brien J., Scherer A. and Dapkus P. Tailoring of the resonant mode properties of optical nanocavities in two-dimensional photonic crystal slab waveguides. Journal of Optics A: Pure and Applied Optics 2001; 3(6) S161-S170.

[36] Frédérick S., Dalacu D., Aers G., Poole P., Lapointe J. and Williams R. Optical characterisation of InAs/InP quantum dot photonic cavity membranes. Physica E: Low-dimensional Systems and Nanostructures 2006; 32 (1-2) 504–507.

[37] Hennessy K., Högerle C., Hu E., Badolato A. and Imamog◎lu A. Tuning photonic nanocavities by atomic force microscope nano-oxidation. Applied Physics Letters 2006; 89 041118.

[38] Akahane Y., Asano T., Song B. and Noda S. Fine-tuned high-Q photonic-crystal nanocavity. Optics Express 2005; 13 (4) 1202–1214.

[39] Srinivasan K. and Painter O. Momentum space design of high-Q photonic crystal optical cavities. Optics Express 2002; 10 (15) 670–684.

[40] Vuckovic J., Loncar M., Mabuchi H. and Scherer A. Optimization of Q-factor in photonic crystal microcavities. IEEE Journal of Quantum Electronics 2002; 38 (7) 850–856.

[41] Pisanello F., De Vittorio M. and Cingolani R. Modal selective tuning in a photonic crystal cavity. Superlattices and Microstructures 2010; 47 (1) 34–38.

[42] Krauss T. F. K. Photonic crystals: Cavities without leaks. Nature Materials 2003; 2 (12) 777-778.

[43] Carbone L., Nobile C., Giorgi M. D., Sala F., Morello G., Pompa P.P., Hytch M., Snoeck E., Fiore A., Franchini I. R., Nadasan M., Silvestre A. F, Chiodo L., Kudera S., Cingolani R., Krahne R. and Manna L. Synthesis and micrometer-scale assembly of colloidal CdSe/CdS nanorods prepared by a seeded growth approach. Nano Letters 2007; 7 2942–2950.

[44] Pisanello F., Qualtieri A., Stomeo T., Cingolani R., Martiradonna L., Bramati A. and De Vittorio M. High-Purcell-factor dipolelike modes at visible wavelengths in H1 photonic crystal cavity. Optics Letters 2010; 35 (10) 1509–1511.

[45] Rico-Garcia J., Lopez-Alonso J. and Alda J. SPIE 2005: Multivariate analysis of photonic crystal microcavities with fabrication defects: Proceeding of SPIE Europe, 9-11 May 2005, Sevilla, Spain.

[46] Andreani L., Panzarini G. and Gérard J. M. Strong-coupling regime for quantum boxes in pillar microcavities: Theory. Physical Review B 1999; 60 13276.

[47] Adawi, A. & Lidzey, D. A design for an optical-nanocavity optimized for use with surface-bound light-emitting materials. New Journal of Physics 2008; 10 065011.

[48] Ganesh N., Zhang W., Mathias P., Chow E., Soares J., Malyarchuk V., Smith A. and Cunningham B. Enhanced fluorescence emission from quantum dots on a photonic crystal surface. Nature Nanotechnology 2007; 2 (8) 515–520.

[49] Skivesen N., Têtu A., Kristensen M., Kjems J., Frandsen L. H. and Borel P. I. Photonic-crystal waveguide biosensor. Optics Express 2007; 15 (6) 3169–3176.

[50] Li M., He F., Liao Q., Liu J., Xu L., Jiang L., Song Y., Wang S. and Zhu D. Ultrasensitive DNA detection using photonic crystals. Angewandte Chemie 2008; 120 (38) 7368–7372.

[51] Fan J., Chee M. and Gunderson K. Highly parallel genomic assays. Nature Reviews Genetics 2006; 7 (8) 632–644.

[52] Schena M., Shalon D., Davis R. and Brown P. Quantitative monitoring of gene expression patterns with a complementary DNA microarray. Science 1995; 270 (5235) 467-470.

[53] Spurgeon S., Jones R., and Ramakrishnan R. High throughput gene expression measurement with real time PCR in a microfluidic dynamic array. PLoS One 2008; 3 e1662.

[54] Purcell E., Torrey H. and Pound R. Resonance Absorption by Nuclear Magnetic Moments in a Solid. Physical Review 1946; 69 (1-2) 37–38.

http://dx.doi.org/10.5772/54846

[55] Noda S., Fujita M. and Asano T. Spontaneous-emission control by photonic crystals and nanocavities. Nature Photonics 2007; 1 449–458.

[56] Painter O., Lee R., Scherer A., Yariv A., O'brien J., Dapkus P. and Kim I. Two-dimensional photonic band-gap defect mode laser. Science 1999; 284 (5421)1819-1821.

[57] Fan S. and Joannopoulos J. Analysis of guided resonances in photonic crystal slabs. Physical Review B 2002; 65 235112.

[58] Crozier K., Lousse V., Kilic O., Kim S., Fan S. and Solgaard O. Air-bridged photonic crystal slabs at visible and near-infrared wavelengths. Physical Review B 2006; 73 115126.

Chapter 11

Photonic Crystals for Optical Sensing: A Review

Benedetto Troia, Antonia Paolicelli,
Francesco De Leonardis and Vittorio M. N. Passaro

Additional information is available at the end of the chapter

http://dx.doi.org/10.5772/53897

1. Introduction

In recent years, photonic sensors have seen a massive development because of the increasing demand of sensing applications in healthcare, defence, security, automotive, aerospace, environment, food quality control, to name a few.

The development and integration of Microfluidic and Photonic technologies, with specific reference to the CMOS-compatible silicon-on-insulator (SOI) technology, allows to enhance sensing performance in terms of sensitivity, limit-of-detection (LOD) and detection multiplexing capability. Photonic sensors have been the subject of intensive research over the last decade especially for detection of a wide variety of biological and chemical agents. In this context, photonic Lab-on-a-chip systems represent the state-of-the-art of photonic sensing since they are expected to exhibit higher sensitivity and selectivity as well as high stability, immunity to electromagnetic interference and product improvements, such as smaller integration sizes and lower costs.

In recent years, rapid advancements in photonic technologies have significantly enhanced sensing performance, particularly in the areas of light-analyte interaction, device miniaturization, multiplexing and fluidic design and integration. This has led to drastic improvements in sensor sensitivity, enhanced limit of detection (LOD), advanced fluidic handling capability, lower sample consumption, faster detection time, and lower overall detection cost per measurement. With future commercialization of photonic biosensors in a Lab-on-a-chip, next generation biosensors are expected to be reliable and portable, able to be fabricated with mass production techniques, to reduce the cost as well as to do multi-parameter analysis, enabling fast and real-time measurements of a large amount of biological or physical parameters within a single, compact sensor chip.

In this context, photonic crystals (PhCs) represent an intriguing solution for achieving high performance in sensing applications. In fact, since a lot of photonic architectures have been widely investigated and employed in photonic sensing (e.g., ring resonator, surface Plasmon resonance (SPR) – based sensors, microdisks, microspheres, to name a few), PhCs exhibit a strong optical confinement of light to a very small volume, enabling the detection of chemical species characterized by nanometer dimensions. In addition, by using advanced chemical surface functionalization techniques and integration with microfluidic systems, very high performance can be achieved in ultra compact sensor chips. For example, the detection of dissolved avidin concentrations as low as 15 nM or 1 μm/ml, has been experimentally achieved by using functionalized slotted PhC cavities with integrated microfluidics (Scullion et al., 2011). Ultra high performance have been experimentally and theoretically demonstrated, such as a LOD less than 20 pM for anti-biotin, corresponding to less than 4.5 fg of bound material on the sensor surface and fewer than 80 molecules in the modal volume of the integrated microcavity (Zlatanovic et al., 2009).

PhC-based sensors have been also proposed as gas sensors in mid infrared (mid-IR), since many gases (e.g., CO_2, CH_4, CO) exhibit absorption lines in mid-IR wavelength region. Other applications reported in literature concern with the detection of temperature, pressure, stress and humidity measurements, to name a few.

From a technological point of view, photonic sensors based on PhCs, including photonic crystal optical fiber (PCFs) and integrated planar photonic crystals, are suitable for multiplexing and label-free detection. For example, a large-scale chip-integrated PhC sensor microarrays has been recently proposed and demonstrated for biosensing on a SOI-based platform (Zou et al., 2012). Standard and CMOS compatible technological processes (i.e., electron-beam lithography, inductively coupled plasma (ICP) etching, plasma enhanced chemical vapor deposition (PECVD)) are generally employed for PhCs fabrication, making these sensors suitable for mass-scale and low cost production. Finally, PCFs can be easily fabricated by stacking tubes and rods of silica glass into a large structure of the pattern of holes required in the final fiber.

In this chapter, a complete review on planar PhC- and PCF-based sensors, is presented. In particular, it will be focused on the choice of materials and sensing applications.

Optical sensing principles will be described in detail, with particular reference to homogeneous and surface sensing, optical absorption, fluorescence, surface Plasmon resonance (SPR) and photonic detection based on non linear effects (e.g., Four Wave Mixing, Raman effect, surface enhanced Raman scattering). In addition, several advanced waveguide structures and microstructured optical fibers (MOFs) will be analyzed. For example, resonant microcavities based on integrated PhCs, slotted resonant cavities, interferometer configurations (e.g., directional couplers and Mach-Zehnder Interferometer (MZI), Sagnac interferometer), active PhC-based sensing devices, to be named. Sensing applications and performance of PhC-based sensors are reviewed and compared with those exhibited by other conventional photonic architectures in literature. The state of the art of PhC- based sensors is analyzed, highlighting on the actual strategic approach of integrating PhC sensors chips with Optofluidic Microsystems (Choi et al., 2006), and on advanced technologies and measurement setups employed in PCF-based sensing.

http://dx.doi.org/10.5772/53897

2. Integrated photonic crystal sensors

Nowadays, integrated PhC-based sensors represent one of the most popular class of photonic sensors, generally employed for physical and chemical/biochemical sensing. In this context, the principal advantages of these intriguing photonic sensor architectures are ultra-high light confinement in very small volumes, high wavelength selectivity, ultra high sensitivity and selectivity in sensing mechanism.

Materials usually employed for sensing PhC planar devices are heteropitaxial layers such as AlGaAs/GaAs, III-nitride compound layers or dielectric layers such as Si_3N_4, TiO_2, SiO_2 and the well-known SOI wafers (Biallo et al., 2006). In addition, organic compounds, and polymers have attracted an increasing interest in the last few years. Finally, porous silicon photonic crystals have been recently proposed for organic vapor sensing, too. However, future integrated photonic sensors are expected to be CMOS-compatible, able to be realized with low cost processes, and suitable for mass-scale production.

In this context, the SOI technological platform represents undoubtedly the most suitable platform for fabricating ultra-compact and ultra-high performance PhC-based integrated sensors.

To this purpose, several types of PhC-based sensor architectures are presented in this section, focusing on employed sensing principles (i.e., refractive index (RI)-based sensor, optical absorption, opto-mechanical, nonlinear effects) and application performance.

2.1. RI-based PhC sensors

Refractive index based sensors represent the most diffused class of PhC sensors. In fact, a large number of advanced architectures (e.g., integrated microcavities and interferometric configurations) employ the refractive index sensing for detection. RI-based PhC sensors present numerous advantages such as minimal sample preparation without fluorescence labeling, real–time detection, high sensitivity and selectivity. In particular, the sensing principle consists in measuring RI changes of a bulk solution (e.g., deionized water, $n_{Water} = 1.33$ or air, $n_{Air} = 1$ @ $\lambda = 1.55$ μm) due to the presence of chemical analytes or gases generally characterized by higher refractive indices. Applications in gaseous and aqueous environment have been studied to detect concentrations of chemical and biological species. In fact, by using these sensors, it is possible to quantify molecule and protein (e.g., streptavidin, DNA, mRNA) surface or volumetric density. Recently, advanced PhC-based sensors properly designed for single-molecule detection have been demonstrated to be able to detect the number of molecules concentrated into a complex solution (Lin et al., 2012).

In this section, two fundamental sensing principles commonly employed in photonic RI sensing are presented, i.e., surface and homogeneous sensing.

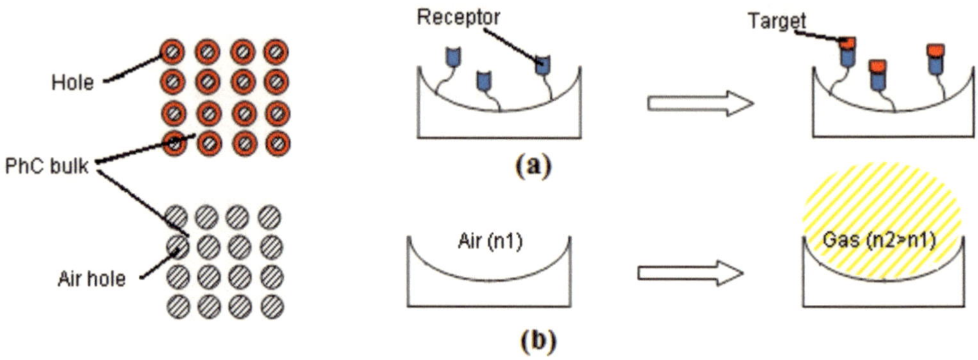

Figure 1. a) Surface sensing and (b) homogeneous sensing. Probe molecules (blue) are deposited on hole surfaces and target molecules (red) are captured by receptors forming an adlayer on the sensor surface.

In Fig. 1(a), the surface sensing principle is sketched. In particular, PhC holes are initially functionalized by receptor molecules properly chosen in order to selectively adsorb target analytes in a complex solution. Consequently, when the device is exposed to a chemical sample, target molecules are immobilized by receptors on the sensor hole inner surfaces. The adsorbed layer characterized by a thickness t_{ad} induces a localized refractive index change around the hole region. Finally, a surface sensitivity S_s is generally defined as follows:

$$S_s = \frac{\partial n_{eff}}{\partial t_{ad}} \tag{1}$$

where n_{eff} is the effective index of the optical mode propagating in the sensor device and t_{ad} has been previously defined.

The remaining sensing mechanism named homogeneous sensing, is schematically sketched in Fig. 1(b). In this case, the effective index of the propagating optical mode changes because of cover RI variations induced by gases or liquid samples properly concentrated in the cover medium, where the photonic sensor is exposed. Consequently, a homogeneous sensitivity S_h can be defined as follows:

$$S_h = \frac{\partial n_{eff}}{\partial n_c} \tag{2}$$

being n_c the cover refractive index.

In the following, main architectures of PhC RI-based sensors are presented, focusing on their operating principles and performance.

The first class of sensors to be discussed are those based on photonic crystal resonant micro-cavities. These devices are fabricated by introducing localized defects (i.e., removing one or more holes) in the periodic hole distribution. In this way, the perfect periodicity of the photonic crystal is compromised and a defective state in the band gap map is introduced, allowing the excitation of resonance modes.

Two parameters, i.e., the quality factor Q and the wavelength sensitivity S_λ, have to be considered for appreciating PhC-cavity-based sensor performances. The Q-factor defines the shape of resonant peaks and consequently the value of the Full Width at Half Maximum (FWHM) and it is expressed as follows:

$$Q = \frac{\omega_0 U(t)}{P(t)}, \; or \;\; Q = \frac{f_0}{\Delta f} \tag{3}$$

where ω_0 is the angular resonant frequency, $U(t)$ is the energy stored in the cavity mode, $P(t)$ is the energy dissipated per cycle (i.e., a single roundtrip in the resonant cavity), f_0 is the resonant frequency and Δf is the peak bandwidth.

In particular, PhC-cavity-based sensors can be interrogated into two distinct modes. The first one is the wavelength interrogation mode and the second one is the intensity interrogation mode. In the first method, the optical readout consists in monitoring the wavelength of the optical signal through an optical spectrum analyzer (OSA), while in the latter one, it is possible to monitor the intensity changes of the output signal by using a photodetector (PD). In this context, the wavelength sensitivity S_λ represents a fundamental parameter for quantifying the sensor performance in case of wavelength interrogation scheme. S_λ is defined according to Eq. (4), as the ratio between the shift of resonant wavelength ($\Delta\lambda$) induced by the change of the background refractive index (Δn). Moreover, it is given in units of nm/RIU (refractive index unit), as:

$$S_\lambda = \frac{\Delta\lambda}{\Delta n} \tag{4}$$

In Fig. 2 a typical example of two-dimensional (2D) PhC microcavity in silicon-on-insulator (SOI) wafer is shown (Liu et al., 2012). As sketched in Fig. 2, air holes are etched only in the upper silicon layer and they can be realized by standard anisotropic etching. The periodic structure is characterized by an hexagonal cell with lattice constant a=515 nm. The radius r of air holes and the thickness of the silicon layer h are chosen to be the ratios r/a=0.33 and h/a=0.427, resulting in r=170 nm and h =220 nm. As it is possible to observe in Fig. 2, the microcavity is obtained by removing seven air holes at the centre of PhC in the ΓK direction. Such microcavity is formally indicated as L7-cavity, because of the number of holes removed in the periodic PhC structure.

Different arrangements of air holes near the cavity centre improve the Q-factor of the micro-cavity presented above. In particular, by shifting three rows of air holes in the ΓM direction spaced from the cavity centre of a distance of 0.02a, 0.014a and 0.017a, it is possible to obtain

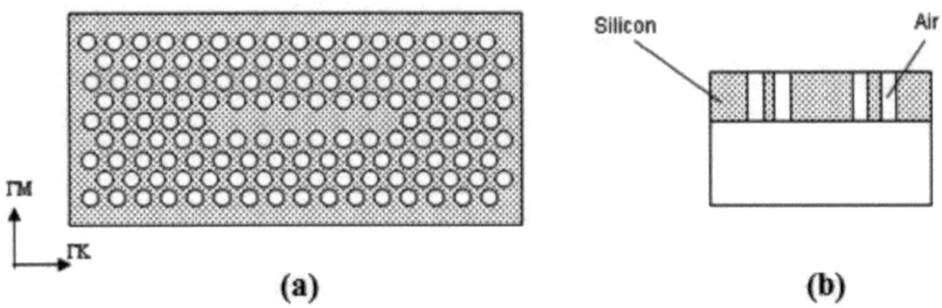

Figure 2. PhC microcavity (a) realized by removing seven holes as line defect (L7 cavity) and cross section (b).

an improvement of *Q*-factor of a factor ~1,000. In addition, by placing three pairs of mini holes into the cavity region it is possible to further increase the performance of PhC resonant cavity. The image of this new arrangement is sketched in Fig. 3. In particular, the pair named C has a radius r_C=0.78r and an outward position shift d_C=0.2a. The second pair centre (i.e., B) is not moved, but the radii of these holes are minimized, resulting in d_B=0 and r_B=0.2r. Finally, the innermost pair of holes have a displacement from original position d_A=0.2a and a radius slightly larger than previous pair to be r_A=0.28r. Under these design conditions, in air-infiltrated case, the cavity achieves a *Q*-factor of 2,600, exhibiting a resonant wavelength around 1550 nm.

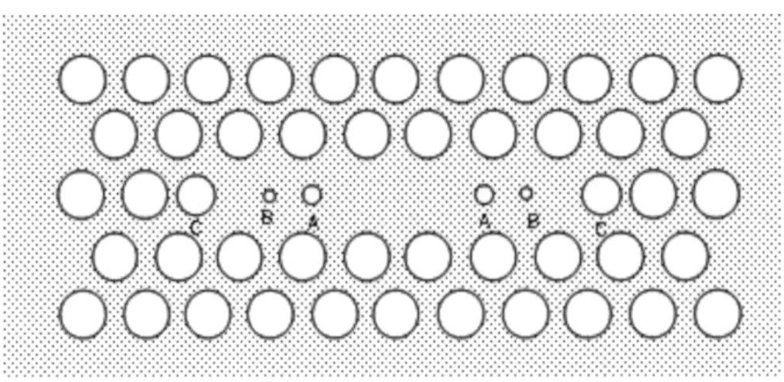

Figure 3. Zoom of cavity region of L7 cavity.

Performances of the sensor proposed have been evaluated in case of water or ethanol infiltration, whose refractive indices are estimated to be 1.332 and 1.359 at $\lambda \approx 1.55$ μm, respectively. The resonant wavelength shift measured in the first case is 22.28 nm and in the latter is equal to 12.65 nm. Finally, the device described until now exhibits a sensitivity as high as 460 nm/RIU, being larger than sensitivities usually achieved by L3 cavity.

An interesting RI-based sensor employing a PhC resonant microcavity is characterized by a cavity region in an air slot. In this sensor, the technological approach employed for realizing the microcavity does not consist in modifying the lattice constant or the hole radii character-

http://dx.doi.org/10.5772/53897

izing the PhC, but in introducing a straight line defect in which a modified waveguide width acts as resonant cavity (Jágerská et al., 2010).

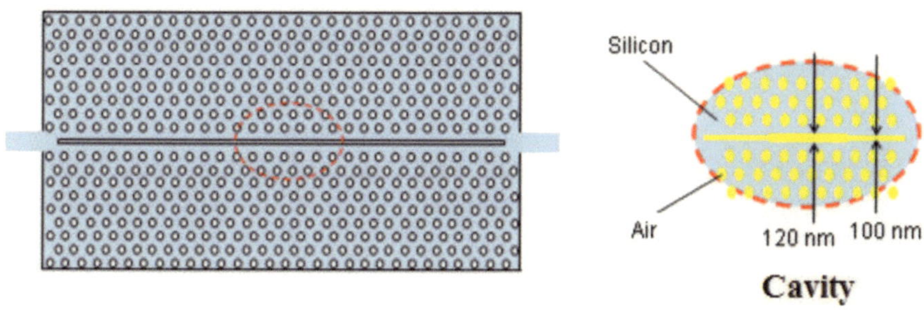

Figure 4. Air-slot PhC cavity with zoom of slit in the slot.

As sketched in Fig. 4, the device consists of a 2D PhC characterized by hexagonal cell and lattice constant $a=510\ nm$. The waveguide region is obtained by removing a row of air holes in the middle of the structure. An air – slot is embedded in this line defect region. The width of the air slot is kept constant at 100 nm, except in the middle. A slit is made by increasing the width of air slot from 100 nm to 120 nm, in the centre. This reduction in slot width results in the formation of reflective barriers for traveling mode, thus in a resonant cavity whose length is $L\ =3a$.

The complete device has been processed on a 220-nm-thick SOI wafer with a 2 μm buried oxide layer. The cavity mode is strongly confined into the cavity region, the effective mode volume is $V_{eff}=0.05\mu m^3$ and the spatial overlapping between cavity mode and air slot is $\Gamma>0.73$.

The sensitivity can be expressed as a function of the spatial mode overlapping Γ, as reported in the following expression:

$$S_\lambda = \frac{\Delta\lambda}{\Delta n} = \Gamma\frac{\lambda_0}{n_{eff}} \tag{5}$$

where $n_{eff}^2 = \Gamma n_{gas}^2 + (1-\Gamma)n_{Si}^2$ is the effective index of the cavity mode, $n_{Si}=3.46$ is the silicon RI, n_{gas} is the gas RI and λ_0 is the resonant wavelength. When air characterized by the refractive index $n_{air}=1$ is infiltrated in PhC holes, the sensor exhibits the resonance frequency $\lambda_0=1570\ nm$, near to cut-off frequency which occurs at 1590 nm. The quality factor has been estimated to be $Q=26{,}000$ and the sensitivity $S_\lambda=570$ nm/RIU.

Sensor performances have been experimentally quantified by exposing the cavity to different gases. Several gases have been used. In particular, air ($n = 1.000265$) as reference gas, nitrogen (N_2, $n = 1.000270$), helium (He, $n = 1.000032$), carbon dioxide (CO_2, $n = 1.000406$), acetylene (C_2H_2, $n = 1.000579$) and propane (C_3H_8, $n = 1.000999$). All refractive indices are given at atmospheric pressure at the resonant wavelength λ_0=1570 nm and room temperature, T=20°C.

Experimental results show a blue shift for He gas and a red shift for CO_2 or C_3H_8 gas characterized by higher RIs. In fact, the resonant wavelength shift is demonstrated to be linearly dependent on refractive index changes (n_{gas} - 1) calculated with respect to the reference gas, i.e. air (n = 1).

The PhC sensor described above is limited by the fact that resonant wavelength shifts are not only influenced by cover refractive index changes, but also by external parameters such as temperature, pressure, adsorbed humidity or progressive oxidation of sensor surface. In order to minimize this effect, it is necessary to test the sensor with O_2-free gases or use an identical sensor architecture acting as a reference one for compensating undesired effects mentioned above.

A PhC sensor based on a ring resonator cavity has been proposed for monitoring the level of seawater salinity between 0% to 40% (Robinson et al., 2012). In particular, a ring resonator is realized by removing a number of silicon rods in the PhC structure characterized by periodic distribution of square cells of silicon rods in air. The sensor architecture is shown in Fig. 5 below.

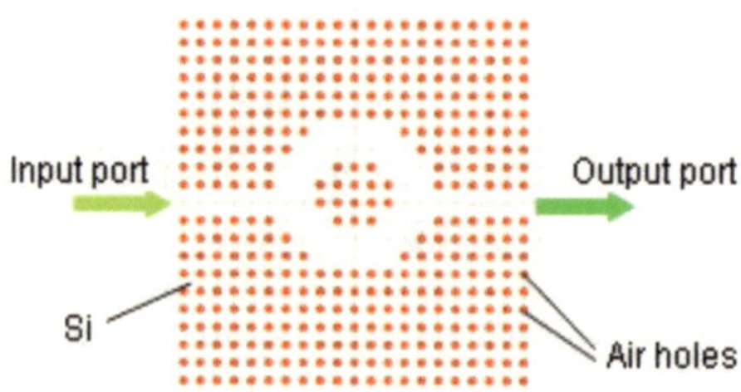

Figure 5. Schematic structure of sensor for seawater sensing.

The lattice constant is a=540 nm and radius of rods is r=0.185a=100 nm. The refractive index of silicon rods is n = 3.46 while the background RI is set to be the seawater RI. An input and output waveguides are placed in horizontal direction with respect to the center of the ring cavity.

Different salinity percentages induce background RI changes, resulting in a detectable variation of the sensor transmission spectrum. In particular, it is possible to adopt both an intensity and a wavelength interrogation scheme, resulting in a more accurate and precise optical readout. The output efficiency of the sensor decreases from 99% to 80% by increasing the salt level in water from 0% to 40%. Moreover, in the range of salinity 0÷40%, the water RI increases from 1.33300 to 1.34031, respectively. According to Eq. (4), changes of the background solution RI produce resonant wavelength shifts. In fact, in the range of interest, i.e., 0÷40% the resonant wavelength shifts down from 1590.55 nm to 1590.05 nm. Finally, the Q-factor changes as a function of the salinity. In fact, by increasing the salt level in water the Q-factor increases.

A PhC sensor based on an air-bridge cavity has been proposed, exhibiting a good sensitivity (Junhua et al., 2011). The photonic crystal sensor is characterized by a triangular (or equivalently hexagonal) array of air holes with lattice constant $a=440\ nm$ and radius $r=0.29a=127.5\ nm$. The microcavity has been realized by decreasing the radius of the central hole ($r_d=0.2a=88\ nm$). The air-bridge is made by removing a portion of buried silicon dioxide that is sandwiched between two silicon layers, as sketched in Fig. 6. The top silicon layer with thickness $t=0.591a$ is separated by 1 μm of SiO_2 from the second silicon layer on the bottom of the same structure.

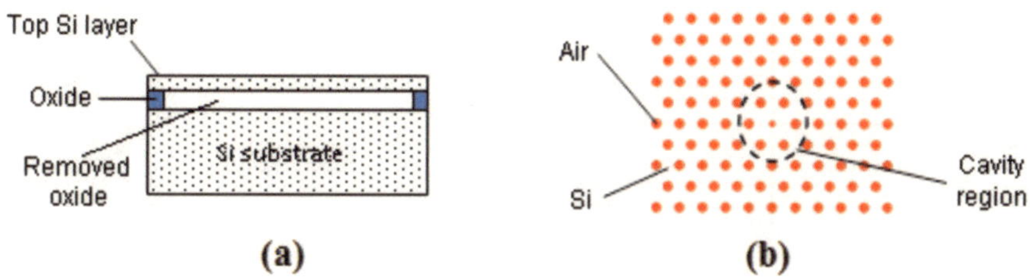

Figure 6. Cross section (a) and top view (b) of the PhC sensor based on air-bridge cavity.

The band gap map evidences a photonic band gap only for transverse electric (TE)-polarized mode in the microcavity. Under these conditions, the sensitivity has been estimated to be $S=570$ nm/RIU, higher than sensitivity evaluated in a simple SOI PhC-based sensor characterized by the same physical characteristics.

In the sensor presented above, the design of defect radius and slab thickness assumes a fundamental role for enhancing sensing performances. In fact, by changing the defect radius it is possible to obtain different band gap maps. Moreover, different defective states, and, consequently, different resonant conditions, can be properly introduced. For example, by increasing the defect hole radius a blue shift of resonant wavelength occurs, exhibiting a higher sensitivity. This effect is justified by the fact that a greater portion of localized field is distributed into the cavity region, resulting in a high overlapping between the resonant optical modes and the chemical/biochemical species to be sensed.

The slab thickness represents a key parameter because it influences performance parameters, i.e., S_λ and Q-factor. In addition, also the field distribution and the energy bandgap are seriously affected by the slab thickness. In fact, for a thinner slab S_λ is larger and Q-factor is smaller because of less confinement and larger overlapping between the electric field intensity and the background. On the contrary, the resonant wavelength shifts to longer wavelengths by increasing the slab thickness.

A very interesting solution proposed for enhancing sensing performance consists in a PhC sensor based on RI sensing principle and characterized by an array of resonant microcavities.

The first device is embedded on monolithic silicon substrate of thickness $t=0.55a=232.65\ nm$ (Yang et al., 2011a). The 2D PhC is characterized by a triangular lattice of holes with lattice constant $a=423\ nm$ and air hole radius $r=0.32a=135.36\ nm$. As shown in Fig. 7, the device consists of a waveguide obtained by removing a row of air holes along the horizontal direction. Such waveguide guides light from the sensor input to the output. The cavity region is adjacent to the waveguide and it is realized by pulling outwards two holes in the opposite direction (parallel to the waveguide). In addition, radii of surrounding holes have been adjusted in order to optimize the cavity dimension. In particular, radius of left and right holes is set to be r_x=0.25a and radius of top and bottom holes is r_y=0.35a. The overall geometry is sketched in Fig. 7.

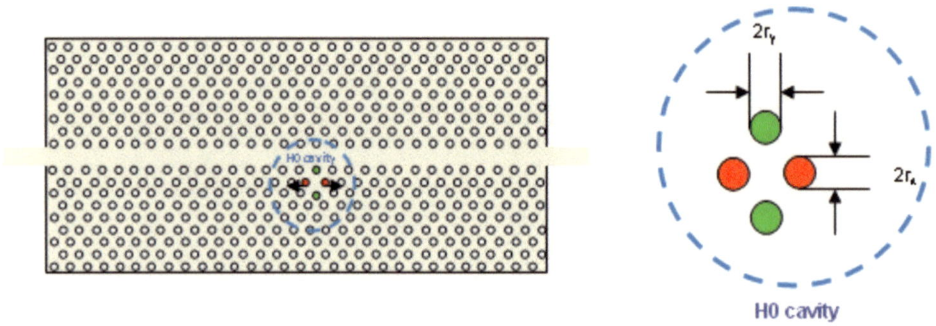

Figure 7. H0 cavity structure with the zoom of hole arrangement.

The hole shift represents a strategic design parameter. In fact, by varying the shift of cavity holes, the resonant wavelength of the transmission spectrum shifts, resulting in changes of the Q- factor, too. For the structure sketched in Fig. 7, it has been demonstrated that the optimal shift is s_x=0.2a, because the maximum Q-factor is obtained at this value. In addition, the best set of radii has been set to r_x=0.32a and r_y=0.28a, in order to obtain a quality factor as high as Q=2,761.

Another fundamental design parameter is the number of functionalized holes around the cavity region for mass sensitivity analysis. Initially, probe receptors are deposited on the inner hole surfaces near the cavity region. When target molecules are infiltrated into holes, the refractive index around the cavity area changes. This phenomenon is due to binding between probe molecules and target objects, resulting in a surface sensing.

It is convenient to introduce a new parameter known as mass sensitivity S_m, that illustrates the dependence of the resonant wavelength shift $\Delta\lambda$ on the number of functionalized holes N:

$$S_m = \frac{\Delta\lambda}{N} \tag{6}$$

As expected, the sensitivity increases by decreasing the number of functionalized holes. The most sensitive holes to RI changes are the innermost ones in the y direction. Consequently, in order to optimize the sensing event, it is necessary to deposit a layer of probe molecules only on the surface of holes mentioned above.

In this way, n sensors are made in cascade, side-coupled with the same waveguide, allowing a multiple and parallel sensing. As sketched in Fig. 8, each H0-cavity is characterized by the same architecture with different sets of geometrical parameters. The transmission spectrum shows n different dips, each one independent from the others. When a binding event occurs, only the correspondent transmission dip is affected by a shift. This allows a multiple and simultaneous sensing of different chemical species.

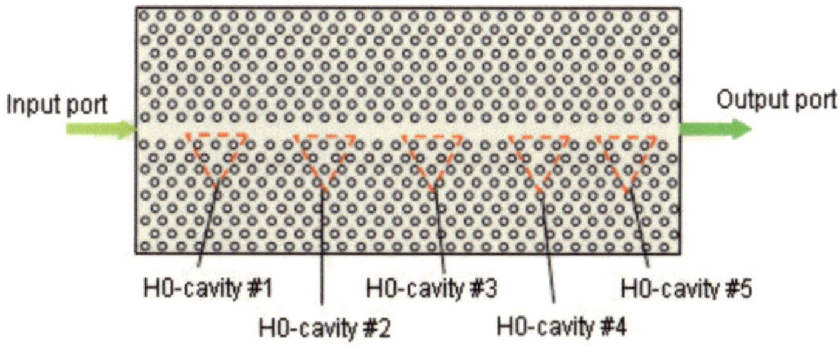

Figure 8. Array of H0 resonant cavities.

In conclusion, this sensor exhibits a wavelength sensitivity of 115.60 nm/RIU. In addition, the sensitivity can be also varied from 84.39 nm/RIU to 161.25 nm/RIU, by adjusting the number of functionalised holes from 2 to 28, respectively. The advantage of this architecture is represented by the low mass limit of detection achieved with small functionalized area, resulting in a good level of optical integration and large degree of multiplexed sensing in aqueous environment.

A RI-based PhC sensor based on the same architecture described above (i.e., a series of cascaded resonant cavities) has been proposed for simultaneous sensing of different species in aqueous environment (Mandal et al., 2009). The architecture consists of arrays of one-dimensional (1D) PhC resonators coupled to a single bus waveguide. Each cavity has a slightly different width with respect to the others, so that everyone can independently detect a different bio-molecular specie in response to changes of surrounding medium RI. The sensing mechanism occurs when bio molecules concentrated in the sensor cavity are captured by receptor molecules previously deposited on the sensor surface. Ring resonators have been designed in order to exhibit different and unique resonant wavelengths, so allowing multiplexed detection with a single waveguide. When target molecules are selectively captured by receptors, the dip in transmission spectrum shows a red shift. Analysis of the magnitude of this red shift provides quantitative information about concentration of target molecules in the sample and, consequently, about their bound mass.

The structure consists of a single mode silicon waveguide designed to be 450 nm wide and 250 nm tall, while resonators have been realized with the cavity region surrounded by 8 air holes at both sides with 200 nm in diameter, being the 1D lattice constant equal to 390 nm. The cavity area of the first sensor has been obtained by shifting outwards the innermost holes of 39 nm from the centre, as shown in Fig. 9.

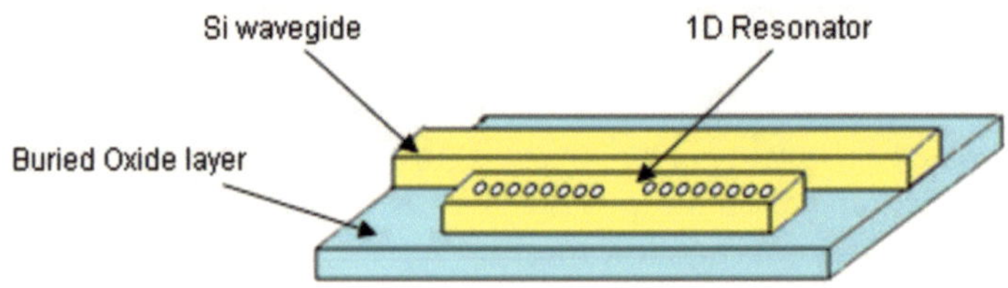

Figure 9. 1D-photonic crystal mycrocavity.

The architecture presented above has been adopted to detect different bio-molecules. The first configuration consists of five resonators, each one to be functionalized by a different probe. In particular, resonators functionalized with glutaraldehyde and streptavidin serve as control for non specific analyte adsorption. The other resonators functionalized with monoclonal antibodies are designed for monitoring and detecting in-*vivo* concentrations of interleukins 4, 6 and 8 (Mandal et al., 2009). The device can detect antibodies in a concentration range 1 μg/ml÷1 mg/ml, suitable for clinical application and medical diagnostics, such as HIV test and drug screening.

The same architecture characterized by analogous arrangement of five resonators near the PhC waveguide, has been designed with a different number of functionalized holes rather than a different size of cavity regions (Mandal et al., 2008). Each resonator has been functionalized by a 50-nm-thick single stranded DNA (ssDNA) monolayer with refractive index $n_{ssDNA}=1.456$. A detection event occurs when the complementary ssDNA hybridizes with the functionalized capture probes, forming a double stranded DNA (dsDNA). Moreover, the sensor sensitivity can be tuned by changing the number of functionalized holes, as reported in the expression below:

$$S_\lambda = \frac{\Delta\lambda}{\Delta m} = \frac{\Delta n}{\Delta m / A} \times \left[\frac{\Delta\lambda}{\Delta n} \times \frac{1}{A} \right] \tag{7}$$

where Δn is the refractive index change due to the binding event, A is the functionalized area of the sensor, $\Delta\lambda$ is the resonant wavelength shift and Δm is the mass of bound target. The sensor sensitivity increases by decreasing the number of functionalized holes N, as already demonstrated for sensor architectures previously analyzed. In this specific case, for only two functionalized holes a sensitivity as high as 3.5 nm/fg is achieved, while for sixteen holes the sensitivity drops to 1 nm/fg.

In this review on RI-based PhC integrated sensors we present also an innovative sensor able to detect in-*vivo*, single particles as small as viruses in aqueous and gaseous environment (Lee et al., 2008). The PhC sensor is characterized by an hexagonal array of cylindrical air holes embedded in a SOI wafer with a lattice constant $a=400\ nm$ and a pore radius $r=120\ nm$. The radius of central hole is $r_d=342.5\ nm$, resulting in a band gap ranging from 1440 nm to 1590 nm for TE modes.

http://dx.doi.org/10.5772/53897

A latex sphere with a refractive index $n = 1.45$ can be trapped into the central hole of the PhC, characterized by a bigger diameter compared to that of surrounding holes and to the diameter of the same sphere to be sensed. In this way, it is possible to detect the presence of the single particle trapped into the biggest hole by observing the resonant wavelength shift in the sensor transmission spectrum. In fact, when the sphere is trapped into the sensor cavity, the resonant peak characterized by a modest Q-factor around 2,000, is red shifted of about 4 nm. Moreover, the red shift proportionally increases as the latex sphere diameter increases, too. Finally, the sensor described until now represents a useful tool in medical and health applications for single molecule detection.

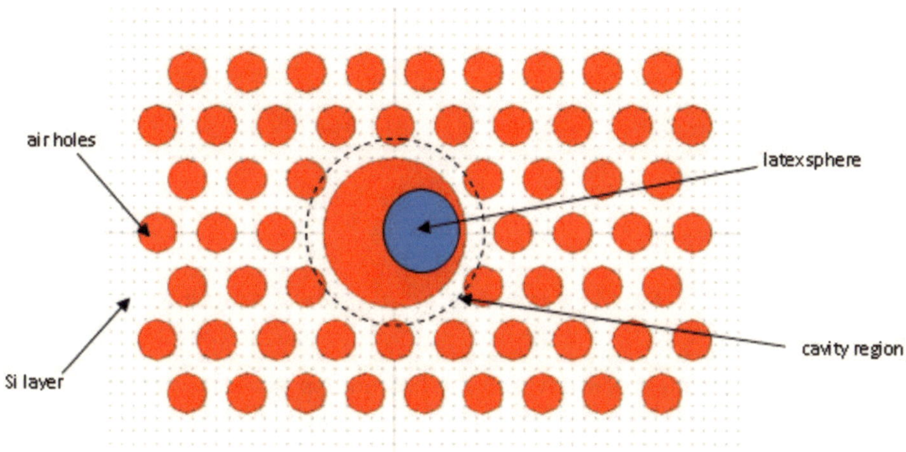

Figure 10. Top view of the PhC device with one latex sphere in the central defect of microcavity.

In several RI-based sensors the PhC waveguide directly acts as sensing element without designing any integrated microcavity, as previously analyzed in other sensor architectures. As before, a PhC waveguide is generally realized by introducing a line defect in the periodic planar structure. In such waveguide structures, only light at wavelengths outside the PhC bandgap can be guided. In particular, propagating modes are confined by Total Internal Reflection (TIR) along vertical direction and by the periodical structure laterally.

Performances of RI-based sensors can be quantified by monitoring the changes of cut-off wavelength (λ_{cutoff}) as a function of the cover RI. In particular, λ_{cutoff} describes the maximum wavelength at which the optical mode can propagate in the PhC waveguide and it depends on the cover medium RI. Consequently, the sensor sensitivity can be defined as the ratio between changes of cut-off wavelength and changes of cover RI, as follows:

$$S = \frac{\partial T}{\partial n_c} = \frac{\Delta \lambda_{cutoff}}{\Delta n_c} \tag{8}$$

where T is the transmission spectrum, Δn_C is the cover medium RI change and $\Delta\lambda_{cutoff}$ is the change of cut-off wavelength.

A RI-based PhC sensor has been proposed for detecting ssDNA, exhibiting a detection limit of 19.8 nM (García-Rupérez et al., 2010). This sensor, potentially able to detect very low analyte concentrations (e.g., proteins, bacteria, DNA) is fabricated in a SOI wafer with the silicon layer thickness of 250 nm and a 3-μm-thick buried silicon dioxide.

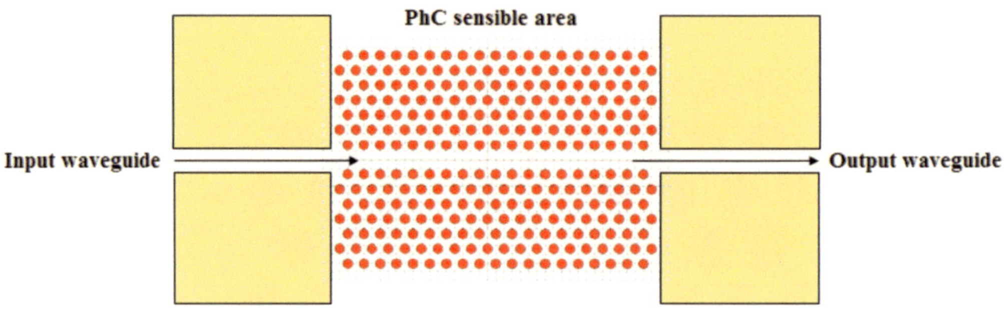

Figure 11. Schematic of the PhC-based DNA sensor characterized by input and output 500-nm-wide single mode waveguides.

The PhC lattice constant is a=390 nm and the hole radius is $r=111nm$. Consequently, the structure exhibits a guided TE mode with its band edge located around $\lambda=1550\ nm$. At the input and output of the PhC waveguide, light is coupled or collected by a 500-nm-wide single mode waveguide, as sketched in Fig. 11.

Sensing operative regime is performed by spectral peaks created by the excitation of multiple-k modes in the slow-wave regime near the band edge. In fact, changes of peak positions are continuously monitored, thus defining the sensor sensitivity according to Eq. (8).

The sensor has been tested with a complementary ssDNA solution bind to the ssDNA probe pre-deposited on the sensor surface. A peak shift of $\Delta\lambda=47.1\ pm$ corresponds to a DNA concentration of 0.5μM in the complex sample.

An important parameter to be properly designed for increasing sensing performance in single line PhC waveguide sensor, is the radius of holes localized at both sides of the line defect (Bougriou et al., 2011). In this context, an integrated sensor based on a PhC waveguide has been proposed. The sensor architecture is characterized by circular air holes in silicon wafer, as sketched in Fig. 12. The triangular lattice structure has a lattice constant a=370 nm and hole radius $r=120\ nm$. The waveguide is obtained by removing an entire row of holes in the horizontal direction, resulting in 9.5 μm long PhC waveguide. In addition, 12 rows of holes are periodically distributed on each side of the line defect. The PhC sensor exhibits a large band gap between 1230 nm and 1720 nm for TE modes and a very small band gap for TM polarization.

http://dx.doi.org/10.5772/53897

Device sensitivity has been evaluated by monitoring the cut-off wavelength shift when the sensor, initially exposed to air cover (n_C=1), is then covered by aqueous solution (i.e. deionized water with n_C=1.33). Consequently, the cut-off wavelength shift is estimated to be 30 nm, due to cover RI change of 0.33 (1.33-1).

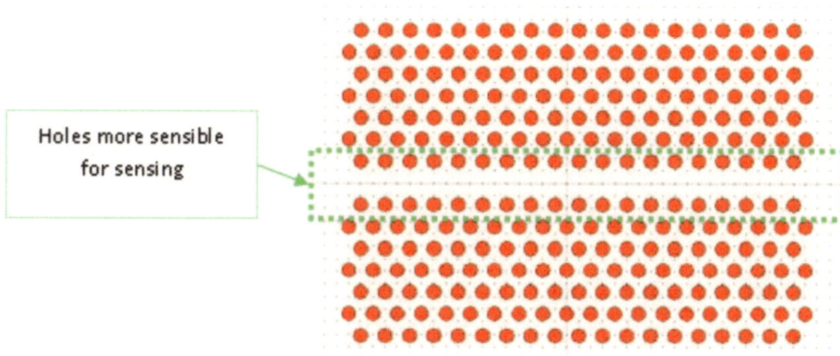

Figure 12. Photonic crystal waveguide sensor with sensible holes.

Sensor sensitivity can be improved by infiltrating the sample to be analyzed only in holes adjacent to the line defect. This arrangement produces a cut-off shift of 20 nm corresponding to a sensitivity of 60 nm/RIU. Experiments have shown that the sensitivity can be further increased by optimizing size holes near the line defect, which are more sensitive than outlying regions. In fact, by increasing the hole size, wavelength cut-off shift of 80 nm and sensitivity as high as S_λ = 240 nm/RIU can be achieved, resulting in an improvement of about 62% with respect to the original sensor.

A RI-based sensor has been also proposed for gas sensing. In particular, the interaction between the slow light mode propagating in the structure and the gas infiltrated in it, is transduced by the waveguide effective refractive index changes, resulting in changes of slow light regime wavelength (Awad et al., 2011). This type of sensor has the advantage of improving the sensing performance because of the enhanced light-matter interaction. In addition, the selectivity of the sensor is ensured because the transmission spectrum changes its amplitude only when the gas is filled in the PhC structure.

As shown in Fig. 13, the sensor consists of an InP air bridge membrane configuration. In particular, a layer of air on the bottom and on the top of the 285-nm-thick PhC slab ensures the device symmetry. The PhC structure is embedded on the InP slab with triangular periodicity, lattice constant a=441 nm and radius of air holes r=0.33a. The waveguide is obtained by removing an entire row of air holes.

Sensing performances have been estimated by exposing the sensor initially covered by air, to Argon (*n = 1.000282)* and Helium *(n = 1.000035)* gas, properly filled in the PhC waveguide. In particular, a shift of 0.6 nm has been detected in case of Helium filled in the waveguide and a shift of 0.05 nm in case of Argon. The sensor exhibits a good tolerance from environmental

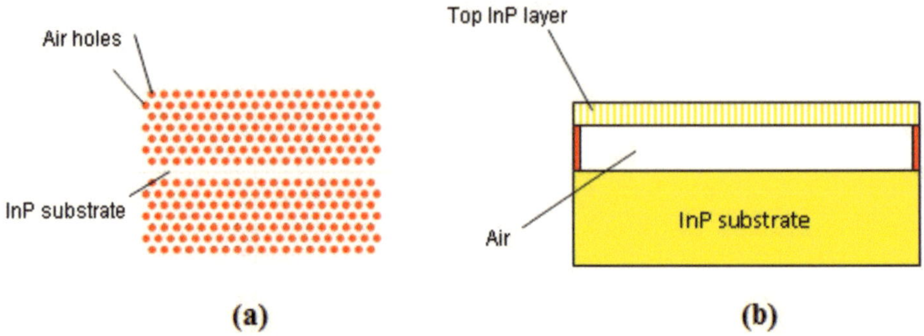

Figure 13. a) Top and (b) cross-sectional view of photonic crystal waveguide sensor for gas sensing.

perturbations and it is able to detect all gases characterized by refractive indices similar to Argon and Helium ones, with the exception of those whose refractive indices are very close to air RI (i.e., $n_{Air} = 1$).

In conclusion, PhC-based sensors integrated in interferometer architectures have been reviewed. Generally, sensing principle characterizing interferometer architectures are based on phase-shift measurement. In particular, if a perturbation occurs only on one arm of the interferometer, the output signal differs from the input signal, showing a variation of its phase or amplitude. In Mach Zehnder configuration it is possible to quantify the amount of phase shift on the active arm by tracking the signal output intensity. The perturbation mentioned above can be associated to the change of cover refractive index in one arm of the MZI, due to the presence of analytes in the sample. This phenomenon can be described by the following relation:

$$\Delta\varphi = \Delta\beta L = \frac{2\pi}{\lambda}\Delta n L \tag{9}$$

where Δn is the refractive index variation, L is the length of the waveguide and λ is the wavelength.

As sketched in Fig. 14, a MZI-based sensor with slot PhC waveguide has been fabricated and experimentally characterized (Chen et al., 2008). The sensor is composed by PhC waveguides in both active and reference arms, rib waveguides for sensor input and output, Y-junctions, electrodes and electrode pads. In particular, PhC waveguides are realized by removing a single row of air hole in a silicon slab. The device is fabricated on a SOI wafer, in which the thickness of the silicon core layer is $t = 215\ nm$, the top cladding medium is air and the bottom layer is a 2-μm-thick buried oxide layer. In the silicon slab, the holes are arranged by hexagonal cells with lattice constant $a = 400$ nm resulting in air hole diameter $d = 0.53a$.

For sensor applications, it is possible to induce localized RI changes by filling the waveguide region with gas- or liquid-phase analyte materials. By this way, the output signal intensity can be varied because of the conjunction effect of static driving voltage supported through

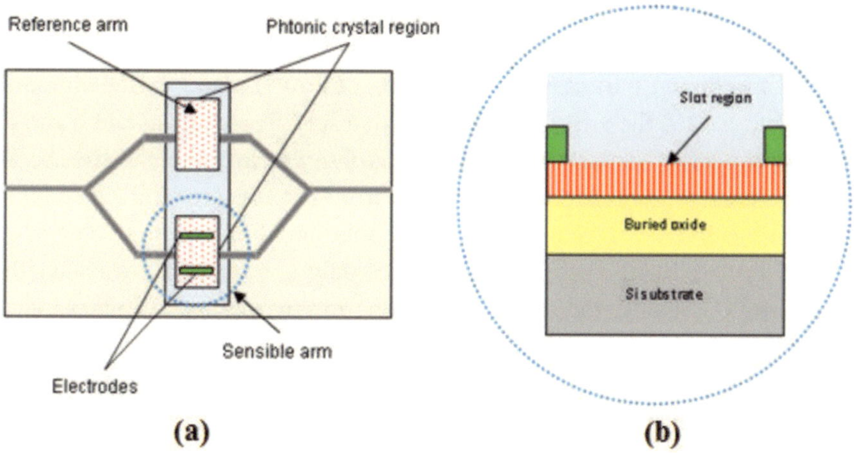

Figure 14. MZI with slot Photonic Crystal waveguide (a) and cross-sectional view (b).

electrodes on the sensor arms and cover RI changes. Finally, experimental comparison evidences how silicon slot PhC waveguide provides 30 times effective index change compared with conventional silicon slotted strip waveguide.

A sensor which combines the optical power splitting characteristics of multi-mode interference (MMI) power splitter and transmission drop resonance characteristics of multiple PhC microcavities arrays, has been recently investigated and proposed (Zou et al., 2012). The device, sketched in Fig. 15, is fabricated on a SOI platform and consists of a 1x4 MMI optical power splitter which splits the input light from a ridge waveguide into four output channels. The MMI has a length and a width of 123 μm and 16 μm, respectively. The input waveguide is 2.5-μm-width and outputs are separated by 1.5 μm. PhC waveguides are line defects with uniform lattice constant $a=400$ nm and diameter holes $d=0.54a$ which is embedded on a silicon slab thickness $t=0.58a=232nm$.

On arms #1 to #3, the edge air holes on the axis of PhC microcavity are shifted outward in the horizontal direction by a distance equal to $0.15a$. On the arm #4, two microcavities spaced of 50 μm, are designed as L13 type. In the first one, edge-air holes are shifted inward by $0.15a$, in the second microcavity edge-air holes are shifted outward by $0.15a$ resulting in zero cross talk.

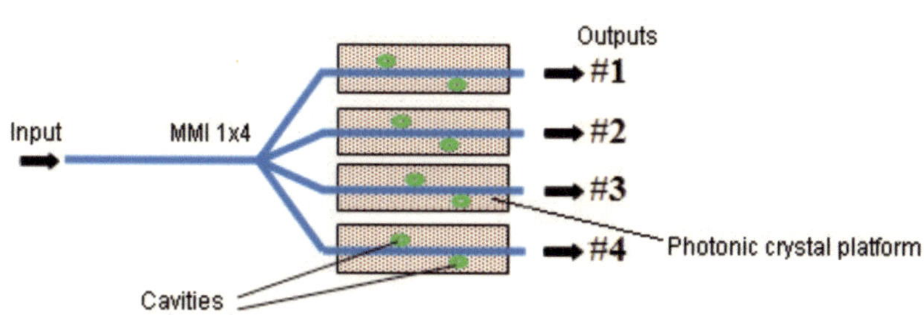

Figure 15. Schematic of 1x4 MMI device.

Each microcavity is coated with different receptor biomolecules, each responsive to its specific conjugate. In fact, by introducing into the sensible area 60 μl of 600 nM of goat anti-rabbit IgG Abs dissolved in PBS, only transmission spectrum of arm #2 changes, showing a resonant wavelength shift. In fact, the arm #2 is printed with a specific conjugate rabbit anti-goat IgG Abs. The arm #4 operates similarly while any shift is observed in remaining arms because arm #1 and #3 are printed with human IL-10 Abs, and arm #4 in the second microcavity is coated with BSA. Consequently, by changing the sample solution, thus by introducing 60 μl of 600 nM of rat anti-human IL-10 Abs dissolved in PBS, only one resonance wavelength shift is observed in arm #1. Finally, the sensor is immune to environmental changes and allows multiple detection, being very suitable for high throughput-screening.

2.2. Photonic crystal sensors based on optical adsorption

A lot of gas and liquid molecules absorb radiation in near- and mid-infrared, being spectroscopically detectable. In particular, when the wavelength of the optical signal matches the natural frequencies or resonances of the irradiated gas or molecule, the energy states of vibrating atoms change in discrete steps. The resonance frequencies or wavelengths depend on the number and mass of atoms in molecules as well as the number and strengths of chemical bonds. If the chemical structure of the molecules is complex, then a range of resonant vibrations characterize the optical absorption of molecules.

Based on these simple principles, the infrared (IR) spectroscopy is the simplest and the most reliable spectroscopic and sensing technique. In particular, the absorption spectroscopy is based on the Beer-Lambert law, defined as follows:

$$I = I_0 \exp(-\alpha L), \quad \alpha = \eta C \tag{10}$$

where I_0 is the intensity of the incident light, α is the absorption coefficient of the chemical specie being linearly dependent to the analyte concentration C via the molar absorptivity η, and L is the interaction length.

In this context, a photonic sensor based on a PhC slot waveguide has been proposed (Chakravarty et al., 2011a). The PhC waveguide is obtained by removing a single row of air holes from the input to the output of the device, resulting in a line defect with uniform lattice constant a. The device is designed for xylene sensing. Consequently the sensing platform is coated with a hydrophobic ~0.8 - μm-thick film of poly-dimethyl siloxane (PDMS). In this way, sensor selectivity is enhanced because adsorption of xylene is ensured by the functionalized coating and the adsorption of water is inhibited by the hydrophobic properties of PDMS. The width of PhC waveguide is $1.3\times\sqrt{3}a$, the hole radius is r=0.25a, the slot width w=0.2a and the waveguide length is L =300 μm. The slow light guided mode propagates at normalized frequency a/λ=0.275 and at this wavelength the constant lattice is set to be a=461 nm.

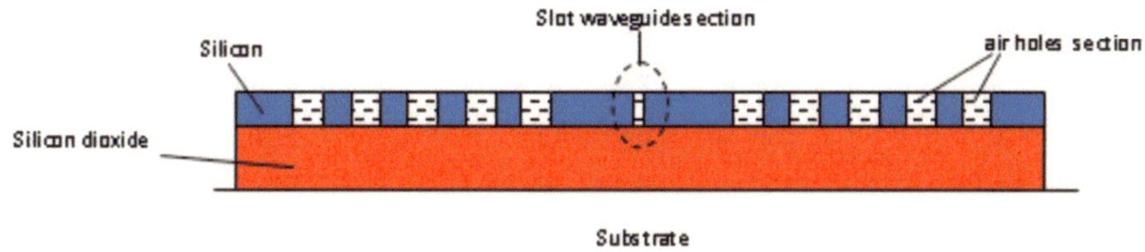

Figure 16. Cross section of PhC slot waveguide.

The whole device, sketched in Fig. 16, has been fabricated on SOI wafer with 230-nm-thick top layer and 3-μm-thick buried oxide. The sensing mechanism takes place by introducing analyte in the device sensible area through tygon tubes.

The absorption spectrum of xylene is characterized by three absorption peaks approximately centered at 1674 nm, 1697 nm and 1725 nm, being the latter peak characterized by a very small intensity. Moreover, when a xylene concentration dissolved in deionized (DI) water is introduced into the device, the intensity of the strongest absorbance peak at 1697 nm changes linearly for small concentrations of xylene (from 100 ppb to 1 ppm), in agreement with the Beer-Lambert law. At higher concentrations the absorbance curve deviates from linearity because the PDMS film reaches saturation and reduces its absorption capacity.

An analogous architecture with the same values of geometrical parameters is used for methane detection (Chakravarty et al., 2011b). The major absorption peak of methane occurs at 1.665 μm.

The experimental response of the sensor proposed follows the linear Beer-Lambert function for low concentrations of methane in nitrogen, being 100 ppm the lowest detectable methane concentration, and exhibits a deviation for higher concentrations, as previously analyzed in case of xylene detection.

Another similar sensor characterized by slow light mode propagation has been proposed for detection of hazardous gases and analytes in aqueous environment (Thévenaz et al., 2012). The PhC waveguide consists of a 180-nm-thick GaInP slab with a triangular lattice of air holes having a lattice constant of 486 nm and a defect line. In various experiments, the size of the first row of holes has been changed in order to firstly modify dispersion properties of the device and secondly to optimize sensing performance. In the first case, the hole size has been increased from 204 nm to 233 nm, in the second case from 224 nm to 253 nm. These two different geometrical arrangements influence the group refractive index n_g. In fact, when a TE-polarized mode is launched in the sensor the group index is measured to be n_g=4.9 and n_g=6.7 for the first and second configuration, respectively. In case of transverse magnetic-(TM)-polarized mode launched into the sensor, n_g remains the same in both configurations, to be equal to 1.5.

The sensor is placed in a gas chamber hermetically closed, filled with acetylene gas at 50 torr. In this way, the absorption coefficient is maximized and the linewidth of the absorption peak is kept narrow. Finally, the sensor confirms a linear dependence of molecular absorption on the group index and evidences how the distribution of the electric field is a very important parameter in gas-sensing measurements. In fact, by considering TM-polarized mode propagating in the sensor, a stronger optical absorption is achieved compared with that obtained when TE-polarized mode propagates in the same sensor. This effect, is due to fact that in TM polarization the electric field inside the lower-index slab material is increased by discontinuity at dielectric interface.

Theoretically, it is possible to define the absorption coefficient as the ratio between the electric field obtained by coupling optical wave and electric dipole and the Poynting vector. An analytical expression for α is reported below:

$$\alpha \approx \frac{\int_{gas} Im\left[\varepsilon\right]\left|E\right|^{2} dV / 2}{\mathrm{Re}\left[\int_{A} E \times H^{*} dA\right]} \tag{11}$$

where H is the magnetic field, ε is the complex electric permittivity and *Re* and *Im* denote the real and imaginary part of a complex number, respectively. In conclusion, the enhanced overlapping between electric field intensity and molecule to be sensed evidences a decisive influence on the absorption enhancement, thus in sensing performance.

Finally, a PhC sensor based on infrared absorption has been proposed for azote oxide (NO_2) detection (Maulina et al., 2011). The sensor consists of 1D PhC characterized by two different defects. In addition, by changing the refractive index and thickness of both defects, it is possible to tune the position of photonic pass band (PBB) in the PhC band gap.

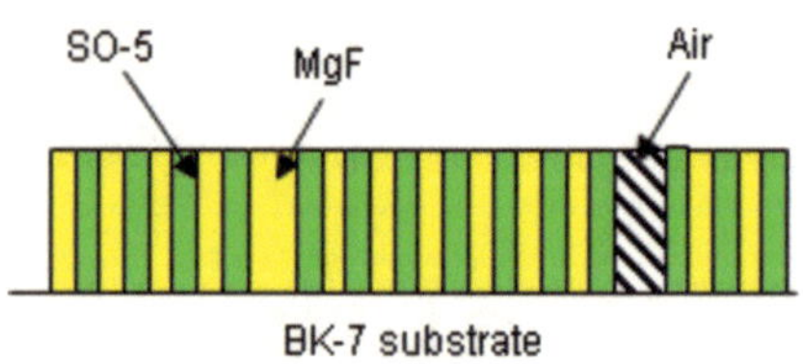

Figure 17. 1D-PhC with two defects for NO_2 sensing.

The two defects are named as regulator and receptor. Changes in the regulator defect influence the wavelength of the transmission spectrum, while changes in receptor defect induce variations in transmittance value. In particular, as plotted in Figure 17, the sensor consists of a first defect sandwiched between 4 and 6 periodic cells, and the second adjacent defect sandwiched between 6 and 2 cells. A cell is characterized by two alternate layers being the first layer (i.e., OS-5) characterized by a high refractive index $n = 2.1$, and the second one magnesium fluoride (MgF_2) with the lower refractive index, $n = 1.38$. In addition, the first defect is OS-5 having twice the thickness of the other layers. The second defect is a void (low air refractive index $n = 1$) to be filled with the sample to be detected. Finally, the whole structure is embedded on a layer of glass material borosilicate crown, known as BK-7 whose refractive index is $n = 1.52$. Experimental results have been performed by absorbing NO_2 in air, combining Beer-Lambert's law and PBB phenomena. The PBB spectra change with respect to different concentrations of absorbed NO_2 gas. Results of spectroscopic measurements evidence a linear rise of transmittance value with increasing the gas concentrations. Finally, the sensor presented until now exhibits an efficiency up to 99 %.

2.3. Integrated photonic crystal sensors based on non linear effect

Recent studies have shown how new PhC sensors based on non linear effects represent a new and intriguing approach for advanced sensing applications. Actually, the main non linear effects investigated in these structures are Kerr nonlinearities (Van Driel, 2003), Raman effect and harmonic generation.

In this section, an original sensor is proposed, consisting in a PhC microcavity in which the Raman effect related to the vibrational excitations mode in silica is excited. By considering the quantum mechanical approach, a photon of the incident field (i.e., the pump wave) is scattered by a molecule of the medium in which the field propagates, resulting in the generation of a photon of lower energy (i.e., the Stokes wave). At the same time, the residual energy is absorbed by molecules via phonons. The Raman shift is then the frequency difference between the incident wave and the scattered one (Stokes wave) in a stimulated Raman interaction.

The device proposed in this section, is based on a PhC cavity fabricated on a 220 nm SOI wafer (Van Leest et al., 2012). Air holes in the silicon slab are arranged in hexagonal cells with lattice constant $a = 430\ nm$ and radius of air holes $r = 0.3a = 129\ nm$.

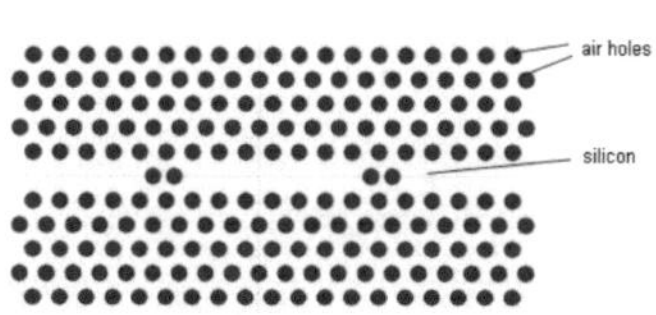

Figure 18. Schematic of L6 cavity to generate Raman effect.

The device proposed is characterized by a L6 PhC cavity, i.e. a PhC structure without six central air holes, as sketched in Fig. 18 above.

A resonant cavity with the overall length $L_{cav} = 2.9\ \mu m$ is obtained by shifting outward the inner holes along horizontal direction by a distance equal to $0.2{*}a = 0.086\ \mu m$. Mathematical modeling of the device suggests this design approach as fundamental in order to ensure the generation of Stokes wave into the cavity, away from the resonant wavelength of about 15,6 THz. This wavelength shift due to pump wave and resonant Stokes wavelength generated into the cavity, is typical for silica.

In conclusion, the sensor investigated evidences intriguing potentialities and sensing performance of new class of PhC sensors based on non linear effect (i.e., Raman effect). In particular, such sensors are expected to be able to detect single particle in aqueous solutions, with very small dimensions comparable to that of virus or proteins.

Several research efforts are still being done in order to comprehend how to employ non linear effects for sensing applications in PhC sensors fabricated on SOI technological platform.

2.4. Opto-mechanical sensors based on PhC

In this section, PhC-based sensors designed for pressure, force, strain and torsion sensing, are discussed. The sensing principle consists in monitoring variations of optical characteristics induced by the physical deformation of the PhC-based device.

In this context, a force and strain sensor is sketched in Fig. 19 (Li et al., 2011).

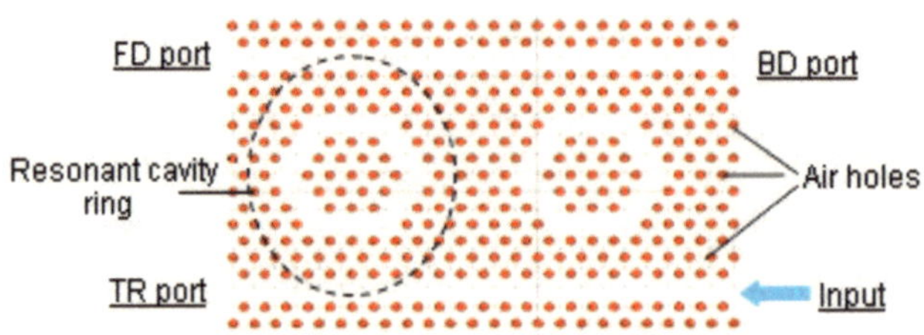

Figure 19. Schematic of DNR channel drop filter.

The architecture presented above is a typical dual-nanoring (DNR) channel drop filter on 2D photonic crystal with hexagonal lattice. A silicon PhC crystal slab of 220 nm thickness is released on a SOI substrate and the ratio between radius of air holes and lattice constant is set to be $r / a = 0.292$.

The nanoring is obtained by removing localized air holes to form an hexagonal defect. The dual-nanoring is made by two aligned nanorings with a centre-to-centre distance $d = 11a$. The structure is sandwiched between two waveguides, so that it is possible to identify four ports in the PhC platform. The first one acts as input port (red arrow) and the other ones are used for transmission (TR) and forward (FD) or backward drop (BD), indicated in Fig. 19 as TR port, FD port and BD port, respectively.

The device characterized by PhC single-nanoring structure shows a photonic band gap map for TM modes, characterized by a band gap in the range of normalized frequency extended from 0.26 to 0.33. The corresponding band gap wavelength ranges from 1242 nm to 1577 nm and a resonant peak displayed in BD port is located at 1553.6 nm revealing a *Q*-factor of 3,884, while FD and TR ports reveal a spectral dip at the same wavelength mentioned above.

Simulations evidence that two rings are always phase-matched and their resonances are not independently. In fact, Li et al. have demonstrated that the wavelength of the resonant peak at the BD port is strongly dependent on the ring size and separation distance *d* between two rings. Consequently, when physical structure is deformed, a variation of the resonant wavelength can be detected. The application of an external force to the device induces a strain linearly proportional to the applied force located at the junction between cantilever and SOI substrate.

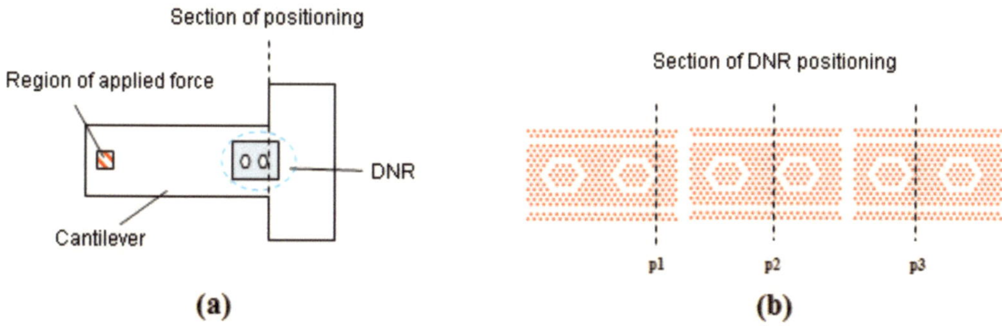

Figure 20. Cantilever architecture (a) and different positioning of DNR on cantilever (b).

In particular, the stronger the applied force, thus the strain induced to the sensor, the bigger the red shift of the resonant wavelength to be detected. As sketched in Fig. 20, three different configurations are adopted by moving the DNR resonator on the cantilever and three different correspondent sensor responses are obtained. For example, in cantilever labelled as type p1, the resonant peak shifts from 1553.6 nm to 1554.9 nm, corresponding to a force variation ranging from 0 to 400 nN. In this configuration, as the load force approaches to 500 nN, the output signal intensity at the BD port is lightly reduced, while the signal intensity at the FD port increases. Consequently the signal intensity at the TR port is also increased. The device exhibits a degraded resonant behaviour due to deformed DNR resonator.

The cantilever labelled as type 2p, maintains the channel drop mechanism up to the load force of 700 nN before losing its resonant behaviour. In the last configuration indicated as type p3, only the second ring is kept within the deformation region and the sensitivity of the DNR is reduced. In fact, a less range of load force is detected through this type of cantilever. The BD behaviour is degraded in case of applied force higher than 100 nN.

A PhC micro-pressure sensor has been fabricated and characterized (Bakhtazad et al., 2010). The device is based on an air-bridged line-defect silicon slab PhC waveguide, as sketched in Fig. 21.

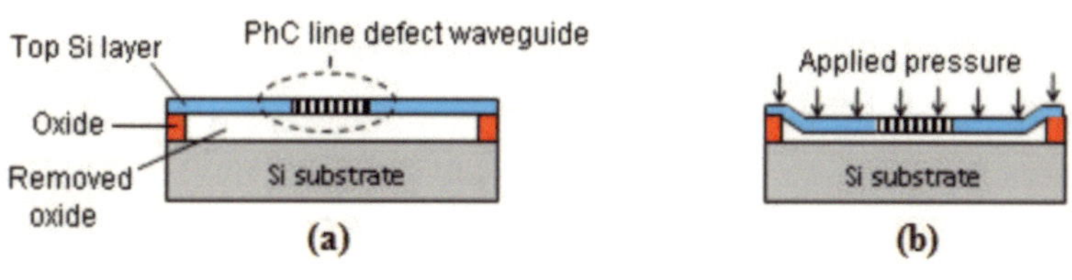

Figure 21. Sensor at rest (a) and sensor with applied pressure (b).

The device is fabricated in SOI technology. In this structure the bridge is obtained by removing a portion of buried oxide layer of 1000 nm thickness, while the top silicon layer is 320-nm-thick.

The PhC structure has hexagonal lattice with lattice constant a=430 nm, diameter holes d=300 nm. A waveguide is embedded by removing a row of holes in ΓK direction. Two 700-nm-thick channel waveguides are placed at input and output section in order to ensure an efficient light coupling. In case of TM polarization, the structure shows a band gap with its centre at about 1550 nm.

The sensor operation is based on the optical field profile changes to the proximity of surrounding material induced by the applied pressure on the top of the sensor. Consequently, changes in the transmission spectrum are directly linked to the magnitude of the applied force. To this purpose, it is possible to defined the device sensitivity as follows:

$$S = \frac{\delta T}{\delta P} = \frac{\delta T}{\delta h} \times \frac{\delta h}{\delta P} \qquad (12)$$

where T is the optical transmittance, P the applied pressure and h is the photonic crystal height over the substrate. The best total sensitivity obtained is S=0.039 (%/MPa) under 1MPa uniform pressure, i.e. a maximum displacement at centre of the rectangular platform of 78 pm.

A PhC optical cavity has been designed for detecting torsion or flexure induced by external magnetic couple excitations (Wu et al., 2012). The design of this sensor is optimized to support low effective mass torsional and flexural mechanical modes. The sensor is characterized by a PhC opto-mechanical cavity in which the central element results to be suspended. This type of sensors have been proposed for magnetic applications and are suitable for probing nano-magnetic torques.

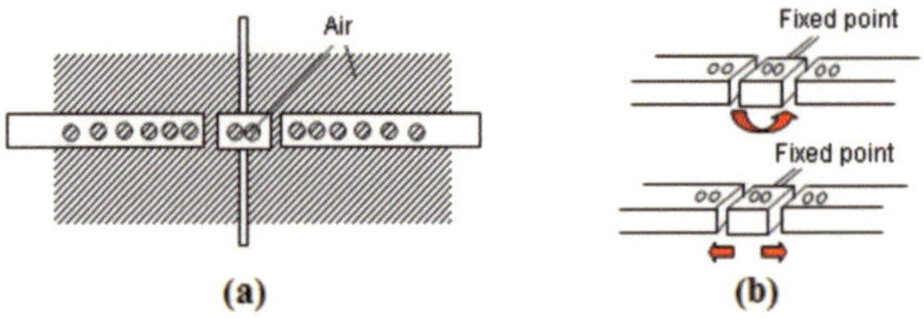

Figure 22. Schematic of the torsion PhC-based sensor (a) and operative configurations (b): right torsional (top) and flexural (bottom) movements of central sensor element.

As sketched in Fig. 22, the overall structure is characterized by air holes with radius r=145 nm, patterned in a free standing silicon nanowire with refractive index n=3.5, thickness t=250 nm and width w=600 nm. The hole spacing varies quadratically over six periods, ranging from a_C=350 nm to a_0=450 nm. The suspended element is disconnected from the two arms of the sensor by a gap of size d=0.58a_0. Central position is defined as $a_2^{'}$=1.12 a_2 where a_2

is the nominal spacing between the first and second hole of the uncut structure. The gap reduces the maximum value of device *Q*-factor. Moreover, the structure can support a *Q*-factor higher than 10^4 by tuning gap size, *d* parameter and gap position.

In presence of an external magnetic field, torsional and flexural resonance may be excited by magnetic moment from nanomagnetic materials attached to the nanocavity paddle. These magnetic resonance affects the suspended central element in the PhC cavity, inducing changes of optical properties. Experimental results show higher *Q*-factor values and lower effective mechanical mass (~100 *fg*).

2.5. Advances in PhC-based sensors: Materials and technologies

In this section, techniques commonly used for PhC fabrication, are briefly discussed. The fabrication of these devices requires a series of technological steps which establish the effective physical characteristics of the photonic crystal. In fact, periodicity, dept and radii of PhC holes can be precisely controlled during technological processes, strongly influencing real performances of the resulting PhC-based sensor. In particular, the geometrical arrangement strongly affects the photonic crystal band gap map, influencing optical mode propagation properties.

In the following, an example of photonic crystal device fabrication is described. In particular, when the photonic crystal is embedded on SOI chip, the first step generally consists in cleaning the device surface with nitric acid in order to remove organic residuals. Consequently, a positive electron beam (e-beam) resist is spun on the top surface of the SOI chip. The photonic crystal is patterned using an e-beam lithography system operating at 100 keV. The resist is developed with a microchem 1:3 methyl-isobytyl ketone:isopropanol (MIBK:IPA) solution for 30 s followed by a 10 s IPA rinse. Successively, the exposed areas are then etched using inductively coupled plasma (ICP) dry etching system or Chlorine based ICP. The remaining resist can be dissolved in a dilute 100:1 HF solution.

The top of device is often covered with a layer of polydimethylsiloxane (PDMS) that is realized by using a soft lithography technique.

Micro fluidic channels, widely used in chemical and biochemical PhC-based sensing architectures, require precise fabrication steps. Initially, a mold for the channels is created by using a 4-inch pure silicon wafer. A 1-μm-thick UV resist layer is spun on the top of silicon, followed by Near-UV lithography for writing the desired patterns of fluidic channels. Pattern-transfer is done by reactive ion etching (RIE) and a depth of approximately 30 μm can be realized by conventional etching. In conclusion, a PDMS layer is poured onto the silicon mold in order to create flow channels.

The fabrication steps can differ for various devices depending on the materials employed, thus the technological platform. For example, a guiding SiN_x layer can be deposited on the substrate by either Plasma Enhanced Chemical Vapor Deposition (PECVD) or Low Pressure Chemical Vapor Deposition (LPCVD) techniques in order to optimize the film uniformity. This layer can be patterned in a photonic crystal structure using either optical or e-beam lithography. The first lithography technique is often used for large area array, the second one is preferred for much smaller device dimensions, being characterized by an higher resolution. The e-beam is well-suited for flexible PhC lattice design. The pattern is transferred from the resist to the SiN_x layer using Fluorine-based dry etching.

Recently, several technological processes have been investigated and tested for fabrication of 3D PhC devices. This new class of devices is not popular as 2D PhCs, because the planar technology is more established and applications of 2D PhCs are well established in optical telecommunication, signal-processing and sensing.

In this context, the first sensor proposed is a PhC hydrogel for sensing of highly toxic mercury ion (Hg^{2+}) in water (Arunbabu et al., 2011). The detection of Hg^{2+} concentration is due to diffraction of visible light from polymerized crystalline colloidal array (PCCA) which consists of highly charged polystyrene particles which are polymerized within the polyacrylamide hydrogel (named crystalline colloidal array or CCA). Different concentrations of analytes in the 3D PhC change the volume of hydrogel resulting in an alteration of the lattice spacing of CCA and, consequently, in a shift of the diffraction wavelength of light. Therefore, the concentration of analytes can be extracted from the wavelength of diffracted light by PCCA. In particular, this sensor employs an urease immobilized PCCA based sensing material for determination of Hg^{2+}. In fact, the presence of Hg^{2+} in a solution in contact with the PCCA inhibits the urea hydrolysis and suppresses the normal production of NH_4^+ and HCO_3^-. The optical result is a red shift diffraction which increases linearly for low concentration of Hg^{2+} and devices to linearity for higher concentration of Hg^{2+}.

The sensor shows reversibility and LOD as low as 1 ppb, i.e., 1 μg/L. In conclusion, the sensor can be used with the same physical principle for detection of Ag^+ and Cu^{2+} ions which are, as Hg^{2+}, the principal inhibitor of urease.

A mechanically robust and highly sensitive sensor, with short response time, characterized by a planar defect in the 3D macroporous array of pH-sensitive hydrogel poly (methacrylic acid) (PMMA), has been proposed for pH detection (Griffete et al., 2011). Two different configurations of the structure fabricated by a Langimur-Blodgett technique and characterized by hexagonal arrangement of spheres have been designed. In particular, one structure is a defect-free (DF) colloidal crystal made from 10 layers of 280 nm diameter particles. The other one is a planar defect-containing (DC) colloidal crystal which consists of a layer of silica particles of 390 nm diameter between two sections of 5 layers of particles of 240 nm diameter. As previously analyzed in case of 2D planar PhC sensors, the defective layer in 3D technology also introduces a change into the band gap map, influencing the optical properties of the device (see Fig. 23).

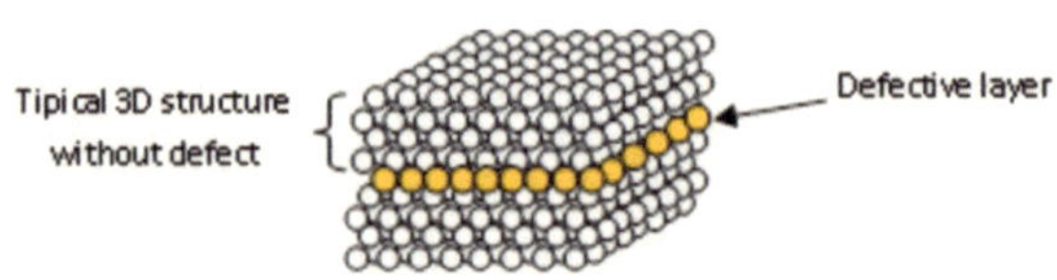

Figure 23. 3D-PhC with defected layer.

In both structures, by increasing the pH of the complex sample a red shift of the diffraction peak can be observed due to the ionization of the ionic gel. The diffraction red shift occurs after the gel is swollen. The presence of defect in the 3D PhC structure enhances both sensitivity and response time of the PhC sensor. In fact, the device characterized by the defect shows a red-shift of $\Delta\lambda = 60\ nm$ that is greater than the wavelength shift $\Delta\lambda = 40\ nm$ obtained with the defect-free device and estimated for the same pH concentration.

A different architecture used for pH sensing (Jiang et al., 2012) is based on poly(vinyl alcohol) (PVA)/poly(acrylic acid) (PAA) photonic crystal materials. This sensor exhibits good durability and adjustability. The pH response is monitored by diffraction wavelength shift. A solvent-assisted method is used to physically cross-link a thermo-reversible PVA hydrogel around CCA and form a gelated crystalline colloidal array photonic crystal material (GCCA). Glutaraldehyde is used to chemically cross-link the PVA hydrogel in order to avoid the collapse of cross-linked GCCA during the procedure for the introduction of environmentally sensitive components. It has been demonstrated that the sensing is better for high concentration of glutaraldheyde solution, because the high cross-link density improves equilibrium hydrogel volume needed for the diffraction shift measuring. The modified Bragg's law reported in Eq. (13), regulates the phenomenon mentioned above:

$$\mathrm{m}\lambda = \sqrt{\frac{8}{3}} D \left(\sum_i n_i^2 V_i - \sin^2\theta \right)^{\frac{1}{2}} \tag{13}$$

where D is the center-to-center distance between the nearest spheres, n_i and V_i are the refractive index and volume of each component, respectively, and θ is the angle between the incident light and the sample normal. As pH increases the hydrogel absorbs water and swell. The center-to-center distance also increases and a red shift of the diffraction peak is generated. The sensor exhibits a wavelength shift of $\Delta\lambda = 96\ nm$ when a solution of pH 7.6 is concentrated in cover medium for 30 min, being the same sensor initially exposed to a solution of pH = 4.8.

In conclusion, a photonic sensor consisting of a glass substrate and a three dimensional photonic crystal realized by using nanoparticles and poly(dimethysiloxane) (PDMS) elastomer has been investigated (Endo et al., 2007). The PhC is generated by infiltrating the opaline lattice of particles with a liquid prepolymer to PMDS in voids. Subsequently the material is thermal cured.

Even in this case, the physical sensing is governed by the Bragg's law that can be written in terms of spacing between planes of crystal (d_{111}) as follows:

$$\mathrm{m}\lambda = 2\mathrm{d}_{111} \left(n_{eff}^2 - \sin^2\theta \right)^{\frac{1}{2}} \tag{14}$$

where m is the order of diffraction and n_{eff} is the mean refractive index of the crystalline lattice.

The first step for the sensor fabrication consists in drying aqueous dispersions of polystyrene (PS) nanoparticles (with 202 nm-diameter) on the glass substrate. After the dry-up process when the PS nanoparticles are spatially ordered, a PMDS solution without any air bubbles is distributed on the top of PhC and all voids between the PS nanoparticles are totally filled. The PDMS is first cured at room temperature, then it is baked at 60°C for 1 hour.

The sensor obtained, exhibits a Bragg reflection peak at 552 nm. When a non-polar organic solvent such as xylene is in contact with the structure, it is possible to see a color change of 3D structure from green to red. The shift increases when concentration of solvent in the solution increases. In addition, the detection limit of this optical chemical sensor is found to be dependent on the polarities of the solvents.

3. Photonic crystal fiber sensors

Photonic crystal fibers (PCFs), also named as micro-structured optical fibers (MOFs), represent nowadays a new and intriguing typology of optical fibers suitable for sensing applications such as measurement of strain, refractive index, pressure, temperature, magnetic field, to name a few. PCF-based sensors are characterized by high sensitivity, small size, robustness, flexibility and ability for remote sensing. Other advantages concern with the possibility to be used even in the presence of unfavorable environmental conditions such as noise, strong electromagnetic fields, high voltages, nuclear radiation, for explosive or chemically corrosive media, and at high temperatures.

Substantially, PCFs are fused-silica optical fibers characterized by a hollow or silica core surrounding by a regular pattern of voids running along the fiber axis, as sketched in Fig. 30 below.

In particular, it is possible to appreciate the difference among PCFs, as in Fig. 24(a-b), with respect to conventional single mode fibers (SMFs), as in Fig. 24(c). In particular, propagation properties in conventional optical fibers and PCFs can be tuned by properly designing geometrical parameters, such as the hole diameter indicated with h, the fiber core diameter d, the pitch x (i.e., the distance between the center points of two consecutive holes), the fiber length L and, obviously, materials.

In PCFs, light can be guided by two different mechanisms, i.e., index-guiding or bandgap-guiding, as a function of the principle of the light confinement (Buczynski, 2004). In particular, in PCFs characterized by a solid core or by a core with a refractive index higher than the micro-structured cladding's one, light is guided as in conventional silica fibers (i.e, doped silica core surrounded by the silica cladding). In fact, light propagates in the high refractive index region by the total internal reflection (TIR) principle at the interface between the core and the low refractive index cladding. In addition, air holes periodically arranged over the fiber cross-section characterize the micro-structured silica cladding, resulting in an effective cladding index. Consequently, the TIR at the core-cladding interface is known as modified TIR and it can occur with very low core-cladding refractive index (RI) contrasts, enabling the fabrication of both core and cladding by the same material.

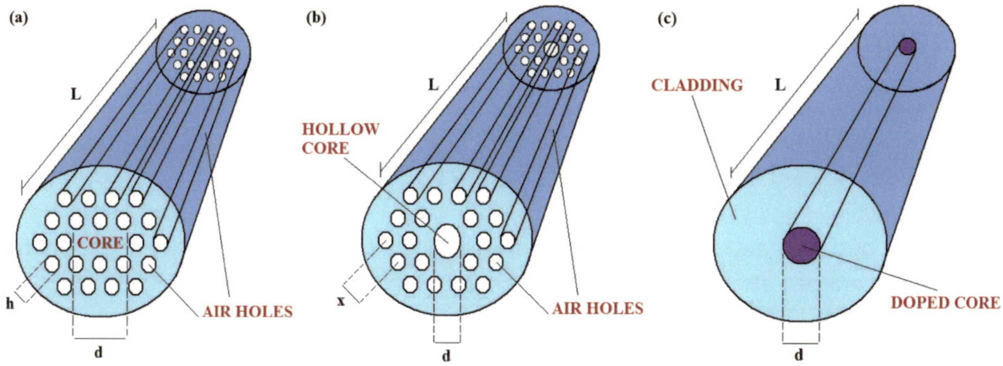

Figure 24. Schematic of (a) solid core PCF, (b) hollow core PCF, and (c) conventional SMF.

On the contrary, by surrounding the fiber low index core (i.e., hollow core) with a photonic crystal structure, it is possible to localize light in the fiber core by the photonic bandgap effect (Barkou et al., 1999). Consequently, only those wavelengths that do not fall within the photonic crystal stop band, can be confined in the core and propagate along the optical fiber. In this way, the TIR principle does not occur and the fiber core can be characterized by a RI lower than the cladding's one, making these optical fibers very suitable for sensing applications. In conclusion, hybrid PCF have also been theoretically and experimentally demonstrated, revealing the possibility of guiding light through simultaneous photonic bandgap-guiding and modified TIR (Xiao et al., 2007).

In both index-guiding and bandgap-guiding PCFs, the sensing mechanism consists in changing fiber optical properties (i.e., cladding effective refractive index) by filling air holes with chemical/biochemical liquids or gases. In this way, the interaction between the propagating light and the analyte to be detected is improved as it is not possible to achieve with standard optical fibers, where the sensible area is realized by removing the cladding from the fibers and directly exposing the fiber core to cover medium where the sample is concentrated.

From a technological point of view, it is possible to highlight the flexibility of the modeling and design of PCFs with respect to conventional SMFs. In fact, in a SMF the only parameter to take into account is the core diameter, while in a PCF there are three physical parameters to be properly set: the core diameter (which for solid core PCF is defined as the diameter of the ring formed by the innermost air holes), the diameter of the air holes of the cladding *d* and the pitch Λ. These three physical parameters, in combination with the choice of the refractive index of the material and the type of lattice, make the fabrication of PCFs very flexible and open up the possibility to manage its properties leading to a freedom of design not possible with common fibers.

In addition, the principal method of fabrication of PCF is the so called multiple thinning (Buczynski, 2004). In summary, the method consists of four fundamental steps: creation of individual capillaries, formation of the preform, drawing of intermediate preform, and finally, drawing of the final fiber. In particular, in the last step extra layers of polymer are usually added to create a coating protecting the fiber mechanically.

PCFs have been widely used in sensing applications because of their ultra-high sensitivity, selectivity and immunity to optical noise and to external interferences. According to an exhaustive review on PCF sensors already published (Pinto et al., 2012), it is convenient to distinguish between physical and biochemical PCF-based sensors. In particular, physical PCF sensors are designed and implemented for measuring and monitoring physical parameters such as temperature, strain, refractive index, pressure, electromagnetic field, vibration, to be named. Moreover, PCF-based chemical and biochemical sensors are usually employed for gas sensing (e.g, acetylene, methane, oxygen), molecular and protein detection, humidity and pH monitoring.

3.1. Photonic crystal fiber sensors for physical sensing

In this paragraph, PCF sensors are investigated in detail, focusing on design criteria and measurement setups usually employed in sensing procedures. To this purpose, by firstly considering the class of PCF physical sensors, an highly sensitive torsion sensor has been experimentally demonstrated by incorporating a segment of novel side-leakage PCF (Chen et al., 2011). In Fig. 25, the cross-section of the fabricated PCF and the relevant experimental setup are sketched.

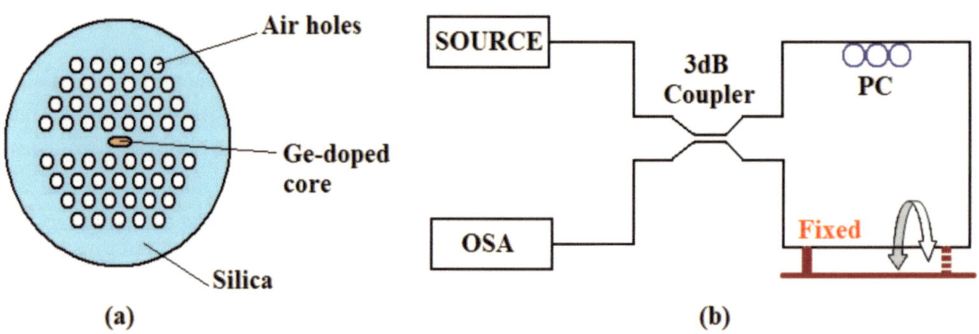

Figure 25. Schematics of the (a) Ge-doped PCF cross-section and (b) of the experimental setup for torsion sensing.

The fiber is characterized by a 125 μm cladding diameter. The elliptical Ge-doped core, adopted to introduce the fiber birefringence, is characterized by the major diameter of 4 μm while the minor one is equal to 2.88 μm. Moreover, the diameter and the pitch of air holes are ~ 5.48 μm and ~ 8.06 μm, respectively. The experimental setup equipped for torsion sensing is based on the Sagnac interferometer. In particular, it is composed by an optical source, a 3dB coupler that splits input light in two distinct optical signals counter-propagating in the Sagnac loop and by a polarization controller (PC), used for the interferometer optimization. In addition, a 14.85-cm-length segment of side-leakage photonic crystal fiber is incorporated in the Sagnac loop. As sketched in Fig. 31, one of the optical fiber extremity is fixed while the other one is not bounded, thus it can be twisted in clockwise and counter-clockwise directions. Finally, an optical spectrum analyzer (OSA) is adopted for monitoring the output spectrum.

When the fiber is twisted, the linear defect can induce different mechanical stresses to the elliptical Ge-doped core. The combination effects of the torsion-induced circular birefringence and the intrinsic birefringence of the PCF fiber, generate an elliptical birefringence. Consequently, the elliptical birefringence is proportional to the torsion angle, and its rotary direction is determined by the torsion direction. In fact, when the PCF is twisted clockwise, the elliptical birefringence is right-rotary, on the contrary it is left-rotary when the fiber is twisted in the opposite direction. Finally, the torsion-induced wavelength shift $\Delta\lambda$ characterizing the sensor transmission spectrum can be estimated by using the following expression:

$$\Delta\lambda = \lambda\eta b_t \Delta\tau \tag{15}$$

where λ is the operative wavelength, η is the circle birefringence ratio of the torsion-induced circle birefringence to the sum of the fiber birefringence, b_t is a constant that described the torsion-induced variation of the circle birefringence, and $\Delta\tau$ is the torsion angle. Moreover, the wavelength shift $\Delta\lambda$ is negative when the fiber is twisted clockwise, whereas it is positive when the fiber is twisted counter-clockwise.

Interesting results have been experimentally demonstrated with the PCF-based sensor described until now. In particular, a maximum torsion sensitivity of about 0.9354 nm/° has been achieved with a torsion angle measurement error due to the temperature effect of about 0.054 ~ 0.178 °/°C.

Generally, PCF sensors are designed to be strain and temperature independent. To this purpose, temperature insensitivity can be achieved by engineering the fiber composition and geometry. Otherwise, it can be contemplated the use of fiber Bragg grating (FBG) or long period grating (LPG) in the measurement setup, but making the sensor architecture complicated and costly (Gong et al., 2010).

A PCF-based modal interferometric torsion sensor has been investigated and experimentally tested according to the experimental setup sketched in Fig. 26 (Nalawade et al., 2012).

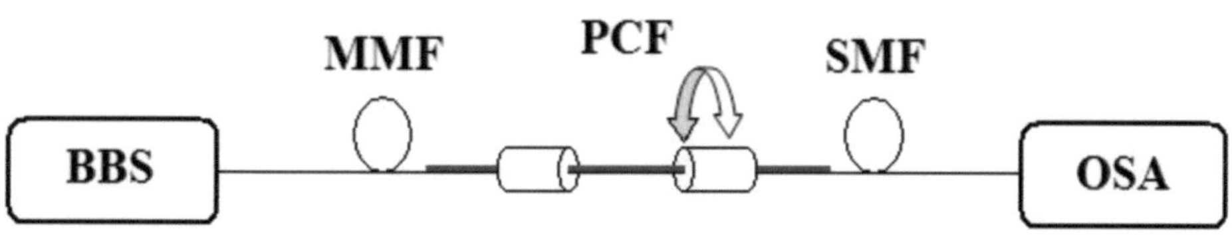

Figure 26. Schematic of the experimental setup (Nalawade et al., 2012).

The measurement setup consists in a broadband source (BBS), a multi-mode optical fiber for guiding the signal to the PCF and a single-mode fiber (SMF) for collecting the signal to the OSA. A torsion sensitivity of about 79.83 pm/° has been achieved in the dynamic range of 180°.

In addition, strain and temperature effects on torsion sensitivity have been demonstrated to be negligible in the range 0÷4500 $\mu\varepsilon$ and 30÷200 °C, revealing very high performance. In conclusion, sensing performance described above and other intriguing experimental results such as a torsion sensitivity of 1 nm/° with a temperature sensitivity of -0.5 pm/°C in the range 30÷100 °C (Zu et al., 2011), suggest PCF-based sensors as good candidates for torsion sensing.

PCFs have been widely used in industry and reservoir engineering for monitoring fundamental parameters such as temperature and pressure. To this purpose, high performance have been theoretically demonstrated by using a PCF-based sensor, as in Fig. 27 (Padidar et al., 2012):

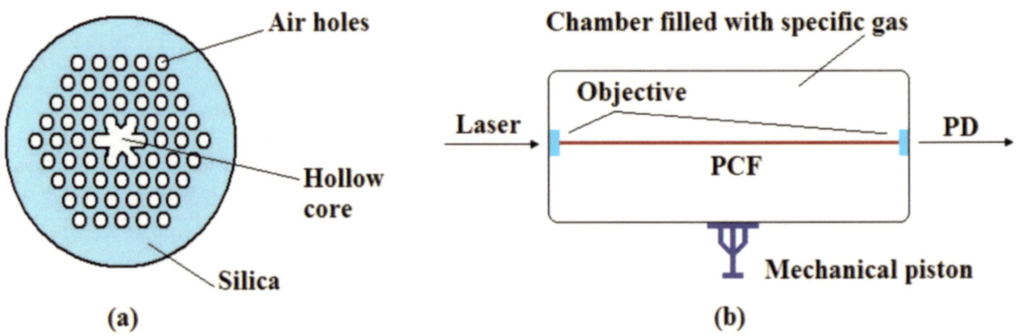

Figure 27. (a) Cross section of the PCF fiber and (b) schematic of the PCF-based pressure/temperature sensor.

As sketched in Fig. 27(a), the PCF cross section is characterized by air holes properly designed and periodically surrounding the inner hollow core. Consequently, this device has been designed for operating by the band-gap guiding principle, making the sensor proposed extremely selective in terms of operative wavelength. According to Fig. 27(b), the PCF described until now is fixed in a chamber filled with a specific gas. Moreover, left fiber tip receives input light at λ = 1.55 μm emitted by a laser source and the remaining tip guides the propagating light at the photo-detector (PD).

The sensor mechanism is based on the transmission peak wavelength shift induced by the temperature/pressure changes. In particular, it is needed to distinguish between the pressure and temperature sensing principle. In fact, when the sensor is used for pressure measurement, a mechanical piston (see Fig. 33b) can move due to pressure of oil well environmental. This causes the chamber gas to be compressed in high pressure or be dispersed in low pressure. Consequently, the refractive index of the specific gas filling PCF holes changes leading to a variation of photonic band gap and fundamental modes of PCF. Moreover, the wavelength sensitivity expressed in nm/RIU and previously defined in Eq. (4) can be applied also in this kind of PCF sensing application, indicating analogous sensing performance.

The mechanical piston is not yet used when the PCF-based sensor operates as temperature sensor. In particular, thermal gradients induce gas RI changes and, consequently, operative wavelength shifts, according to Eq. (4). Obviously, the filling gas has to be properly chosen such as its thermo-optic coefficient is high enough for ensuring appreciable sensitivity. This approach is used also for making the temperature effects negligible when the sensor is used for pressure monitoring. In fact, the filling gas is chosen such as its thermo-optic coefficient has opposite sign with respect to the silica's one. In this way, silica and gas RI changes due to temperature influences can be properly compensated, resulting in the possibility of completely removing temperature effects.

In conclusion, the optimized PCF-based sensor architecture, characterized by a PCF with a length of 0.25-mm, exhibits a sensitivity of about 480 nm/RIU.

Several measurement setups based on PCFs have been widely investigated for optimizing pressure sensing performance. In particular, one of the most popular is the polarimetric measurement. This principle consists in monitoring the light intensity at the sensor output, modulated by the effect of applied pressure. In particular, by using input and output linear polarizers it is possible to control the passage of only certain orientations of plane polarized light, properly set by the input polarizer. Polarization changes induced by pressure influence, alter the current state of light polarization proportionally to the pressure strength. Consequently, if the state of polarization at the sensor output is not equal to that of the output polarizer, propagation of light is inhibited resulting in a light intensity reduction.

In this contest, an intensity measurement of pressure variation has been demonstrated, resulting in a device sensitivity of 2.34×10^{-6} MPa^{-1} (Gahir et al., 2007). In addition, a high birefringence photonic crystal fiber (Hi-Bi PCF) has been designed and fabricated for realizing an hydrostatic pressure sensor based on a bidirectional modal interferometer, as sketched in Fig. 28 (Favero et al., 2010).

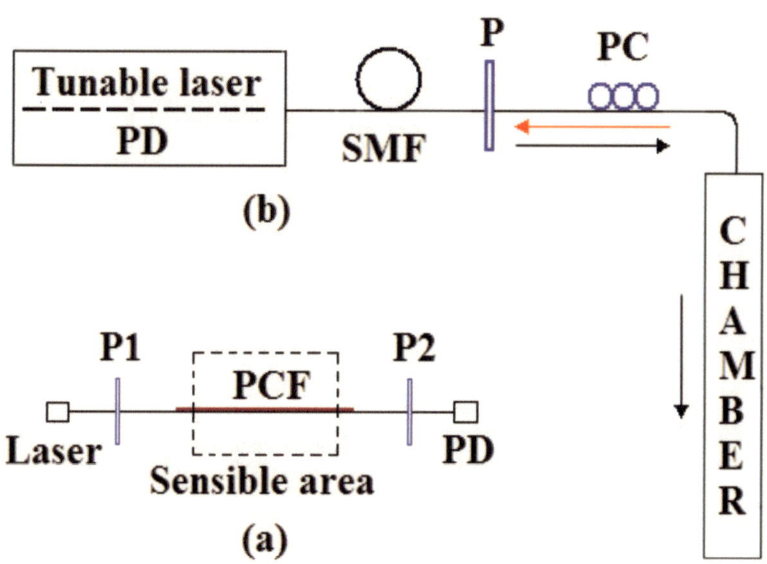

Figure 28. (a) Unidirectional and (b) bidirectional polarimetric measurement scheme used for pressure sensing.

In Fig. 28, it is possible to appreciate the difference between a conventional unidirectional polarimetric measurement scheme, Fig. 28(a), and a bidirectional one, Fig. 28(b). In particular, in the latter one only one polarizer (P) is used in combination with a polarization controller (PC). In the scheme proposed by Favero et al., two orthogonally polarized modes of a high birefringence PCF generate fringes over the optical spectrum of a BBS. In particular, the phase difference between the two modes, indicated with φ, depends on the hydrostatic pressure P, the sensor length L, and the wavelength λ, according to the following expression, being B the phase modal birefringence:

$$\varphi(\lambda,P,L)=\frac{4\pi L}{\lambda}B(\lambda,P) \tag{16}$$

In conclusion, the wavelength measurement of pressure has been numerically and experimentally demonstrated, revealing a sensitivity of 3.38 nm/MPa with an operating limit of 92 MPa.

Polarimetric measurement sensor schemes have been also adopted for temperature sensing. To this purpose, a Hi-Bi PCF has been used for a polarimetric interrogation revealing a sensitivity of 0.136 rad/°C at the operative wavelength of 1310 nm (Ju et al., 2006).

PCFs represent an intriguing and efficient alternative to conventional electrical (E-Field) and magnetic field (H-Field) sensors such as antennas, metal connections and conductive electrodes. The most important advantages of PCF-based E- and H-Field sensors over conventional ones are their immunity to electromagnetic interferences (apart from the sensor head), dielectric isolation between the sensor and the interrogation system and the capability to be used in harmful and remote locations.

To this purpose, an all-fiber sensor based on a nematic liquid crystal infiltrated photonic crystal fiber has been demonstrated as a directional E-Field sensor (Mathews et al., 2011a). In particular, a 1mm-long infiltrated polarization maintaining photonic crystal fiber (PMPCF) is used as sensor head. The length of the infiltrated section of the PCF, subjected to electric field, is initially optimized to have a monolithically varying polarized transmission response with electric field intensity change at 1550 nm. On selective infiltration of the two large holes characterizing the PMPCF, the birefringence of the fiber is set by the refractive indices of the nematic liquid crystal mixture, which vary as its molecules re-orientate on the application of electric field.

The transmitted intensity of the linearly polarized light with the direction of polarization at 45° with respect to the PCF polarization axis, is given by the following expression:

$$I=\frac{I_0}{2}\left\{1+\sin\left(\Phi_0+\left(\frac{\pi\Delta E}{E_\pi}\right)\right)\right\} \tag{17}$$

where Φ_0 and I_0 are the inherent phase retardances due to the infiltrated PMPCF and the light intensity in the absence of field, respectively. The field induced phase retardance can be expressed as follows:

$$\Phi_E = \left(\pi \Delta E / E_\pi\right) \tag{18}$$

where E_π is the sensor characterization term, which is inversely proportional to the infiltration length. Finally, interesting performance have been experimentally demonstrated such as an angular sensitivity of the PMPCF orientation with respect to the electric field as -0.07 dB/degree at a fixed electric field intensity of 3.67 kV_{rms}/mm. The sensor sensitivity to E-fields oriented parallel to the PCF polarization axis is the highest.

A polarimetric sensing scheme with selectively liquid-core infiltrated Hi-Bi PCF has been demonstrated for E-field sensing, exhibiting a sensitivity of ~ 2 dB per kV_{rms}/mm (Mathews et al., 2011b).

In conclusion, an H-Field sensor based on Terfenol coated PCF has been fabricated and experimentally tested (Quintero et al., 2011). In this context, the Faraday effect is generally used for detecting and monitoring electric field current using optical fibers. In particular, the light polarization rotation can be expressed as a function of the magnetic flux B, the interaction length d of the propagating optical field and the magnetic field to be sensed, and of the Verdet's constant υ as in the following equation:

$$\beta = vBd \tag{19}$$

It is intuitive that, for enhancing the light polarization rotation induced by the H-Field, it is possible to increase the fiber length or using fiber materials characterized by an higher Verdet's constant. The first approach is generally avoided because of the necessity of realizing optical compact sensors preventing, at the same time, high propagation losses. The second approach can be obtained by using some soft glasses for fiber optic fabrication, with the disadvantage of being mechanically fragile and temperature sensible. To this purpose, an intriguing alternative is represented by the use of Terfenol particles into optical fibers because of their high magnetostriction properties. In particular, the H-Field sensors proposed by Quintero et al., consists in a HiBi PCF made by a magnetostrictive composite using Terfenol particles with size of 250 μm and a cycloaliphatic epoxy resin with a 30 % volume fraction of Terfenol-D.

In Fig. 29, the measurement setup employed for H-Field sensing is sketched.

The proposed PCF sensor is based on a modal interferometer, where the phase difference between the two orthogonally polarized fiber modes along the optical path generates fringes over a broadband propagation spectrum. By exposing the sensor head to a magnetic field, the magnetostrictive composite changes size, resulting in a conversion of the magnetic energy into a mechanical strain. In particular, the composite deformation causes changes of the cavity length and of the effective RI of the propagating light. Consequently, the number of fringes as

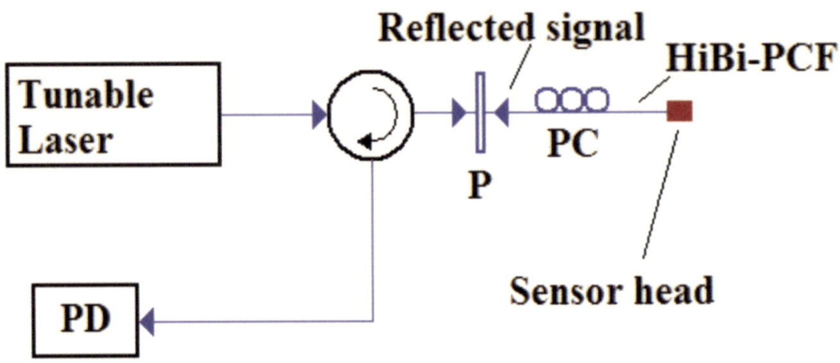

Figure 29. Schematic of the measurement setup of the magnetic field sensor.

well as the distance between consecutive fringes will change, both depending on the cavity length and effective index. In conclusion, experimental results evidence how the optical spectrum shifts towards longer wavelengths as the magnetic field increases, exhibiting a sensitivity of 0.006 nm/mT over a range extended from 0 to 300 mT, with a resolution of about ±1 mT.

Nowadays, strain monitoring represents a very important sensing approach in several application areas such as aeronautics, metallurgy, health monitoring of complex structures, to name a few. Recently, a novel strain sensor has been experimentally demonstrated (Hu et al., 2012). In particular, the sensor is based on a modified PCF-based MZI characterized by three collapsed regions, as sketched in Fig. 30.

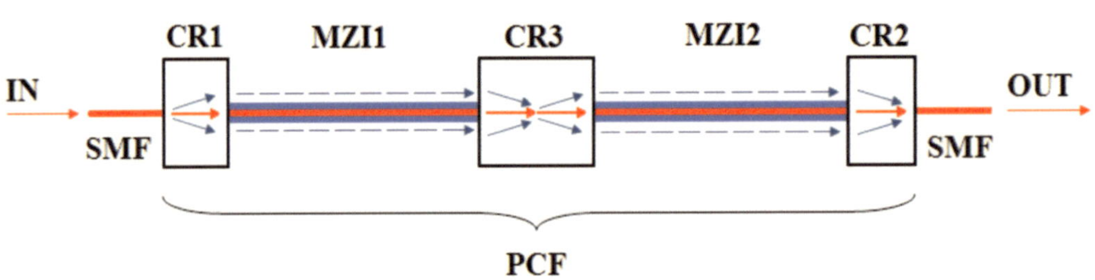

Figure 30. Schematic of the PCF-based MZI strain sensor.

The main advantage of using the modified PCF-based MZI configuration over the conventional MZI scheme, consists in a significantly enhanced extinction ratio of the transmission spectrum, resulting in an increased measurement accuracy. In collapsed regions (CR), realized by collapsing air holes with heat-treatment (Magi et al., 2005), the PCF is not a SMF because there is not any core-cladding structure. A part of the optical beam coming from the core of the lead-in SMF will be coupled into cladding modes in CR1. Then, the two beams propagating along the core and cladding of the PCF will combine and interfere in the collapsed regions CR3 and CR2, successively. Therefore, the modified PCF-MZI is actually a combined one with two cascaded MZIs (indicated with MZI1 and MZI2). Consequently, the interference takes place two times resulting in a higher extinction ratio at the lead-out SMF with respect to the conventional PCF-MZI.

In particular, at the end of the second MZI (i.e., MZI2), the transmission spectrum can be expressed as follows:

$$T_{MZI2} = kT_{MZI1}^2 \tag{20}$$

where k is the factor that describes the insertion loss of the transmission light at CR3 and T_{MZI1} is the total intensity of the transmission from MZI1, equal to:

$$T_{MZI1} = E_0^2 + \sum_{i=1}^{n} E_i^2 + 2\sum_{i=1}^{n}\left(E_0 E_i \cos\frac{2\pi\left(n_0 - n_i\right)L}{\lambda}\right) + 2\sum_{i=1}^{n-1}\sum_{j=i+1}^{n}\left(E_i E_j \cos\frac{2\pi\left(n_i - n_j\right)L}{\lambda}\right) \tag{21}$$

where E_0, E_i, and E_j are the magnitudes of electric field of the core mode and the ith- and j^{th}-order cladding mode of the PCF in MZI1, respectively. Moreover, n_0, n_i, and n_j are the effective indices of the core mode, the ith- and j^{th}-order cladding mode of the PCF in MZI1, respectively. Finally, L is the physical length of the MZI1 and MZI2 (overall PCF length ~ 9.2 cm, CR1 length = CR2 length = 135 μm and CR3 length = 291 μm), and λ is the operating wavelength of the optical source.

When an axial strain is applied on the total length of the PCF, the physical length of each cavity will change, and the effective RI for each mode of the PCF will change due to the photoelastic effect, too. Consequently, the phase differences of MZI1 and MZI2 change due to the applied strain and a wavelength shift of the interference patters can be observed.

In conclusion, the sensor described above exhibits a sensitivity as high as 11.22 dB/mε over a range of 1.28 mε and high-temperature stability (i.e., 0.0015 nm/°C and 0.009 dBm/°C).

A birefringent interferometer configured by a polarization-maintaining photonic crystal fiber (PM-PCF), has been proposed for temperature-insensitive strain measurement (Han, 2009). The strain sensor exhibits a sensitivity of 1.3 pm/με in a strain range extended from 0 to 1600 με and a LOD for strain measurement as low as 2.1 με. In conclusion, the measured temperature sensitivity is -0.3 pm/°C.

3.2. Photonic crystal fiber for biochemical sensing

The review on PCF sensors is completed by focusing on the class of chemical and biochemical PCF-based sensors.

PCFs are very suitable for chemical and biochemical sensing because of several unique features. In particular, in a micro-structured optical fiber the hollow core and air holes characterizing the cladding section, can be properly filled with liquid solutions or gases by using micropumps or particular syringes. In addition, by functionalizing inner walls of voids and hollow core, it is possible to selectively immobilize chemical analytes into the optical device enhancing, in this way, the light-matter interaction. The great overlapping between the propagating optical signal and the analyte to be detected, can be further enhanced by designing PCFs with long interaction lengths. Another important feature of PCFs is their flexibility that allows to employ these sensors for advanced chemical remote sensing.

In the following, different sensing principles adopted in PCF-based sensing are described, focusing on sensor architectures and technologies used for PCFs fabrication.

To this purpose, resonant chemical and biochemical sensors based on low-RI-contrast liquid-core Bragg fibers have been experimentally demonstrated revealing ultra high performance (Qu & Skorobogatiy, 2011). In Fig. 31, the cross section of the Bragg fiber designed and fabricated for chemical and biochemical sensing, is sketched. In particular, the Bragg grating has been realized surrounding the hollow core by a periodic sequence of high and low refractive index layers. In this case, a water filled core ($n_{Water} = 1.33$) is surrounded by a periodic multilayer of polymethyl methacrylate (PMMA) and polystyrene (PS), whose RIs are $n_{PMMA} = 1.487$ and $n_{PS} = 1.581$ at the operative wavelength $\lambda = 650$ nm, respectively.

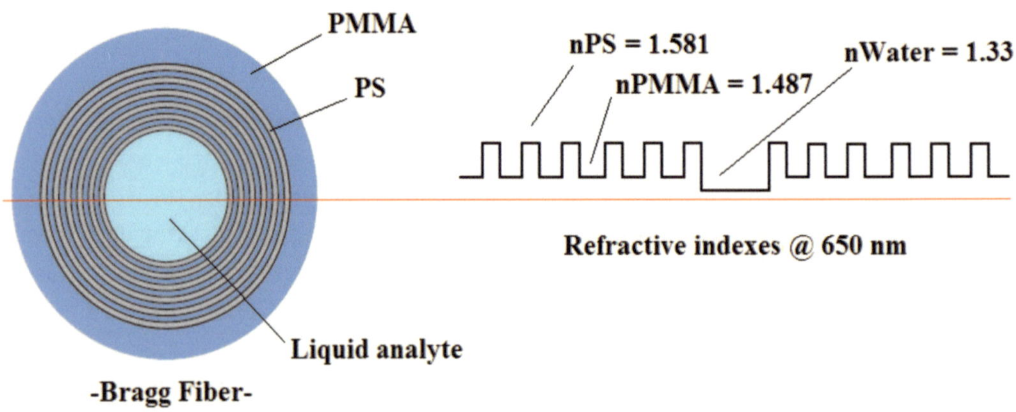

Figure 31. Cross section of the PMMA/Ps Bragg fiber designed for chemical and biochemical sensing.

The Bragg grating is properly designed in order to exhibit a central Bragg center wavelength λ_C depending on optical and geometrical properties of the periodic multilayer, as reported in the equation below:

$$\frac{\lambda_c}{2} = \left[d_h \left(n_h^2 - n_c^2 \right)^{\frac{1}{2}} + d_l \left(n_l^2 - n_c^2 \right)^{\frac{1}{2}} \right] \tag{22}$$

where d_h and d_l are the thickness of the low (n_l) and high (n_h) index layer in the Bragg reflector, and n_c is the refractive index of the liquid filled core. In the sensor proposed, $d_h = 0.13$ μm, $d_l =$

http://dx.doi.org/10.5772/53897

0.37 μm, the number of bi-layers in the Bragg reflector is approximately 25, while refractive indices have been previously indicated in Fig. 31.

The PCF-based sensor has been experimentally tested by filling the hollow core with sodium chloride (NaCl) solutions of different concentrations. In particular, refractive indices of several NaCl solutions as a function of NaCl weight concentrations (wt.%) have been evaluated for accurate calculation (i.e., wt% = 0, 5, 10, 15, 20, 25 and corresponding RIs 1.333, 1.342, 1.351, 1.359, 1.368, 1.378).

As the NaCl concentration increases in water solution, the overall refractive index n_c proportionally increases, too. Consequently, by observing Eq. (22), it is evident that the Bragg center wavelength λ_c decreases, resulting in a measurable wavelength shift in the transmission spectra of the Bragg fiber. The sensor described until now exhibits an experimental sensitivity of ~1400 nm/RIU, defined as in Eq. (4) for homogeneous sensing. In addition, the sensor has been tested also for surface sensing by detecting changes in thicknesses of thin layers deposited directly on the inner surface of the fiber core. In particular, by coating a thin layer, the localized refractive index near fiber inner surface changes, resulting in the modification of resonance guidance of the Bragg fiber, thus to the resonant wavelength shift in the Bragg fiber transmission. Moreover, if d_a is the thickness of the coated layer, the surface wavelength sensitivity can be calculated as in Eq. (23):

$$S_{\lambda,S} = \frac{\partial \lambda_c}{\partial d_a} \tag{23}$$

The presence of a 3.8-μm thick layer of sucrose solution leads to a 3.5 nm red shift of the transmission spectrum with respect to the initial position referred to a water-filled fiber without a sucrose layer. In conclusion, a moderate sensitivity $S_{\lambda,S} \approx 0.9$ nm/μm results due to poor overlap between core guided modes and the coated layer.

A PCF has been demonstrated as chemical sensor by selectively coating the fiber core with thin film containing fluorescent probe (Peng et al., 2009). The Sol-Gel method has been applied for chemical sensor functionalization. The acetylcholinesterase sensor has been experimentally tested for monitoring organophosphorus pesticide residue, revealing interesting performance. In particular, in organic pesticide parathion (PIC) and paraoxonase (Paraoxon) determination, the linear measurements ranges could arrive to $1\times10^{-9} \div 1\times10^{-3}$ mol/L with a detection limit up to 10^{-10} mol/L. In conclusion, authors suggest the PCF sensor for several application areas, such as biological/chemical research, clinical medicine, environmental protection, food inspection and preventive war biochemical fields.

A PCF interferometer operating in reflection mode has been proposed for humidity detection (Mathew et al., 2010). Generally, hygroscopic materials are required for this application field including meteorological services, air-conditioning, electronic processing, to name a few. The innovative aspect that characterized the PCF sensor proposed above, consists in the use of all-glass fiber optic based device, without using polymers or particular hygroscopic coatings.

As sketched in Fig 32, the PCF interferometer consists in a BBS, an optical spectrum analyzer (OSA), a SMF spliced to a stub of a pure silica PCF characterized by four rings of air holes arranged in a hexagonal pattern. During splicing, air holes of the PCF are completely closed resulting in a ~300-μm long collapsed region. The end section of the PCF represents the sensor head to be exposed to humidity in a climate chamber. In particular, the PCF end facet is cleaved so that the PCF behaves as a mirror. In this case, air holes are left open allowing humidity to fill in resulting in optical properties changes.

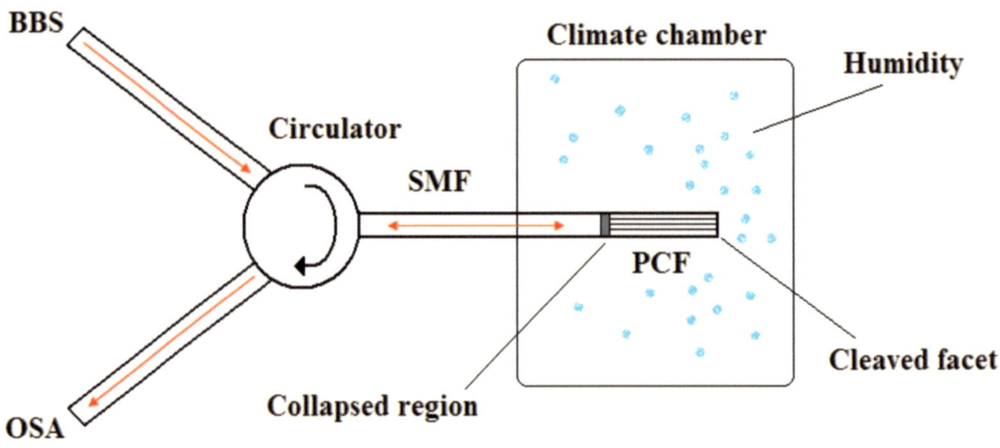

Figure 32. Schematic of the humidity sensor based on PCF interferometer.

The fundamental SMF mode excited by the BBS propagates to the collapse region where the excitation of two core modes occurs. These modes propagate to the cleaved PCF end facet, thus they are reflected to the collapse region where they recombine forming again a SMF fundamental mode. The recombination leads to an interference pattern whose interference peaks shift as a function of the relative humidity (RH) values in the climate chamber (i.e., the adsorption and desorption of H_2O molecules at the air-glass interface within the PCF holes). The sensitivity of the PCF interferometer sensor has been experimentally demonstrated to be about 5.6 pm/%RH in the range extended from 40 to 70% RH. Moreover, the shift of the interference pattern is most significant above 70% RH, exhibiting a sensitivity as high as ~ 24 pm/%RH.

The chemical functionalization of PCF sensors represent an efficient sensing technique for the selective detection of particular analyte in chemical liquid samples properly injected in holes or hollow core, depending on the particular PCF cross section. However, this sensing approach generally requires the repetition of chemical treatments at every new measure process and the change of the functionalizing chemistry as a function of the particular analyte to be sensed by the PCF sensor. In this context, label-free PCF biosensor, in which biomolecules are unlabeled or unmodified have achieved considerable attention. Moreover, label-free sensing allows to preserve chemical properties of the specie to be detected, resulting in the possibility of executing *in vivo* analysis in addition to common *in vitro* ones.

http://dx.doi.org/10.5772/53897

In this context, a novel PCF-based low-index sensor has been theoretically investigated, revealing ultra high performance (Sun et al., 2011). In Fig. 33, two examples of PCF cross sections are sketched. Both PCFs are characterized by a two distinct cores. In particular, in PCF named Fiber (a) there is a pure silica core and the other core is obtained by filling air holes with liquid analyte (i.e., water with RI $n_{Water} = 1.33$). In Fiber (b), the solid core is the same as that previously described, while the other core is made with an enlarged analyte-filled hole.

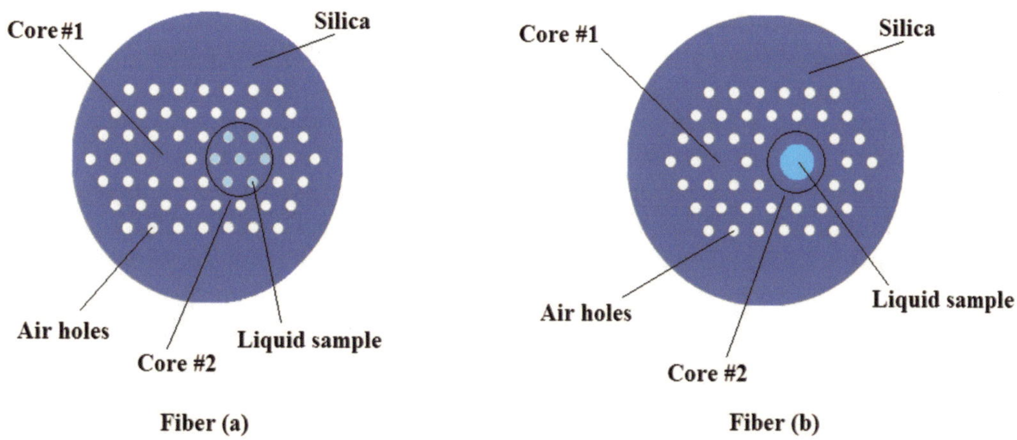

Figure 33. Cross sections of the analyte-filled micro-structured dual-core PCF (a), and dual-core PCF with an enlarge analyte-filled hole (b).

The resonant wavelength of the propagating optical mode can change as a function of the liquid analyte refractive index. In particular, the sensor sensitivity can be estimated according to Eq. (4). Moreover, performance results calculated by semi-vectorial beam propagation method, are listed in Table 1.

Numerical results		Fiber (a)		Fiber (b)
Analyte refractive index	1.33	1.35	1.42	1.4
Sensitivity, S (nm/RIU)	8500	8750	12750	10750
Detection limit, DL (RIU)	2.02×10^{-6}	1.54×10^{-6}	2.94×10^{-7}	4.75×10^{-6}

Table 1. Sensing performance of microstructured optical fiber simulated by semi-vectorial beam propagation method.

In conclusion, ultra-sensitive microstructured optical fiber refractive index sensor described until now, are able to detect liquid analyte characterized by a refractive index lower than that of the background, thus extending their regime of operation to low index liquid sample such as water.

An all-solid twin-core photonic bandgap fiber has been also designed and simulated for refractive index sensing (Yuan et al., 2010). In particular, two solid cores are separated by a analyte-filled single hole acting as a microfluidic channel. By operating in the bandgap guiding regime the proposed sensor is capable of measuring low refractive indices around that of water (i.e., $n_{Water} = 1.33$), revealing a theoretical sensitivity as high as 70,000 nm/RIU.

A refractive index sensor based on a PCF interferometer has been recently designed and experimentally tested, revealing very interesting performance (Wang & Tang, 2012). In particular, the sensor configuration is constituted by a single MZI realized by fusion splicing a short section of PCF between two standard SMFs. The operation principle is analogous to that already described in the sensor configuration based on two cascade PCF-based MZIs (Hu et al., 2012). In particular, the excitation and recombination of cladding modes in collapsed regions lead to a transmission spectrum characterized by sinusoidal interference pattern which shifts differently when the cladding/core surface of the PCF is immersed with different RI of the surrounding medium. Interesting performance have been experimentally demonstrated by using wavelength-shift interrogation. In particular, two PCF sensor configurations, i.e. sensing length for 3.5 and 5 cm, have been exposed to different concentrations of sucrose solution revealing resolutions ranging in $1.62\times10^{-4} \div 8.88\times10^{-4}$ RIU for the 3-cm sensor long and $1.02\times10^{-4} \div 9.04\times10^{-4}$ RIU for the 5-cm sensor long. Sensing performance reported above have been achieved for refractive indices in the range $1.333 \div 1.422$, suggesting these sensors as suitable for biochemical sensing and environmental monitoring applications.

Recently, research efforts concerning PhC-based sensors are oriented to the optimization of light-matter interaction, with the aim to increase sensing performance. In this context, novel PCF are proposed for advanced chemical and biochemical sensing. In particular, a index-guided PCF characterized by a hollow high index ring defect at the center of its cross-section, has been proposed and theoretically investigated (Park et al., 2011). The PCF cross section is sketched in Fig. 34.

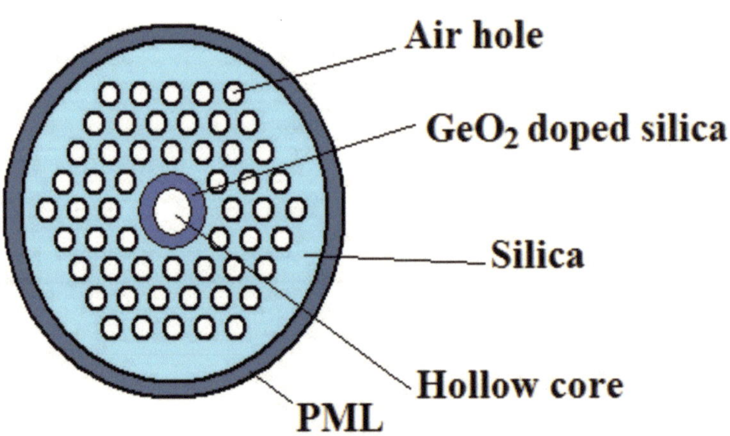

Figure 34. Cross section of the PCF characterized by a hollow GeO_2-doped high index ring defect.

As shown in Fig. 34, the PCF cross section is characterized by a periodic distribution of air holes surrounding the high index GeO_2-doped silica ring defect. A perfect matched layer (PML) is used for numerical analysis. In particular, theoretical simulations reveal an high mode intensity distribution inside the ring defect, i.e., the high refractive index region. Finally, the proposed PCF designed with optimal parameters, exhibits a relative sensitivity as high as 5.09%, and a confinement loss as low as 1.25 dB/m, suggesting this hollow core architecture as suitable for biochemical sensing, such as for the characterization of gas species (Dicaire et al., 2010). In this context, a PCF spliced to a standard optical fiber, has been arranged in an interferometer sensing scheme for detecting volatile organic compounds (VOCs) (Villatoro et al., 2009). In particular, the PCF consists of five rings of air holes arranged in a hexagonal pattern and guides light by means of the internal reflection effect. The PCF voids have been experimentally infiltrated with vapors of methyl alcohol (CH_3OH), acetonitrile (CH_3CN), isopropanol (C_3H_6OH), or tetrahydrofuran (THF), without using any permeable material. The sensing mechanism, as previously analyzed for interferometer sensing architectures, consists in the variation of the effective cladding index induced by the presence of VOCs in air holes. In this way, the reflection spectrum characterized by a regular interference pattern, is affected by interference peak shifts proportional to refractive index changes. The proposed sensor based on PCF interferometer has been fabricated and experimentally characterized, revealing interesting performance. In particular, detection limits can be estimated by associating the maximum shifts observed in the interference pattern of the reflection spectrum with the maximum volume that can be housed in air voids, i.e., ~520 picoliter. Consequently, with THF the amount of VOC detected in this low volume lies in the $\sim 4\times10^{-10}$ mole range, for acetonitrile this value is $\sim 10.5\times10^{-10}$ moles. Finally, the sensor proposed is able to detect in the few hundreds or thousands of picomoles (10^{-12}) range for VOCs, as previously presented. In this way, the sensor can be used for advanced biochemical applications, such as trace chemical or gas detection.

The review on PCF biochemical sensors includes also the so called long-period fiber grating (LPG). In particular, a LPG is a one dimension (1D) periodic structure formed by introducing periodic modulation of the refractive index along the optical fiber. LPG resonantly couples light from the fundamental core mode to some co-propagating cladding modes and leads to dips in the transmission spectrum. LPGs have been widely used for sensing purposes, such as strain, temperature and biochemical detection (Massaro, 2012). In this context, long-period gratings in photonic crystal fibers (PCF-LPG) have been experimentally demonstrated and used for label-free detection of biomolecules (Rindorf et al., 2006). In Fig. 35, a schematic of the cross section of the PCF-LPG is shown.

The sensor proposed has been functionalized with poly-L-lysine (PLL) in order to selectively immobilize charged DNA molecules on hole surfaces (see Fig. 41b). In particular, the PLL and DNA layers are characterized by refractive indices in the range 1.45÷1.48, thus closer to that of silica (1.453 @ 850 nm) than that of H_2O (1.328 @ 850 nm, 25°C).

A deep in the transmission spectrum of the PCF-LPG sensor can be experimentally appreciated at the resonant wavelength λ_R of the LPG. Moreover, the resonant wavelength can be expressed as follows:

$$\lambda_R = \Lambda_G\left(n_{co}^{eff}(\lambda) - n_{cl}^{eff}(\lambda)\right) \tag{24}$$

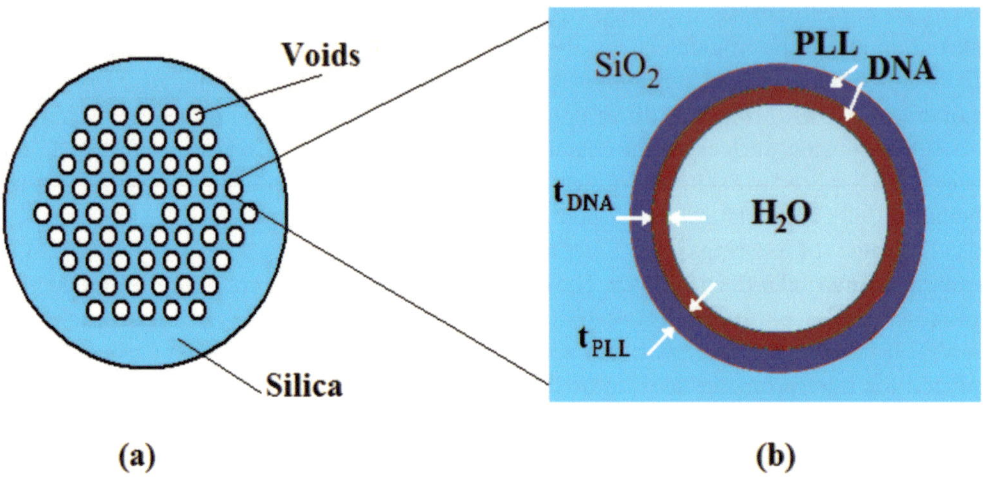

Figure 35. Cross section of the PCF used as sensitive biochemical sensor (a). Single void hole filled with water and functionalized with poly-L-lysine (PLL) of thickness t_{PLL} on which the DNA layer of thickness t_{DNA} is adsorbed (b).

where, Λ_G is the period of the LPG and $n_{co}^{eff}(\lambda)$ and $n_{cl}^{eff}(\lambda)$ are the effective indexes of core and cladding modes as a function of the free-space wavelength, respectively.

The resonant wavelength of the LPG when air is inside holes, is measured to be $\lambda_{Air} = 753.6$ nm. Successively, voids have been filled with phosphate buffered saline solution (PBS) resulting in a shifted resonant wavelength $\lambda_{PBS} = 842.5$. Moreover, the immobilization of PLL shifts the resonant wavelength to $\lambda_{PLL} = 849.2$ nm, and when DNA is adsorbed on the functionalized surface the new λ_R is measured to be $\lambda_{DNA} = 851.4$ nm. The average thicknesses of PLL and the double-strained DNA monolayer are estimated to be $t_{PLL} = 4.79$ nm and t_{DNA} 1.65 nm, respectively.

In conclusion, the PCF-LPG is able to detect a cladding effective refractive index change of approximately 10^{-4} RIU, exhibiting a wavelength sensitivity as high as 1.4 nm/nm (i.e., the shift in resonance wavelength in nm per nm thickness of biomolecular layer).

3.2.1. Photonic crystal fiber sensors based on nonlinear effects for biochemical sensing

Non linear effects, such as Four Wave Mixing (FWM) and surface enhanced Raman scattering (SERS) have been widely used for sensing applications.

In this context, a microstructured optical fiber (MOF) has been proposed for label-free selective biosensing of streptavidin (Ott et al., 2008). In particular, the nonlinear biosensor is based on the change in the degenerate FWM gain spectrum induced by the selective adsorption of streptavidin antigen biomolecules on the walls of the holes, properly functionalized with α-streptavidin, thus the antibody in the biochemical ligand.

In degenerate FWM, two pump signals at the same frequency ω, generate two new signals in the MOF, i.e. Stokes and anti-Stokes signals characterized by frequencies ω_S and ω_{aS}, respectively. These new signals are generated symmetrically around the pump ω because of the principle of the energy conservation. In particular, it results that $\omega_S = \omega - \Omega$, and $\omega_{aS} = \omega + \Omega$. Moreover, the gain of the degenerate FWM can be expressed as in the following expression:

$$g(\Omega) = \sqrt{(\gamma P_0)^2 - \left(\frac{\kappa}{2}\right)^2}, \ \kappa = 2\gamma P_0 + \Delta\beta \tag{25}$$

where $\gamma = n_2\omega_0/(cA_{eff})$ is the nonlinear parameter, c is the speed of light in vacuum, ω_0 is the pump frequency, n_2 is the nonlinear refractive index of the material used for MOF fabrication (i.e., $n_2 = 2.6\times10^{-20}$ m^2/W for silica), P_0 is the peak power of the pump and A_{eff} is the effective area of the guided mode. Finally, $\Delta\beta$ is the linear phase-mismatch expressed as follows:

$$\Delta\beta = \beta(\omega_S) + \beta(\omega_{aS}) - 2\beta(\omega) \tag{26}$$

where $\beta(\omega)$ is the linear propagation constant at the frequency ω.

In the proposed sensor, each hole of the silica PCF is properly functionalized by forming a layer of α-streptavidin with a thickness of 40 nm on the inner hole surfaces, as it has been analogously described for PLL surface functionalization, previously sketched in Fig. 35. The functionalized sensor is designed in order to exhibit precise Stokes and anti-Stokes signals around the pump frequency ω. When streptavidin biomolecules are adsorbed onto the functionalized surface, a bio-molecular adlayer of thickness $t_{bio} = 5$ nm is formed. Consequently, the hole diameter is reduced resulting in a change of the effective area of the guided mode, A_{eff}. This effect causes a change in the degenerate FWM gain, resulting in shifts of Stokes and anti-Stokes signals around the pump signal ω. In conclusion, by tracking the Stokes and anti-Stokes frequency shifts it is possible to detect adsorbed biomolecules with ultra high performance. In fact, the nonlinear sensor described until now exhibits a wavelength sensitivity of ~ 10.4 nm/nm, which is a factor of 7.5 higher than that achieved by Rindorf et al., previously reported to be 1.4 nm/nm.

Nowadays, one of the most important feature often required from optical biochemical sensors is the molecular specificity in addition to high sensitivity, low cost, easy fabrication, label-free, short-time detection, reusability, compactness, flexibility, to be named.

To this purpose, Raman spectroscopy represents a powerful optical technique due to its unique molecular specificity. In fact, Raman signal carries the specific vibrational information of the molecules to be sensed. The main drawback of this technique is represented by the weak of the Raman signal, especially in case of very low concentration of molecules in a complex liquid sample. In order to enhance the Raman signal, SERS has been widely used because of the possibility of amplifying the Raman signal by orders of magnitude due to the strong enhancement of the electromagnetic field by the Surface Plasmon Resonance (SPR) of the metallic nanostructures and the surface chemical enhancement. In particular, a surface plasmon is a localized electromagnetic wave that propagates along the metal-dielectric interface and exponentially decays into both media. Surface plasmons can be excited due to the resonant transfer of the incident photon energy and momentum to collectively oscillating electrons in a noble metal (e.g., silver, gold).

In this context, liquid-core PCFs have been experimentally investigated and theoretically analyzed for biochemical sensing of various molecules such as Rhodamine B, Rhodamine 6G (R6G), human insulin, and tryptophan, revealing excellent performance (Yang et al., 2011b). Several biomolecules have been also detected at low concentrations (i.e., 10^{-6} M ÷ 10^{-7} M) by LCPCF based on SERS, such as Prostate Specific Antigen (PSA) and alpha-synuclein, which are indicators of prostate cancer and Parkinson's disease, respectively (Shi et al., 2008).

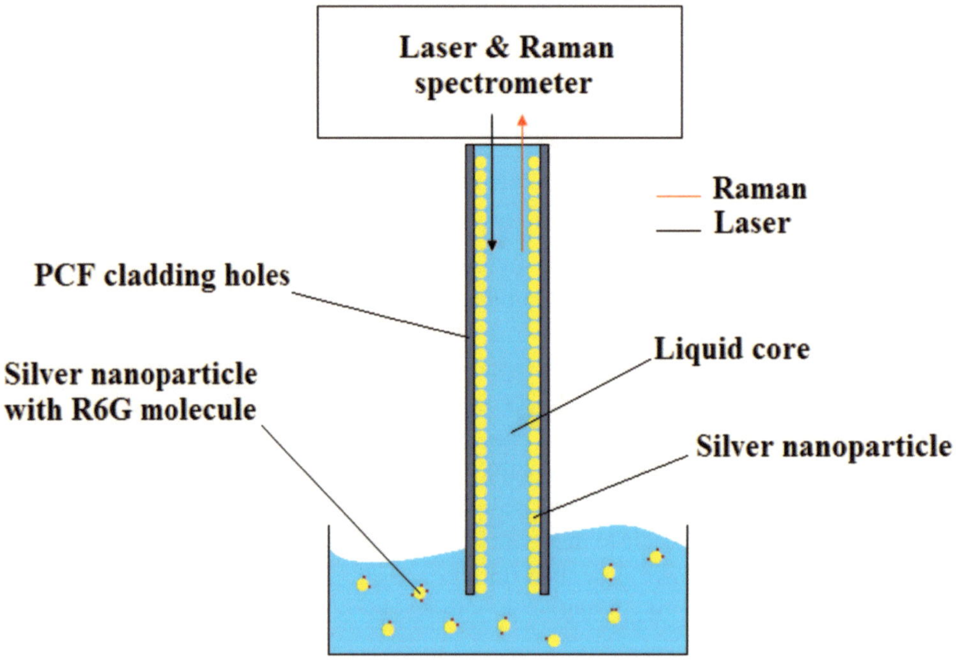

Figure 36. Schematic of the inner wall-coated LCPCF sensor used for R6G detection (Yang et al., 2011b).

In Fig. 36, the schematic of the functionalized inner wall-coated of LCPCF employed for R6G detection is sketched. The optical signal emitted by the laser source propagates along the LCPCF and excites the SPR at the surface of silver nanoparticles (SNPs). Some of these metal particles have attached on their surfaces R6G molecules to be detected. Consequently, the enhanced Raman signal containing detailed molecule vibrational information, counter-propagates to the Raman spectrometer in order to be analysed.

Interesting performance have been experimentally demonstrated. In fact, by using SNPs as the SERS substrate and R6G as a test molecule, the lowest detectable concentration that has been achieved is 10^{-10} M.

Finally, a PCF sensor based on SERS on silver nanoparticle colloid has been experimentally demonstrated for the detection of 4-mercaptobenzoic acid (4-MBA) molecules (Xie et al., 2008). Experiments have been done by mixing 200 μL of 0.01 mM 4-MBA aqueous solution with 100 μL of silver colloid and by filling the mixture solution into air holes of PCF through a particular syringe setup.

3.2.2. *Photonic crystal fiber sensors based on surface plasmon resonance*

Optical sensors based on SPR have been widely used in biological sensing because of their high sensitivity, and label-free detection.

Recently, the theoretical investigation of a PCF based on SPR has been demonstrated as efficient biosensor operating in aqueous environment (Akowuah et al., 2012). The cross section of the PCF SPR sensor proposed is sketched in Fig. 37. In particular, the proposed sensor consists of two metalized microfluidic slots, air holes for light guidance and a small central air hole for facilitating phase matching between guided and Plasmon modes. Extra air holes have been inserted between the main air holes for reducing propagation losses and ensuring efficient coupling between the core guided and plasma modes.

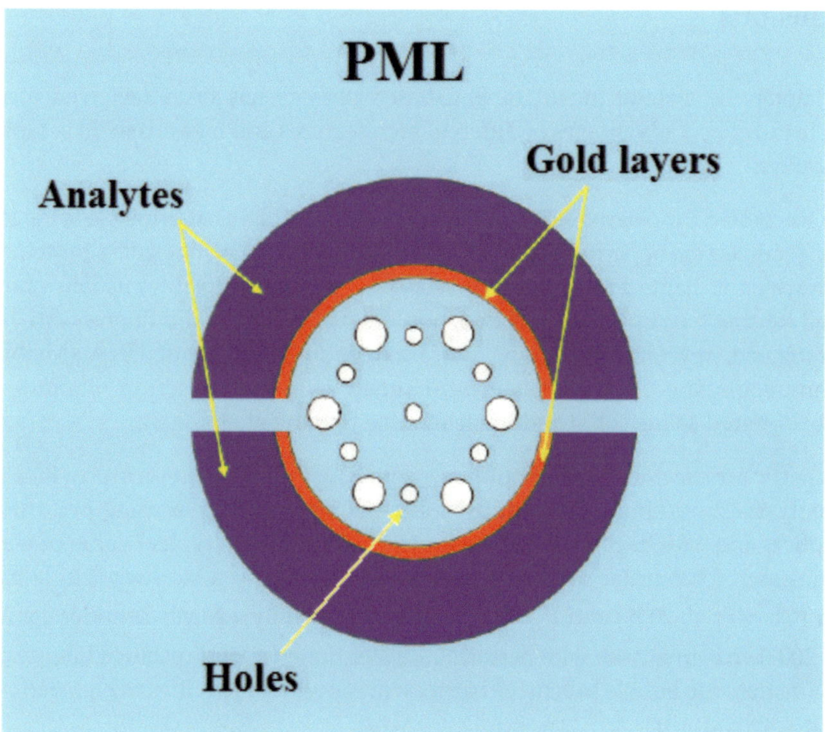

Figure 37. Schematic of the PCF based on SPR simulated for biochemical detection in aqueous environment.

The sensor has been simulated and optimized by using the full-vectorial Finite Element Method (FEM) with perfect matched layer (PML). In particular, when the sensor operates in the wavelength interrogation mode, changes in the analyte refractive index are detected by measuring the displacement of a plasmonic peak, and the wavelength sensitivity is defined as in Eq. (4), in which λ is equal to λ_p representing the wavelength corresponding to the SP resonance condition, in this case. Consequently, by changing the analyte refractive index from 1.33 (pure water) to 1.34, the optimized PCF SPR sensor exhibits a wavelength sensitivity as high as 4000 nm/RIU and a limit of detection as low as 2.5×10^{-5} RIU.

Nowadays, PCF SPR sensors are investigating for biochemical sensing because of their ultra-high performance. For example, a PCF-based refractive index sensor employing the SPR as sensing principle, has been recently theoretically investigated. In particular, sensing performance have been estimated to be S = 1700 nm/RIU (refractive index sensitivity), and LOD = 5.9×10^{-5} RIU, in aqueous environment (Peng et al., 2012).

In conclusion, PCF SPR sensors are also easy to fabricate. In fact, according to Fig. 37, deposition of metal layers inside of the microfluidic slots can be performed either with the high-pressure chemical vapor deposition technique or electroless plating techniques.

4. Conclusions

In this chapter the state-of-the-art of PhC-based sensors has been reviewed focusing on principal features and advantages of different architectures and measurement setups presented in literature.

In case of integrated PhC-based sensors, the most common physical principle is the RI sensing employed for detection of particles, gases, chemical and biological molecules, proteins, viruses, salinity in water, to name a few. In addition to other sensing principles usually employed in PhC-based sensing (i.e., optical absorption, nonlinear effect), SPR and fluorescence have been also investigated, revealing intriguing performance in detection of DNA (Mathias et al., 2010), immunoglubine G (IgG), Goat anti-Human IgG, bovine serum albumin (BSA) in phosphate buffered saline (PBS) and Cysteammine (Huang et al., 2008).

Interferometric architectures and photonic sensors based on resonant microcavities represent undoubtedly the most efficient integrated solutions for ultra high sensing performance and simple optical and CMOS-compatible readouts. In this context, typical value of wavelength sensitivities are of the order of S_λ=298 nm/RIU achieved by a slot-waveguide-based ring resonator (Claes et al. 2009) and S_λ=26 nm/RIU exhibited by a Mach-Zehnder configuration (Lu et al., 2009). In comparison with performance mentioned above, resonant PhC cavities have been demonstrated to be able to achieve higher wavelength sensitivities as high as S_λ=460 nm/RIU and S_λ=570 nm/RIU.

The principal disadvantage characterizing PhC-based sensors is represented by the rigorous control of technological processes to be executed during design and fabrication. These

http://dx.doi.org/10.5772/53897

requirements are necessary because of nanometric sizes of defects and holes in PhC periodic structure and the very high sensor operating sensitivity with respect to geometrical and physical changes. Anyways, the well-known SOI technological platform identifies a cheap and standard facility for the mass-scale production of PhC-based sensors.

Photonic crystal fibers represent a very efficient sensing solution for industrial, medical and environmental applications.

In fact, due to their fabrication simplicity, high fabrication tolerances and other advantages such as immunity to external effects, PCFs are usually employed for remote sensing in dangerous and harmful environments. Strategic approaches for improving sensing performance concern with PCF technological optimization, selection of suitable materials to be employed for sensing purposes (e.g., sol-gel, polymers), and the development of efficient and innovate measurement setups for improving readout capabilities.

Finally, photonic crystal technology surely represents a research field to be further investigated for its incredible potential in applications such as optical signal processing, telecommunications and, last but not the least, advanced optical sensing.

Author details

Benedetto Troia, Antonia Paolicelli, Francesco De Leonardis and Vittorio M. N. Passaro

Photonics Research Group, Dipartimento di Elettrotecnica ed Elettronica, Politecnico di Bari, Italy

References

[1] Akowuah, E.K.; Gorman, T; Ademgil, H.; Haxha, S.; Robinson, G. & Oliver, J. (2012). A Novel Compact Photonic Crystal Fibre Surface Plasmon Resonance Biosensor for an Aqueous Environment. Photonic Crystals - Innovative Systems, Lasers and Waveguides, A. Massaro (Ed.), 978-953-51-0416-2, InTech.

[2] Arunbabu, D.; Sannigrahi, A. & Jana, T. (2011). Photonic crystal hydrogel material for the sensing of toxic mercury ions (Hg2+) in water. Soft Matter, (February 2011), 7(6), 2592-2599.

[3] Awad, H.; Hasan, I.; Mnaymneh, K.; Hall, T.J. & Andonovic, I. (2011). Gas sensing using slow light in photonic crystal waveguide. 7th Workshop on Fibre and Optical Passive Components (WFOPC), 978-1-4577-0861-9, (July 2011), 1-3.

[4] Bakhtazad, A.; Sabarinathan, J. & Hutter, J.L. (2010). Mechanical Sensitivity Enhancement of Silicon Based Photonic Crystal Micro-Pressure Sensor. International Symposium on Optomechatronic Technologies (ISOT), 978-1-4244-7682-4, (October 2010), 1-5.

[5] Barkou, E. S.; Broeng, J. & Bjarklev, A. (1999). Dispersion properties of photonic bandgap guiding fibers. OFC/IOOC ®99. Technical Digest, (February 1999), 4, 117-119.

[6] Biallo, D.; D'Orazio, A.; De Sario, M.; Marrocco, V. & Petruzzelli, V. (2006). Proceedings of 2006 International Conference on Transparent Optical Networks (ICTON), (June 2006), 2, 44-48.

[7] Bougriou, F.; Bouchemat, T.; Bouchemat, M. & Paraire N. (2011). High sensitivity of sensors based on two-dimensional photonic crystal. Electronics, Communications and Photonics Conference (SIECPC), 2011 Saudi International. 978-1-4577-0069-9, (April 2011), 1-4.

[8] Buczynski, R. (2004). Photonic Crystal Fibers, Acta Physica Polonica A, 106(2), 141-167.

[9] Chakravarty, S.; Lai, W.-C.; Zou, Y.; Lin, C.; Wang, X. & Chen, R.T. (2011a). Silicon-nanomembrane-based photonic crystal nanostructures for chip-integrated open sensor systems. Proceeding of SPIE, (November 2011), art. 819802, 8198.

[10] Chakravarty, S.; Lai, W.-C.; Wang, X.; Lin, C. & Chen, R.T. (2011b). Photonic crystal slot waveguide spectrometer for detection of methane. Proceedings of SPIE, (January 2011), art. 79410K, 7941.

[11] Chen, W.; Lou, S.; Wang, Liwen; Zou, H.; Lu, Wenliang & Jian, S. (2011). Highly Sensitive Torsion Sensor Based on Sagnac Interferometer Using Side-Leakage Photonic Crystal Fiber. IEEE Photonics Technology Letters, (November 2011), 23(21) 1639-1641.

[12] Chen, X. & Chen, R.T. (2008). Sensitivity-enhanced silicon slot photonic crystal waveguides. 5th IEEE International Conference on Group IV Photonics, 978-1-4244-1768-1, (September 2008), 395-397.

[13] Choi, C.J. & Cunningham, B.T. (2006). Single-step fabrication and characterization of photonic crystal biosensors with polymer microfluidic channels. Lab on a Chip, (August 2006), 6(10), 1373-1380.

[14] Claes, T.; Molera, J.G.; Schacht, E.; Baets, R. & Bienstman, P. (2009). Label-Free Biosensing With Slot-Waveguide-Based Ring Resonator in Silicon on Insulator. IEEE Photonics Journal, (September 2009), 1(3), 197-204.

[15] Dicaire, I.; Beugnot, J.-C. & Thevenaz, L. (2010). Optimized conditions for gas light interaction in photonic crystal fibres. Proceedings of SPIE, (September 2010), art. 76530L, 7653.

[16] Endo, T.; Yanagida, Y. & Hatsuzawa, T. (2007). Photonic crystal based optical chemical sensor for environmental monitoring. 7th IEEE International Conference on Nanotechnology (IEEE-NANO 2007), 1-4244-0608-0, 947-950.

[17] Favero, F.C.; Quintero, S.M.M; Martelli, C; Braga, A.M.B.; Silva, V.V.; Carvalho I.C.S.; Llerena, R.W.A. & Valente, L.C.G. (2010). Hydrostatic Pressure Sensing with High Birefringence Photonic Crystal Fibers. Sensors, (November 2010), 10(11), 9699-9711.

[18] Gahir, H.K. & Khanna, D. (2007). Design and development of a temperature-compensated fiber optic polarimetric pressure sensor based on photonic crystal fiber at 1550 nm. Applied Optics, (March 2007), 46(8), 1184-1189.

[19] García-Rupérez, J.; Toccafondo, V.; Bañuls, M.J.; Griol, A.; Castelló, J.G.; Peransi-Llopis, S. & Maquieira, A. (2010). Single strand DNA hybridization sensing using photonic crystal waveguide based sensor. 7th IEEE International Conference on Group IV Photonics, 978-1-4244-6346-6, (September 2010), 180-182.

[20] Gong, H.P.; Chan, C. C.; Zu, P.; Chen, L. H. & Dong, X. Y. (2010). Curvature measurement by using low-birefringence photonic crystal fiber based Sagnac loop. Optics Communication, 283(16), 3142–3144.

[21] Griffete, N.; Frederich, H.; Maître, A.; Chehimi, M.M.; Ravaine, S. & Mangeney, C. (2011), Photonic crystal pH sensor containing a planar defect for fast and enhanced response. Journal of Materials Chemistry, (July 2011), 21(34), 13052-13055.

[22] Han, Y.G. (2009). Temperature-insensitive strain measurement using a birefringent interferometer based on a polarization-maintaining photonic crystal fiber. Applied Physics B, 95(2), 383-387.

[23] Huang, H; Zhang, J.; Ji, X.; Zhou, J.; Bao, Minhang & Huang, Y. (2008). Photonic Crystal Biosensors Based on Surface Plasmons. 9th International Conference on Solid-State and Integrated-Circuit Technology (ICSICT), 978-1-4244-2186-2, (October 2008), 2565-2568.

[24] Hu, L.M.; Chan, C.C.; Dong, X.Y.; Wang, Y.P.; Zu, P.; Wong, W.C.; Qian, W.W.& Li, T. (2012). Photonic Crystal Fiber Strain Sensor Based on Modified Mach-Zehnder Interferometer. IEEE Photonics Journal, (February 2012), 4(1), 114-118.

[25] Jágerská, J.; Zhang, H.; Diao, Z.; Le Thomas, N. & Houdré, R. (2010). Refractive index sensing with an air-slot photonic crystal nanocavity. Optics Letters, (August 2010), 35(15), 2523-2525.

[26] Jiang, H.; Zhu, Y.; Chen, C.; Shen, J.; Bao, H.; Peng, L.; Yang, X. & Li, C. (2012). Photonic crystal pH and metal cation sensors based on poly(vinyl alcohol) hydrogel. New Journal of Chemistry, (February 2012), 36(4), 1051-1056.

[27] Ju, J.; Wang, W.; Jin, W. & Demokan, M.S. (2006). Temperature sensitivity of a two-mode photonic crystal fiber interferometric sensor. IEEE Photonic Technology Letters, (October 2006), 18(20), 2168-2170.

[28] Junhua, L.; Qiang, K.; Chunxia, W.; Baoqing, S.; Yiyang, X. & Hongda, C. (2011). Design of a photonic crystal microcavity for biosensing. Journal of Semiconductors, (March 2011), art. 034008, 32(3).

[29] Lee, M.; Miller, B.L. & Fauchet, P.M. (2008). Two-dimensional photonic crystal microcavity sensor for single particle detection. Lasers and Electro-Optics, 2008 and 2008 Conference on Quantum Electronics and Laser Science, 978-1-55752-859-9, (May 2008), 1-2.

[30] Li, B.; Hsiao, F.L. & Lee, C. (2011). Computational Characterization of a Photonic Crystal Cantilever Sensor Using a Hexagonal Dual-Nanoring-Based Channel Drop Filter. IEEE Transaction of Nanotechnology, (July 2011), 10(4), 789-796.

[31] Lin, H.; Yi., Z. & Hu, J. (2012). Double resonance 1-D photonic crystal cavities for single-molecule mid-infrared photothermal spectroscopy: theory and design. Optics Letters, (April 2012), 37(8), 1304-1306.

[32] Liu, Y. & Salemink, H.W.M. (2012). Photonic crystal-based all-optical on-chip sensor. Optics Express, (August 2012), 20(18), 19912-19920.

[33] Lu, P.; Men, L.; Sooley, K. & Chen, Q. (2009). Tapered fiber Mach-Zehnder Interferoemter for simultaneous measurement of refractive index and temperature. Applied Physics Letters, (April 2009), art. 131110, 94(13).

[34] Magi, E.C.; Nguyen, H.C. & Eggleton, B.J. (2005). Air-hole collapse and mode transitions in microstructured fiber photonics wires. Optics Express, (January 2005), 13(2), 453-459.

[35] Mandal, S. & Erickson, D. (2008). Nanoscale optofluidic sensor arrays. Optical Express, (February 2008), 16(3), 1623-1631.

[36] Mandal, S.; Goddard, J.M. & Erickson, D. (2009). A multiplexed optofluidic biomolecular sensor for low mass detection. Lab on a Chip, (July 2009), 9(20), 2924-2932.

[37] Massaro, A. (2012). Photonic Crystals – Introduction, Applications and Theory. A. Massaro(Ed.), 978-953-51-0431-5, InTech.

[38] Mathew, J.; Semenova, Y.; Rajan, G. & Farrell, G. (2010). Humidity Sensor Based on a Photonic Crystal Fiber Interferometer. Electronics Letters, (September 2010), 46(19), 1341-1343.

[39] Mathews, S.; Farrell, G. & Semenova, Y. (2011a). Directional electric field sensitivity of a liquid crystal infiltrated photonic crystal fiber. IEEE Photonics Technology Letters, (April 2011), 23(7), 408-410.

[40] Mathews, S; Farrell, G. & Semenova, Y. (2011b). All-fiber polarimetric electric field sensing using liquid crystal infiltrated photonic crystal fibers. Sensors and Actuators A: Physical, (May 2011), 167(1), 54-59.

[41] Mathias, P.C.; Jones, S.I.; Wu, H.-Y., Yang, F; Ganesh, N.; Gonzalez, D.O.; Bollero, G.; Vodkin, L.O. & Cunningham, B.T. (2010). Improved Sensitivity of DNA Microarrays Using Photonic Crystal Enhanced Fluorescence. Analytical Chemistry, (August 2010), 82(16), 6854-6861.

[42] Maulina, W.; Rahmat, M.; Rustami, E.; Azis, M.; Budiarti, D.R.; Miftah, D.Y.N.; Maniur, A.; Tumanggor, A.; Sukmawati, N.; Alatas, H.; Seminar, K.B. & Yuwono, A.S. (2011), Fabrication and Characterization of NO2 Gas Sensor Based on One Dimensional Photonic Crystal for Measurement of Air Pollution Index. 2nd International Conference

on Instrumentation, Communication, Information Technology and Biomedical Engineering (ICICI-BME). 978-1-4577-1166-4, (November 2011), (2), 352-355.

[43] Nalawade, S.M.; Harnol, S.S. & Thakur, H.V. (2012). Temperature and Strain Independent Modal Interferometric Torsion Sensor Using Photonic Crystal Fiber. IEEE Sensors Journal, (August 2012), 12(8), 2614-2615.

[44] Ott, J.R.; Heuck, M.; Agger C., Rasmussen, P.D. & Bang, O. (2008). Label-free and selective nonlinear fiber-optical biosensing. Optics Express, (December 2008), 16(25), 20834-20847.

[45] Padidar, S.; Ahmadi, V. & Ebnali-Heidari, M. (2012). Design of High Sensitive Pressure and Temperature Sensor Using Photonic Crystal Fiber for Downhole Application. IEEE Photonics Journal, (October 2012), 4(5), 1590-1599.

[46] Park, J.; Lee, S.; Kim, S. & Oh, K. (2011). Enhancement of chemical sensing capability in a photonic crystal fiber with a hollow high index ring defect at the center. Optics Express, (January 2011), 19(3), 1921-1929.

[47] Peng, Y. & Cheng, Y. (2009). The Research of photonic-crystal fiber sensor. Proceedings of SPIE, art. 73810G, 7381.

[48] Peng, Y.; Hou, J.; Huang, Z.; Zhang, B. & Lu, Q. (2012). Design of the photonic crystal fiber-based surface plasmon resonance sensors. Chinese Optics Letters, (June 2012), art. S10607, 10(s1).

[49] Pinto, A.M.R. & Lopez-Amo, M. (2012). Photonic Crystal Fibers for Sensing Applications. Journal of Sensors, art. ID 598178, (February 2012), 2012, 1-21.

[50] Qu, H. & Skorobogatiy, M. (2012). Resonant bio- and chemical sensors using low-refractive-index-contrast liquid-core Bragg fibers. Sensors and Actuators B: Chemical, (January 2012), 161(1), 261-268.

[51] Quintero, S.M.M.; Martelli, C.; Braga, A.M.B.; Valente, L.C.G. & Kato, C.C. (2011). Magnetic Field Measurements Based on Terfenol Coated Photonic Crystal Fibers. Sensors, (November 2011), 11(12), 11103-11111.

[52] Rindorf, L.; Jensen, B.J.; Dufva, M.; Pedersen, H.L.; Hoiby, P.E. & Bang, O. (2006). Photonic crystal fiber long-period gratings for biochemical sensing. Optics Express, (September 2006), 14(18), 8224-8231.

[53] Robinson, S. & Nakkeeran, R. (2012). Photonic Crystal based sensor for sensing the salinity of seawater. IEEE – International Conference On Advances In Engineering, Science and Management (ICAESM), 978-81-909042-2-3, (March 2012), 495-499.

[54] Scullion, M.G.; Di Falco, A. & Krauss, T.F. (2011). Slotted photonic crystal cavities with integrated microfluidics for biosensing applications. Biosensors and Bioelectronics, (June 2011), 27(1), 101-105.

[55] Shi, C; Zhang, Y.; Gu, C.; Seballos, L. & Zhang, J.Z. (2008). Low Concentration Biomolecular Detection Using Liquid Core Photonic Crystal Fiber (LCPCF) SERS Sensor. Proceedings of SPIE, art. 685204, 6852.

[56] Sun, B; Chen, M.-Y.; Zhang Y.-K.; Yang J.-C.; Yao, J.-Q & Cui, H.-X. (2011). Microstructured-core photonic-crystal fiber for ultra-sensing refractive index sensing. Optics Express, (February 2011), 19(5), 4091-4100.

[57] Thévenaz, L.; Dicaire, I.; Chin, S. & De Rossi, A. (2012). Gas Sensing using Material and Structural Slow Light System. Optical Sensors, OSA Technical Digest (online) (Optical Society of America, 2012), (June 2012), paper STu2F.1

[58] Van Driel, H.M. (2003). Nonlinear Optics in Photonic Crystal. Proceedings of 2003 5th International Conference on Transparent Optical Networks (ICTON), 0-7803-7816-4, (July 2003), 1, 56-59.

[59] Van Leest, T.; Heldens, J.; Dan Der Gaag, B. & Caro, J. (2012). Photonic crystal cavities for resonant evanescent field trapping of single bacteria. Proceedings of the SPIE, (June 2012), art. 84270T, 8427.

[60] Villatoro, J.; Kreuzer, M.P.; Jha, R.; Minkovich, V. P.; Finazzi, V.; Badenes, G. & Pruneri, V. (2009). Photonic crystal fiber interferometer for chemical vapor detection with high selectivity. Optics Express, (February 2009), 17(3), 1447-1453.

[61] Wang, J.-N. & Tang, J.-L. (2012). Photonic Crystal Fiber Mach-Zehnder Interferometer for Refractive Index Sensing. Sensors, (March 2012), 12(3), 2983-2995.

[62] Wu, M.; Hryciw, A.C.; Khanaliloo, B.; Freeman, M.R.; Davis, J.P. & Barclay, P.E. (2012). Photonic crystal paddle nanocavities for optomechanical torsion sensing. CLEO: Science and Innovations, OSA Technical Digest (online) (Optical Society of America, 2012), paper CW1M.7.

[63] Xiao, L.; Jin, W. & Demokan, M.S. (2007). Photonic crystal fibers confining light by both index-guiding and bandgap-guiding: hybrid PCFs. Optics Express, (November 2007), 15(24), 15637-15647.

[64] Xie, Z.-G.; Lu, Y.-H., Wang, P.; Lin, K.-Q.; Yan, J. & Ming, H. (2008). Photonic Crystal Fibre SERS Sensors Based on Silver Nanoparticle Colloid. Chinese Physics Letters, (September 2008), 25(12), 4473-4475.

[65] Yang, D.; Tian, H. & Ji, Y. (2011a). Nanoscale photonic crystal sensor arrays on monolithic substrates using side-coupled resonant cavity arrays. Optics Express, (October 2011), 19(21), 20023-20034.

[66] Yang, X.; Shi C.; Newhouse, R.; Zhang, J.Z. & Gu, C. (2011b). Hollow-Core Photonic Crystal Fibers for Surface-Enhanced Raman Scattering Probes. International Journal of Optics, art. ID 754610, (February 2011), 2011, 1-11.

[67] Yuan, W.; Town, E. G. & Bang, O. (2010). Refractive index Sensing in an All-solid Twin-Core Photonic Bandgap Filter. IEEE Sensors Journal, (July 2010), 10(7), 1192-1199.

[68] Zlatanovic, S.; Mirkarimi, L.W.; Sigalas, M.M.; Bynum, M.A.; Chow, E.; Robotti, K. M.; Burr, G.W.; Esener, S. & Grot, A. (2009). Photonic crystal microcavity sensor for ultracompact monitoring of reaction kinetics and protein detection, Sensors and Actuators B: Chemical, (August 2009), 141(1), 13-19.

[69] Zou, Y.; Chakravarty, S.; Lai, W.-C.; Lin, C.-Y. & Chen, R.T. (2012). Methods to array photonic crystal microcavities for high throughput high sensitivity biosensing on a silicon-chip based platform, Lab Chip, (July 2012), 12(13), 2309-2312.

[70] Zu, P.; Chan, C.C.; Jin, Y.; Gong, T.; Zhang, Y.; Chen, L. H. & Dong, X. (2011). A temperature-insensitive twist sensor by using low-birefringence photonic-crystal-fiber-based Sagnac interferometer. IEEE Photonics Technology Letters, (July 2011), 23(13), 920–922.

Chapter 12

Silicon Photonic Crystals Towards Optical Integration

Zhi-Yuan Li, Chen Wang and Lin Gan

Additional information is available at the end of the chapter

http://dx.doi.org/10.5772/55041

1. Introduction

During the past two decades, there have been great interests in developing ways to manipulate photons at nanoscale, realizing optical integrations, developing smaller, faster, and more efficient optoelectronic devices for the purpose of next-generation optoelectronic technology. Great progresses have been made in exploring photonic crystals (PCs) [1,2], plasmonic structures [3,4], and other nanophotonic devices for applications. However, plasmonic structures always involve some metal cells and are subject to strong energy dissipation and absorption loss in optical frequencies. Since silicon has a large refraction index and low loss in the infrared wavelength, it becomes an important optical material that has been widely used for integrated photonics applications. Meanwhile, silicon dominates microelectronics and this makes the silicon-based optical devices have the advantage to integrate with electronic devices.

Among all the semiconductor-based optical devices, a class of integrated optical devices that are built in the platform of periodically patterned silicon structures (namely, silicon PCs) are now attracting much attention [5–7]. Analogous to real crystal, electromagnetic (EM) wave is strongly modulated in PC by means of periodic Bragg scattering. Photonic band gaps (PBGs), which can prevent light from propagation in certain direction for a certain range of wavelengths, are formed similarly to electron band gaps. If we introduce a line defect or a point defect in PC, a defect state will take place within the PBG, where light is strongly localized around the defect. These defects can serve as a high efficient waveguide channel or as a microcavity with a high-quality (high-*Q*) factor. The mismatch of the PBGs spatial inversion symmetry breaking and could lead to the optical isolation in any device where the forward and backward transmissivity of light is very much different.In addition, the transmission bands also provide remarkable dispersion properties due to strong Bragg scattering, and negative refraction, self-collimation, superprism and many other anomalous transport behaviors [8–13] can be achieved by engineering the unit cell geometry of PCs. In this review, we briefly introduce the theoretical background of the light propagation in PC and show our recent results on design, fabrication, and characterization of several basic integrated optical devices in the platform of infrared silicon PC slab.

2. Theoretical and numerical tools

In this section, we briefly introduce the theoretical background and the numerical methods for our study of PC. The propagation of electromagnetic waves in PC is governed by the Maxwell equations [14]. For the sake of simplicity, we only consider a nonmagnetic linear system. In particular, the dielectric constant ε is independent of frequency and we neglect any absorption of electromagnetic waves by the material. Furthermore, there are no free charges or currents in our system. With all of the assumptions, the magnetic field within the PC satisfies the following equation that directly originates from the Maxwell equations [15]:

$$\nabla \times \left\{ \frac{1}{\varepsilon(\mathbf{r})} \nabla \times \mathbf{H}(\mathbf{r}) \right\} = \frac{\omega^2}{c^2} \mathbf{H}(\mathbf{r}). \tag{1}$$

Since our system involves a periodic dielectric function ε, we can apply Bloch-Floquet theorem to our situation, which means that the solutions can be expressed as:

$$\boldsymbol{H}_k(\boldsymbol{r}) = \boldsymbol{u}_k(\boldsymbol{r}) e^{i(\mathbf{k}\cdot\mathbf{r}-\omega t)}, \tag{2}$$

where k is the Bloch wave vector and $\boldsymbol{u}_k(\boldsymbol{r})$ is a periodic function of position. This type of solutions are periodic as a function of k.

Several theoretical methods have been developed to handle different problems for PC structures, such as the plane-wave expansion method (PWEM) [15–18], transfer matrix method (TMM) [19], finite-difference time-domain (FDTD) method [20,21], and multiple scattering method [22,23]. Each method exhibits its own benefits and drawbacks. The PWEM is the earliest method applied to PC [15], and has shown its great power in the discovery of three-dimensional (3D) diamond-lattice PCs that have a complete PBG [15]. However, this method has a severe limitation in that it can only deal with the photonic band structures. The TMM is an efficient approach that was designed particularly to calculate the transmission spectra of PC, but it can also be used to solve the photonic band structures. This approach works based on the finite-difference scheme in the real space [19]. Later on, a plane-wave based transfer-matrix method (PWTMM) was developed by Li et al. [24–26]. This method works on the plane-wave space and uses plane wave functions (representing Bragg waves) to describe both the EM fields and dielectric functions. This approach can handle a broad range of general PC problems. In addition to the regular solutions of photonic band structures and transmission/reflection/absorption spectra, this approach can efficiently solve the Bloch wave scattering at the interface between semi-infinite PC structures, because it can also work in the Bloch mode space [27]. So far, other methods have not developed such a peculiar capability. For this reason, it can calculate efficiently modal coupling with multimode PC waveguides [28], transmission efficiency through general two-dimensional (2D) and 3D waveguide bends [29], and band diagrams and field profiles of PC surface states [30]. On the other hand, this method has adopted advanced numerical and mathematical analysis tools to enhance numerical convergence and accuracy, and it has shown its superior power in dealing with some metal PC structures compared with other methods [31]. Recently, this method was also extended to solve nonlinear optical problems in ferroelectric PC structures [32,33]. The FDTD method is a very popular and universal approach in numerical simulations of various PC problems. In addition to the regular band structures and optical spectra calculations, this technique can govern the EM field evolution with time in arbitrary PC structures with infinite or finite structural domain. The reason is that the technique works in the time domain and directly solves the Maxwell equations.

Many free software packages and commercial software packages have been developed worldwide and they are widely used in numerical simulations and solutions of different PC problems. Our group has also developed homemade codes based on several methods including the PWEM, PWTMM, FDTD, and multiple scattering method. In addition, we also utilize publicly available free software packages as they are more numerically economic or have better numerical efficiency. In our case, we use MIT Photonic-Bands (MPB) package [16] to compute the photonic band structures and use MEEP, a free FDTD simulation software package developed at MIT [34] to calculate transmission spectra and model electromagnetic wave transport features in the 2D PC structures and devices.

3. Sample fabrication and optical characterization

After the discovery of PC, many novel devices have been proposed to control light and implement specific functionality of information processing. 2D air-bridged silicon PC slab [Fig. 1(a)] is an excellent platform to fabricate PC integrated optical devices. This system involves a silicon membrane suspended in air, which confines light by high index contrast in the vertical direction, while the periodic structures in the slab give a strong in-plane confinement of light through PBGs. In most cases, a typical PC structure is a kind of periodic array of air holes etched in a silicon-on-insulator (SOI) wafer by microfabrication techniques. The SOI wafer has a Si/SiO_2/Si structure. In our case, it has a 220 nm thick silicon top layer and a 3 μm buried silica layer on top of a 0.5 mm thick single crystal silicon wafer. We directly use focused ion beam (FIB) lithography to drill air holes in the silicon membrane or use electron-beam lithography (EBL) to define PC patterns in a thin film of polymethylmethacrylate (PMMA), and then transfer the patterns into the silicon membrane by inductively coupled plasma (ICP) etching under the atmosphere of SF_6 and C_4F_8 gases. Figures 1(b) and (c) are the top-view scanning electron microscope (SEM) picture of our PC structures fabricated by FIB. By utilizing state-of-the-art microfabrication techniques, the optical properties of the periodic array of air holes can be easily and accurately controlled. For instance, one can change the diameters of

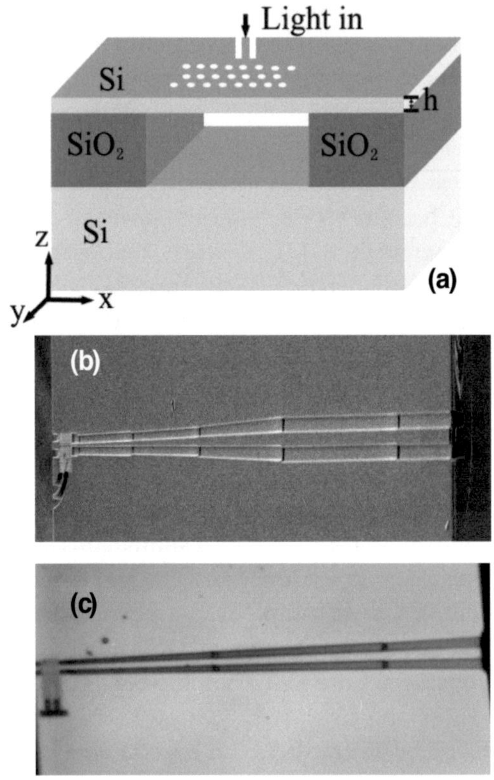

Figure 1. a) Schematic view of a 2D air-bridged PC structures with an input silicon waveguide. The whole structures are fabricated in SOI wafer. The air-bridged structures are formed by HF wet etching; (b) and (c) are the top-view SEM and optical microscopy image of a practical PC sample used in experiment. The long adiabatically tapering ridge waveguide connected with the PC structure can be clearly visualized.

certain holes or omit to etch one or several holes at certain places. These procedures can allow for engineering of the linear and point defect characteristics. After the air holes PC structures get done, we use HF acid wet etching to remove the buried oxide layer under the silicon membrane PC structures to form air-bridged structures. Usually, wide silicon wire waveguides (also with the air-bridged geometry) close to the interface of PC structures are used as the input and output infrared light beam channels, as shown in Fig. 1. These wire waveguides are further connected with long adiabatically tapering ridge silicon waveguides (each about 0.2 mm long) to allow easy coupling with external infrared signals from single mode optical fibers. As a result, a typical PC sample has a total length of about 0.5 mm and the input and output ends are carefully polished to enhance the coupling efficiency of input and output infrared signals.

The transmission spectra of a PC structure effectively reflect its optical properties. To get this important physical quantity, we have set up an experimental apparatus that involves several functional components. The overall measurement setup is schematically illustrated in Fig. 2(a), while a picture of the corresponding real system is displayed in Fig. 2(b). As shown in

Fig. 2(a), the PC samples are placed in the center of the stage, with its two sides connected with the input and output optical fibers. The input optical signal comes from a continuous wave tunable semiconductor laser with the wavelength ranging from 1500 to 1640 nm, launched into one facet of the ridge waveguide via a single-mode lensed fiber. Power meter is used to detect the optical signals transmitted through the PC structures and emitted from the output side. The measurement is made with TE polarization (electric field parallel to the slab plane) since it has a complete band gap in silicon PC slabs. The measurement data are normalized by a ridge waveguide on silicon with the same length and width to yield the final transmission spectra for a specific PC structure.

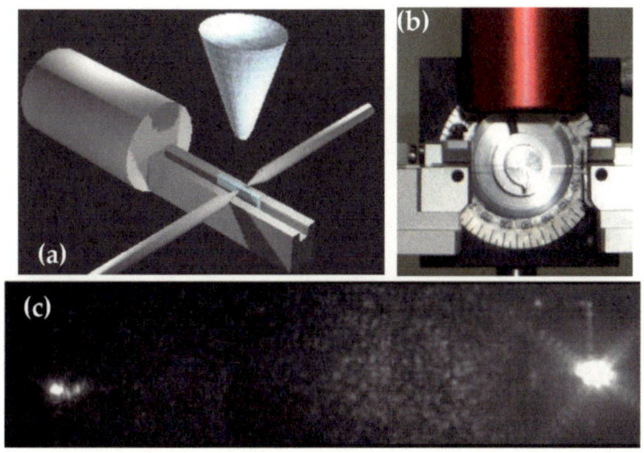

Figure 2. a) Schematic view and (b) experimental setup for the optical characterization of infrared 2D silicon PC slab structures; (c) typical optical microscopy picture recorded by the CCD camera for the PC sample as displayed in Fig. 1(b).

In addition to the measurement of transmission spectra, our experimental setup can offer another big power: it allows for easy and convenient direct monitoring of the transport path of infrared light through PC devices. As depicted in Fig. 2(a), a charge-coupled device (CCD) camera is mounted above the sample and it can *in situ* monitors the transport property by imaging the roughness induced scattering infrared light from the surface of the PC structures. The long-focus microscope objective connected with the CCD camera is shown in the upper part of Fig. 2(b). The ray trace can be directly visualized by the camera to yield images at a personal computer monitor, and this gives the researcher a rough but direct estimate abouthow much the infrared signal has gone into the PC structures. The idea is simple: if the infrared light is coupled into the PC sample with a sufficiently high efficiency, infrared light can transport along the input ridge waveguide, PC devices, and the output ridge waveguide. Significant scattering of infrared signal off the sample can take place and is collected by the CCD camera and visualized *in situ* by the monitor. The strongest scattering occurs at the discontinuity interface, including the end facets of input and output ridge waveguides and the connection section between ridge waveguides and PC structures. Even within the PC structures, remarkable scattering still takes place because of the inevitable roughness on the surface of silicon slab and within the inner walls of air holes. On the other hand, if the infrared signal coupled into the PC samples is weak, then the overall scattering light that can be collected by the CCD camera is also very weak, and no picture can be visualized in the monitor. Figure 2(c) shows a typical microscopy picture recorded by the CCD camera. Such an experimental setup is very convenient to adjust the precise position of the input and output optical fibers, so that they have a precise point-to-point alignment with about 250 nm thick PC samples to allow for high-efficiency optical coupling. The infrared signal transporting through the PC samples usually has a power level of micro-walt when the input signal from semiconductor laser is on the power level of milli-walt. The signal has been already sufficiently strong to allow for transmission spectrum measurement and CCD camera monitoring with high signal-to-noise ratio.

4. PC band–gap devices: waveguides and cavities

4.1. PC waveguides

PC waveguides are one of the most important elements in PC integrated optics, because they offer efficient channels for light propagating at wavelength scale and connect different devices in integrated optical circuits. Usually, PC waveguides are formed by removing one row of holes in a PC structure. The line defects can generate defect states within the complete PBG and serve as waveguide channels for light to propagate efficiently and freely in PC structures. Among many 2D PC slab structures, the triangular lattice of air holes has a relatively large band gap for TE-like electromagnetic modes, where the magnetic field points in the perpendicular direction while the electric field is dominantly within the lateral plane of the slab. In addition, the structures allow for easy fabrication by standard planar nanofabrication technologies such as FIB and EBL, and have good mechanical stability. For these reasons, they are widely and dominantly adopted in designing and exploring PC based integrated optical devices [35,36].

Figure 3(a) shows the calculated TE-like mode photonic band structures of a particular 2D triangular lattice PC slab, where a wide complete PBG is clearly seen. In the region, light propagation inside the PC is prohibited. When removing one or several rows of air holes in the PC structures, some allowed modes (defect states) appear within the PBG [Fig. 3(b)], and they can be used to create waveguides or cavities. In most works, single-mode or multi-mode optical waveguides are usually made along the Γ–K direction in the triangular lattice PC. It has been well established in plenty of literatures that the number of waveguide modes as well as the width of the transmission windows can be controlled by tuning the core width of the line defects. However, waveguides along other directions in the triangular lattice PC were rarely discussed. Just like the Γ–K direction, waveguides along the Γ–M direction should also be able to guide confined modes due to the existence of a complete PBG in the 2D triangular lattice PC. By removing a line of diamond areas, we can obtain a cluster-like waveguide along the Γ–M direction as depicted in Fig. 4(a). This kind of waveguide is called Γ–M waveguide [37].

Figure 4(b) is the SEM picture of the original Γ–M waveguide, where the air holes remain to locate at the original lattice site and the radius of all air holes remains the same. The air holes

http://dx.doi.org/10.5772/55041

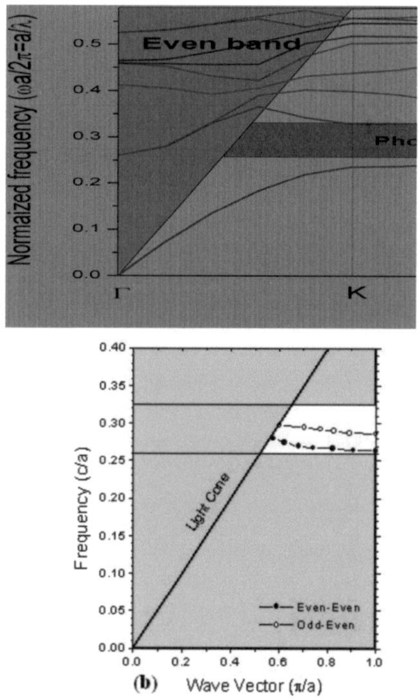

Figure 3. a) Photonic band structures for air holes triangular lattice PC slab; (b) band diagrams for a PC W1 waveguide, where one row of air holes is removed along the Γ–K direction. The upper and lower bands correspond to the even-symmetric and odd-symmetric guided mode, respectively.

are directly drilled by FIB and the lattice constant is 430 nm. According to our simulation and experiment results, we find that the width of the propagation modes for the original Γ–M waveguide [Fig. 4(b)] is only 22 nm. Then we optimize the geometry to improve its transmission characteristics. We shrink the radius r_1 of the air holes in the two nearest-neighboring rows around the waveguide and enlarge the radius r_2 of the air holes in the two second-nearest-neighboring rows, as shown in Fig. 4(c). The key point is to generate a transport pathway with walls as smooth as possible. According to our simulations, the parameters corresponding to an optimized waveguide are that $r_1 = 50$ nm and $r_2 = 170$ nm, while the radius of the original air holes is $r_0 = 120$ nm. Figures 5(a) and (b) are the calculated dispersion relations of the original and optimized Γ–M waveguide, respectively. It's shown that the optimized waveguides have a high pass band that is much broader than the original waveguide. We can also obtain the same conclusion from the measured transmission spectra in Figs. 5(c) and (d). Besides, the intensities of the transmission spectra are much higher than the original one. As the Γ–M waveguide is perpendicular to the usual Γ–K waveguide, it offers an alternative to construct a waveguide interconnection beyond the usual scheme of Γ–K with Γ–K waveguides. A high-performance wide-band Γ–M waveguide should be of great help to build integrated-optical devices, such as interconnection networks, channel-drop filters, and wave division multiplexers, with more flexible geometrical configurations in 2D PC slabs.

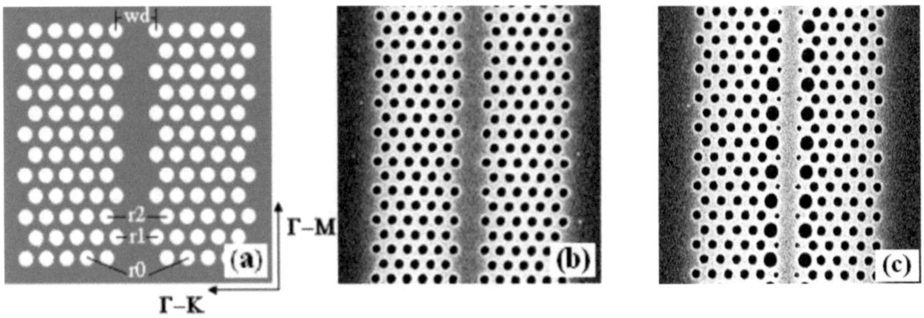

Figure 4. a) Schematic of Γ–M waveguide constructed in a triangular-lattice PC slab. The width of the waveguide w_d, as well as the radius of air holes in the first and second row r_1 and r_2, are the three crucial parameters to optimize the width of the transmission windows; (b) and (c) are SEM pictures of original and optimized Γ–M waveguides [37].

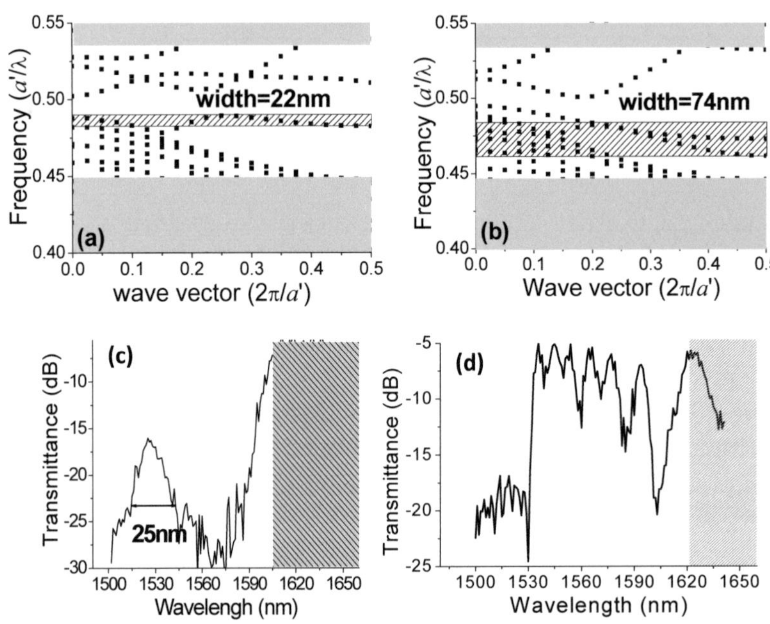

Figure 5. Calculated modal dispersion relation of (a) the original Γ–M waveguide and (b) an optimized Γ–M waveguide. The band width of the waveguide modes (within the dashed boxes) is 22 nm in the original waveguide, which has parameters: lattice constant a = 430 nm, hole radius r_0=r_1=r_2=120 nm, and waveguide width w_0=2a. After optimization by the following parameters as a = 430 nm, r_1 = 50 nm, r_2 = 170 nm, and w_d = 0.65w_0, the waveguide band width is significantly broadened to 74 nm; (c) and (d) are the corresponding measured transmission spectra of the original and optimized waveguides [37].

Based on the design of an optimized Γ–M waveguide, we combine the Γ–K waveguide and Γ –M waveguide together to form a 90° waveguide bend as schematically depicted in Fig. 6(a) [38]. This is the first design of a 90° waveguide bend in the 2D triangular-lattice PC. The whole waveguide bend system is composed of two Γ–K waveguides as the input and output ports and the center Γ−M waveguide as the connection component. In other words, the structure involves two consecutive segments of 90° waveguide bends. In constructing the waveguide bend, we have used the optimized Γ−M waveguide discussed in the above and depicted in Fig. 4(c). The guided modes in the optimized Γ−M waveguide have better phase-matching and mode profile matching with the normal Γ−K waveguides. In the same time, we modify the bend corner geometry by fabricating smaller air holes in the corner to make the guided modes between the two kinds of waveguides matching better. Based on a serial of simulation and experiment tests, we find the best values for r_1 and r_2. Figure 6(b) shows the SEM picture of the waveguide bend with r_1 = 50 nm and r_2 = 150 nm. After optimization, we get 70 nm pass band width while the transmission efficiency of a single bend is 45%. The proposed 90° waveguide bends can help to construct integrated optical circuits with more flexible and diversified infrastructures.

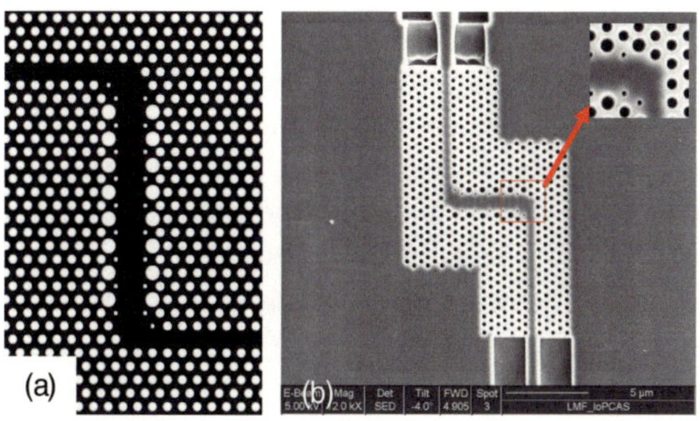

Figure 6. a) Schematic geometry of 90° waveguide bends in a triangular lattice PC slab with optimized Γ−M waveguide; (b) SEM picture of a practical sample of 90° waveguide bends with optimized bend corner geometry [38].

4.2. Coupled–cavity waveguide

Moreover, we have designed an air-bridged silicon PC coupled-cavity waveguides (PCCCWs) [39] and mapped its near-field optical distributions at different wavelengths around 1550 nm with the scanning near-field optical microscopy (SNOM) technology. For PCCCWs, the eigenmodes usually have relatively narrow bandwidth with slow group velocity in the whole band range. Previously, slow light propagation in such specific PCCCWs had still not been experimentally studied via SNOM technique. Figures 7(b)–7(f) show the calculated optical field distribution profiles at different wavelengths with a simulation model schematized in Fig. 7(a). We fabricated the PCCCW in the SOI wafers with FIB system. Figure 8(a) displays the SEM image of the element composed of the central PCCCW (encircled by a red square), two identical W1 PC waveguides, and the input/output ridge waveguides. Figure 8(b)–8(f) displays the near-field optical intensity distribution patterns of the PCCCW at different wavelengths. The scanning area is 12 μm × 15 μm with the incident light propagating upwards from the bottom of the image. Straight yellow lines in Figs. 8(b)–8(f) are used to label the position for showing the cross-sectional profiles of the field distribution patterns.

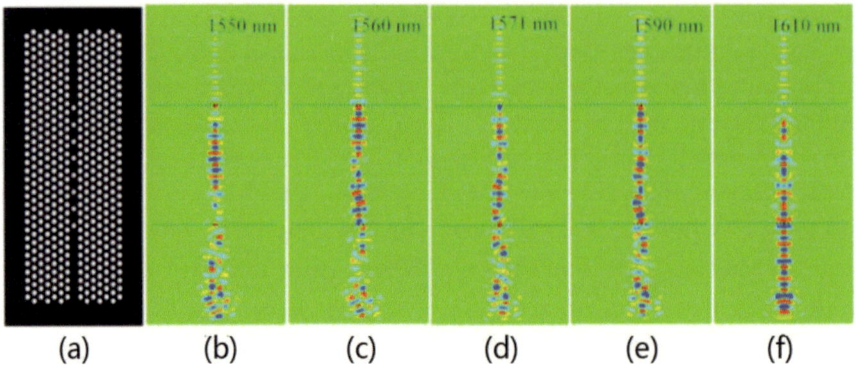

Figure 7. Simulation model (a) and calculated optical field distributions at (b) 1550 nm; (c) 1560 nm; (d) 1571 nm; (e) 1590 nm; and (f) 1610 nm [39].

The optical intensity distribution patterns are different at 1550 and 1610 nm, even though both of them mainly appear as a single narrow line along the central PCCCW region with a full width at half maximum (FWHM) of about 350 nm. Precisely speaking, the pattern demonstrates a little bit shoulder as a result of mode superposition at 1550 nm, since it comprises two eigenmodes. At 1610 nm, the calculated result consists well with the experimental one in Fig. 8(f), which presents a single line along the whole waveguide. The pattern appears bright and wide with obvious interference nodes in the PCCCW section. The simulated field distribution profiles in the W1 PC waveguide sections agree well with the detected ones at all these wavelengths, which show a snake-like/single-line profile in the input/output W1 PC waveguide except that of a snake-like profile in the output W1 PC waveguide at 1560 nm. In addition, the simulated field distribution patterns of the snake-like profile in the PCCCW section appear deviating greatly from the detected ones at 1550, 1571, and 1590 nm. However, if we calculate the optical field distribution patterns at 1550, 1560, and 1571 nm with the even-to-odd amplitude ratios of 1:4, 1:1, 1:4, and 1:6, respectively, we can find the simulated results are consistent with the experimental patterns evolving from single-line, to snake-like, and then to double-line structures for the PCCCW section. Combination of the near-field optical detection and theoretical simulation shows that SNOM is an efficient tool to study the optical propagation in the PCCCW and can help to design slow light elements.

4.3. High–*Q* cavity

Quantum information processing and quantum state manipulation have received great attentions because of their potential revolutionary impact on future network communication. Optical cavities, which can be used to store information, are considered to be one of the most important devices in the quantum communication application. The generation and teleportation of qubits require sufficiently high value of Q/V, where V stands for the mode volume of the optical cavities. As a result, high-Q optical cavities show great potential application in quantum information. Among all the optical cavities, 2D PC slab cavities are the best choice because of their simultaneous high-Q and small mode volume characteristics. It has been

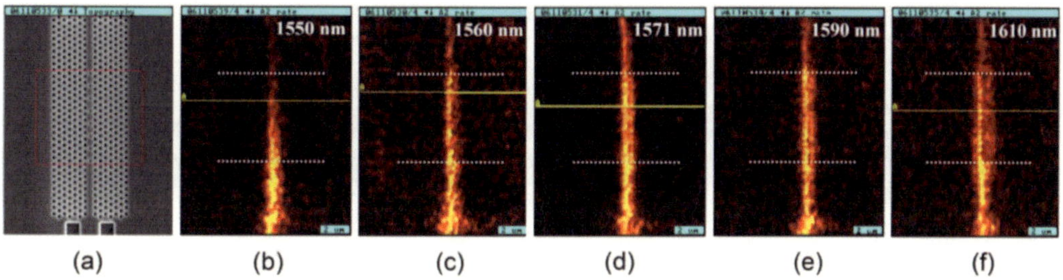

Figure 8. a) SEM topographic image, and the near-field optical intensity distributions at (b) 1550 nm; (c) 1560 nm; (d) 1571 nm; (e) 1590 nm; and (f) 1610 nm. The white dotted lines in each optical picture denote the interface between the W1 PC waveguide and PCCCW. All pictures were obtained for the same scanning area of 12 μm × 15 μm [39].

reported that Rabi splitting can be observed experimentally when the quantum dots are introduced into the PC cavities with high *Q*/*V* [40,41]. Moreover, due to the development of nanofabrication technique, multiple high-*Q* PC slab cavities can be fabricated at the same time on a single slab by the EBL and ICP etching technique. Once atoms or quantum dots are embedded into the high-*Q* PC cavities, various quantum phenomena can be demonstrated on chip. Recently, our works on high-*Q* silicon PC microcavities have achieved great progress after extensive exploration and delicate improvement of nanofabrication techniques and sample processing techniques have been made [42].

We focus on studying the L3 PC microcavities formed by removing three cylindrical air holes in the Γ–K direction in a triangular lattice [Fig. 9(a)]. The lattice constant is 430 nm, the radius of cylindrical air hole is 120 nm and the thickness of silicon slab is 235 nm. The FDTD calculation results indicate that the L3 PC microcavity possesses a quality factor of about 5300 [Fig. 9(c)]. After trying hundreds of simulations, we find that the positions of air holes at the edges of the microcavities affect the *Q* factor dramatically. The electric field pattern of the cavity mode can be tuned to be Gaussian-type by displacing the six air holes outwardly at the edges of the microcavities, and this can increase the quality factor significantly [41,42]. The optimal displacement is found to be 73, 10 and 73 nm for the first, second and third air holes at both edges of the microcavities, which is depicted in Fig. 9(b). The maximum quality factor of 127,323 [Fig. 9(d)] can be achieved, which is 20 times larger compared with the unadjusted one.

Based on the optimal parameters, we successfully fabricate the designed high-*Q* planar L3 PC microcavities in SOI wafer by implementing EBL and ICP [42]. As can be seen in Fig. 10(a), the L3 microcavity is side-coupled to a W1 waveguide with the barrier of three rows of air holes. The samples are measured by our home-made fiber coupling system as described in the above section. When the incident wavelength is off-resonant, light cannot couple with the microcavity, leading to strong output. While, at resonance most energy is tunneled into the microcavity, resulting in weak output. For the case of high-*Q* microcavity, a sharp transmission dip is expected in the transmission spectrum. The lattice constant, radius of cylindrical air hole and the thickness of silicon slab are 430, 120 and 235 nm, respectively. Limited by the fabrication accuracy of 10 nm, the displacement is adjusted to be 80, 20 and 80 nm for the six air holes at both edges of the microcavities. Figure 10(a) shows the enlarged

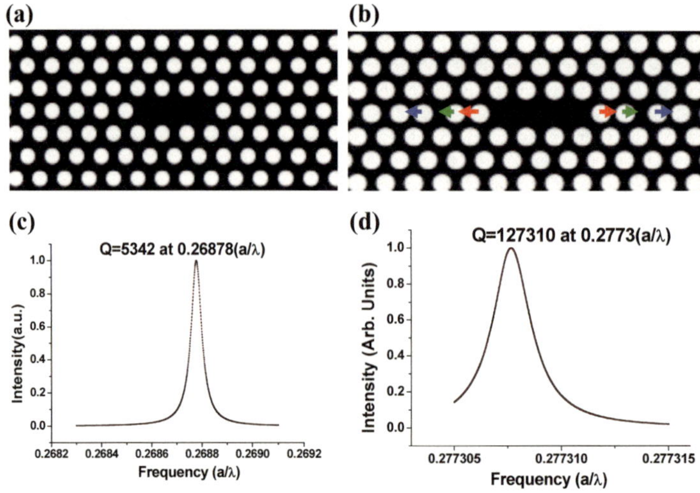

Figure 9. Schematics of (a) the original PC L3 nanocavity and (b) the optimized nanocavity; (c), (d) show radiation spectra of the original PC L3 nanocavity and the optimized nanocavity, respectively [42].

view of the cavity region. A sharp and narrow transmission dip is observed at the 1567.35 nm in the measured transmission spectrum [Fig. 10(c)]. For the purpose of extracting the quality factor accurately, we finely tune the wavelength between 1565 and 1570 nm. The measured spectrum is illustrated in Fig. 10(d) and the *Q*-factor as large as 71,243 is obtained. Nevertheless, there are some deviation between the simulation and experiment. For example, the resonant wavelength is 16.75 nm red-shifted from the simulated result and the maximum quality factor is significantly less than the calculated value of 127,323 [Fig. 10(b)]. We believe that the deviation is caused by the imperfection of the cylindrical air holes and the actual radius is not exactly the same as the value in simulation. The success of fabricating high-*Q* silicon PC slab microcavities enables us to investigate various interesting quantum phenomena, such as strong coupling between light and quantum system, quantum information processing technique, single photon source, all-solid quantum manipulation and high-quality biochemistry sensing devices.

4.4. Channel drop filters

Channel drop filters are key components for extraction of light trapped in a point-defect cavity to a neighboring waveguide and they sit on the basis of wave-division multiplexers and demultiplexers. They have great applications in a wide variety of fields, such as photonic integrated circuits, telecommunications, and quantum informatics. Based on the simulation and experiment experiences about PC waveguides, we design and fabricate an ultra compact three ports filter in 2D air-bridged silicon PC slab by closing the bus waveguide for 100% reflection feedback. Figure 11(a) shows the SEM picture of the three ports filter structure [43]. This filter was fabricated by EBL and ICP techniques. The lattice constant of the PC and the radius of the air hole are 430 and 145 nm, respectively. Port 3 is the input waveguide channel, while ports 1 and 2 are two output waveguide channels, respectively. They are formed by

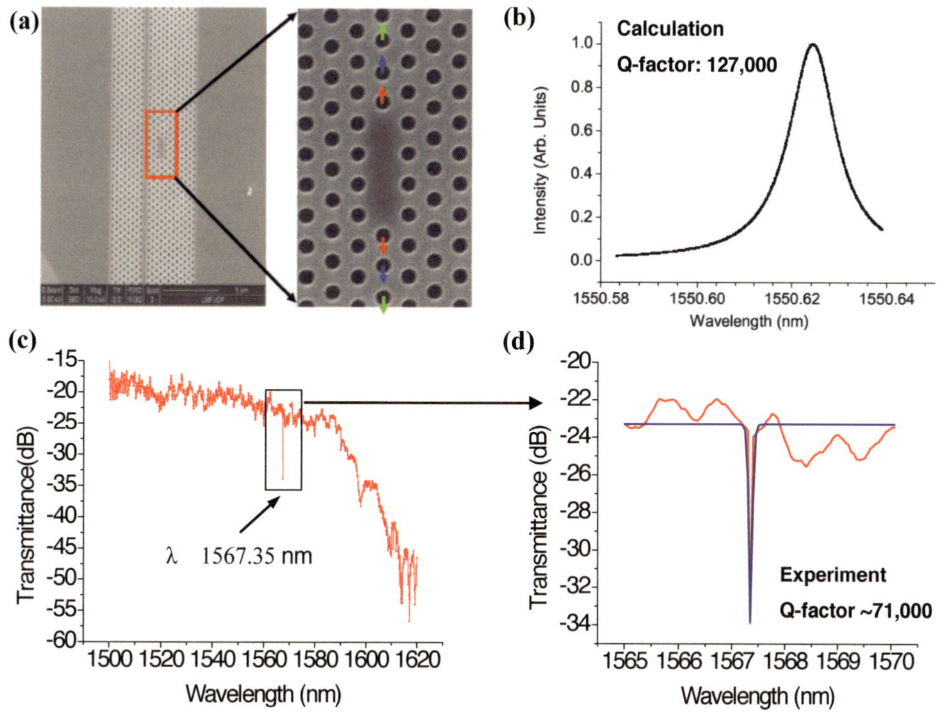

Figure 10. a) SEM pictures of one of the fabricated samples, including the L3 nanocavtiy with displaced air holes; (b) radiation spectra calculated by FDTD method; (c) and (d) show transmission spectra of one of the fabricated samples. The maximum *Q* value of up to 71000 is obtained [42].

missing one row of air holes along the Γ–K direction of the triangular-lattice PC, the so called W1 waveguide. C1 and C2 are two point-defect cavities. The distance between the center of the defect cavity and the neighboring waveguide is 3 rows of holes in the *y* direction. As seen in Fig. 11(a), the C1 cavity consists of three missing air holes, and the two air holes at the cavity edges are shifted outward by 10 nm apart from the regular positions. Similarly, those of the C2 cavity are shifted by 20 nm. The slight shift of air holes is conducive to confine light inside the cavity and leads to a higher quality factor. Meanwhile, the different shifts of the two cavities make the resonant wavelengths slightly different. The experiment results [Figs. 11(b) and (c)] show that the resonant wavelengths of C1 and C2 are 1529.5 and 1531 nm, respectively. The wavelength spacing of the two cavities is about 1.5 nm and might be further reduced by continuously changing the size of the cavity. The full widths at half maximum of the peaks are 1.5 and 1.4 nm and the corresponding quality factors are about 1020 and 1090, respectively. To estimate the drop efficiency, a reference straight wave-guide of the same parameters is positioned near the three-port filter. By keeping the same intensity of input light, the transmission intensities of the reference waveguide and port 1 are 0.330 and 0.158 μw, respectively, when the input wavelength is set at 1529.5 nm. The drop efficiency of port 1 is roughly estimated to be 48% and a similar result has been obtained at port 2.

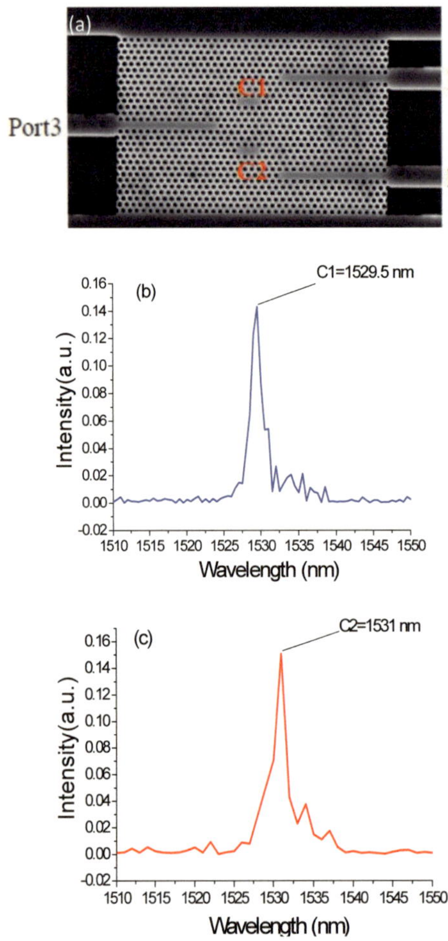

Figure 11. a) SEM image of the three-port filter; (b) and (c) the measured transmission spectra at ports 1 and 2, respectively [43].

It has been well known that structure is the kernel of filter design. Usually, the regulation of microcavity resonant frequency is obtained by changing the size of the cavities. We have proposed a new way to design multi-channel filters by changing the shape of the air holes [44]. When the shape of the air holes changes from circle to ellipse, two parameters, the ellipticity and the orientation angle of the ellipse, in addition to its size can be further explored and they can have a great influence on localized cavity modes. Therefore, we can use this for some special purpose.

http://dx.doi.org/10.5772/55041

Figure 12(a) schematically shows a one-channel PC filter. A horizontal channel (W1 waveguide) serves as the input signal channel, which is created by removing a single line of air holes along the Γ–K direction. A cavity is formed by removing three air holes along the Γ–K direction that is rotated 60° from the W1 waveguide. It is located four rows away from the major channel and is connected with the major channel through an indirect side coupling. Another single-mode waveguide is formed parallel to the cavity and serves as the output signal channel. Figure 12(b) shows an enlarged picture of the filter in the region around the cavity. One of its axes is oriented counterclockwise by an angle θ to the x axis, namely, the input light propagation direction. The sizes of the axes parallel and perpendicular to this orientation are a and b, respectively. Now we have great structural freedom to tune the optical properties of the new PC filter by changing the parameters of θ, a and b. To show this point, we design and fabricate a four-channel PC filter by using different cavity parameters as described in Table 1. The SEM picture of the fabricated four-channel filter is displayed in Fig. 13(a). Four cavities are located on the two sides of the central linear W1 waveguide. They are engineered by leaving several air holes unetched in the Γ–K orientation. The input signal propagates upwards from the bottom input ridge waveguide. Each cavity is coupled with another W1 waveguide that is connected to a ridge waveguide, which serves as the output signal channel.

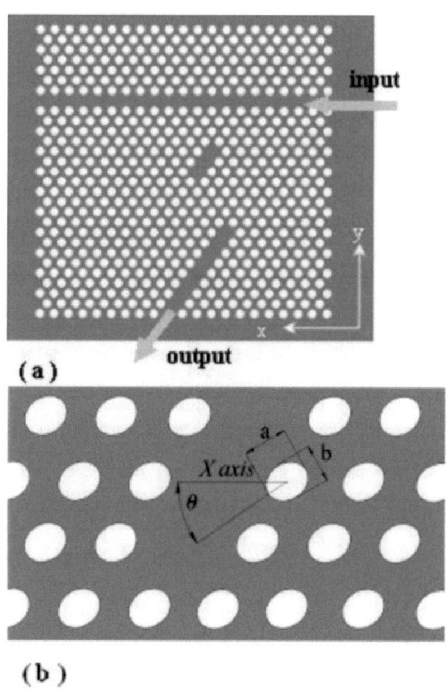

Figure 12. a) Schematic view of a one-channel PC filter, the major channel lies in the x direction, and the cavity and output side channel are parallel to the Γ–K direction of the triangular lattice; (b) enlarged view of the filter around the cavity. The air holes have a general elliptical shape with one of its axes oriented counterclockwise by an angle θ with respect to the x axis. The two axes are of size a and b, respectively [44].

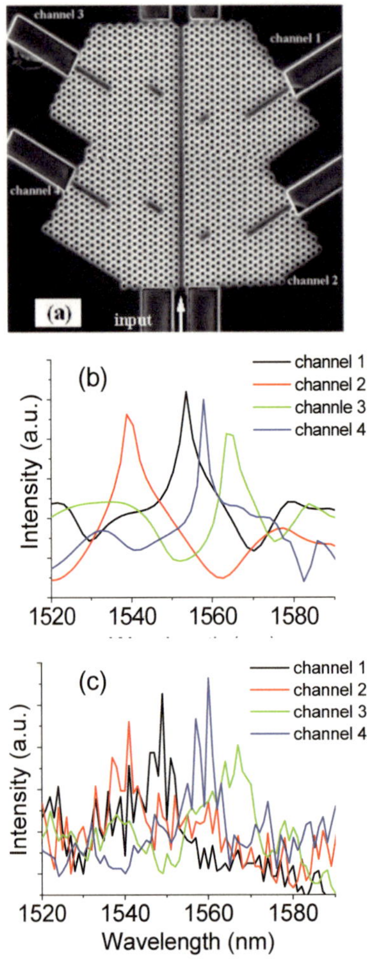

Figure 13. a) SEM image of the four-channel filter; (b) Simulation results of transmission spectra for the four-channel filter. (c) Experiment results of transmission spectra for the same filter [44].

channel	lattice constant /nm	number of missing air holes in cavities	long axis *a* /nm	shot axis *b* /nm	angle θ	theoretical resonant peak /nm	measured resonant peak /nm	deviation nm
1	420	2	240	200	0	1553	1549	4
2	430	2	260	240	0	1539	1541	2
3	420	3	240	220	0	1563	1567	4
4	430	3	280	240	0	1558	1560	2

Table 1. Structural parameters in the four-channel filter

The simulation and experimental results of the transmission spectra for the four channels are displayed in Figs. 13(b) and (c). Although significant noise exists, a resonant peak can be clearly found for each channel. The peaks are located at 1549, 1541, 1567, and 1560 nm for channels 1, 2, 3, and 4, respectively. The results confirm that the air-hole shape has a great influence on the functionality of the PC filter devices. The elliptical air holes can induce a fine tuning of the resonant wavelength by changing the ellipticity of the elliptical air holes.

As described in the preceding section, a high-performance wide-band Γ−M waveguide can be formed by modifying the radii of the air holes along the pathway. The combination of Γ−M waveguides and Γ−K waveguides can offer a more flexible way to interconnect and couple between different devices. We have proposed a type of PC filter using these two kinds of waveguides [45]. The Γ−M waveguide and Γ−K waveguide are used as the input and output signal channels respectively, and they are connected via cavity resonance. Figure 14(a) shows the SEM picture of a four-channel filter structure. We change the size of the cavities by moving the end points of the cavity [marked with "a, b" and "c, d" as shown in the inset of Fig. 14(b)] to change the resonant frequency. Table 2 gives the detailed parameters of the four cavities. The experiment demonstrates that the four resonant peaks are at the wavelengths of 1543, 1545, 1548 and 1551 nm, as shown in Fig. 14(b). In spite of the slight shift in the resonant peak toward higher frequency, which we believe is induced by the uncertainties in the fabrication, the experimental results are in fairly good agreement with the simulation results, where the maximum relative deviation of resonant wavelength is within 2 nm. These results clearly demonstrate that the designed Γ−M waveguides can act together with the usual Γ−K waveguide to construct high-performance multichannel filers with more structural flexibility. In our experiment, we also use the CCD camera to directly monitor the transport of infrared signal within the channel-drop filter. The situation of on-resonance and off-resonance can be clearly visualized and distinguished from the CCD camera images. One typical case is shown in Fig. 14(c).

5. PC band–engineering devices for anomalous transport control

In previous sections, we discuss several PC devices, including waveguides, cavities, and channel-drop filters that are built on the silicon 2D PC platform. These devices work on defects that are brought into PBG and they can be considered as PBG materials. As we have mentioned, PC structures possess another important feature: photonic pass bands. In this section we show several example devices that implement the dispersion and refraction properties of PCs at their transmission bands.

Let's first make an overview of how band dispersion engineering works. Figure 15(a) shows the typical photonic bands structure of an air-bridge PC slab structure composed of a square-lattice array of air holes etched in silicon slab. The areas labeled in Fig. 15(a) show the unconventional light propagation, self-collimation and negative refraction, in PC. One effective way to understand and exploit desirable light propagation properties in PC is using the equifrequency surface (EFS) contours, as shown in Figs. 15(b) and 15(c). Figure 15(b) shows the EFS contours of the first TE-like band. The EFS contours in the red line frame are flat, meaning that

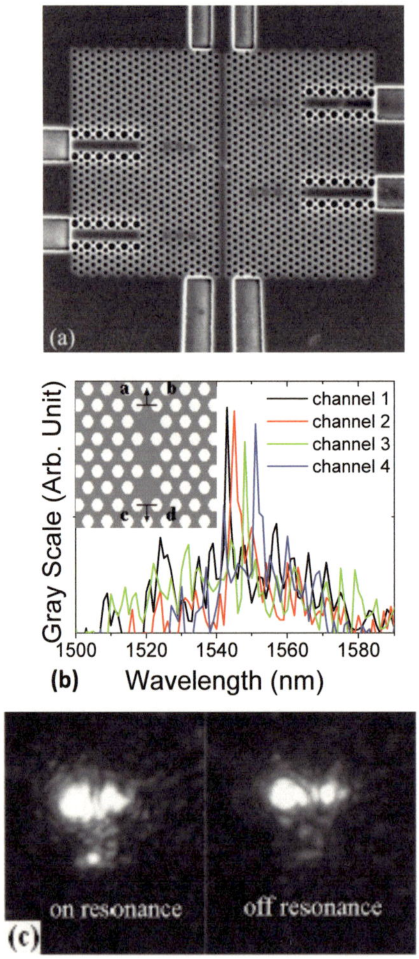

Figure 14. a) SEM image of the fabricated four-channel filter. Four cavities are located on the two sides of the input waveguide; (b) experimental transmission spectra of the four channel filter in linear scale. The inset picture illustrates two groups of end points (air-hole centers) of the cavity marked with "a, b" and "c, d." Black arrows indicate the moving direction of these air holes; (c) infrared CCD camera imaging of the output signal observed in experiment for one channel of the sample. A bright spot appears at the end of the output channel when the input wavelength coincides with the resonant wavelength and disappears when it is at off-resonance [45].

this is the self-collimation region. The reason is that the group velocity, which is parallel to the gradient of the EFS, is pointing in the same direction for all the modes located within the region. As a result, if light propagates along the Γ–M direction, it does not suffer any diffraction in the PC. $\Gamma=(0,\ 0)(\pi\ /\ a)$ and $M=(1,\ 1)(\pi\ /\ a)$ are high-symmetric points in the first Brillouin zone for square lattice. This kind of PC structures can be used as the channelless waveguide in integrated optic devices. Figure 15(c) shows EFS contours of the second band. The EFS contours are roughly circular around the direction at the reduced frequency $(a\ /\ \lambda)$ range 0.28–0.31, as indicated in the red solid line frame. When the frequency increases, the EFS contours move toward the Γ point, which indicates the existence of negative refraction in the region. If the

channel	end points moving distance /nm	theoretical resonant peak /nm	measured resonant peak /nm	theoretical distance from channel 1 /nm	experimental distance from channel 1 /nm	deviation /nm
1	0	1550	1543	-	-	-
2	5	1551.5	1545	1.5	2	0.5
3	10	1553	1548	3	5	2
4	15	1556	1551	6	8	2

Table 2. Structural parameters in the four channel Γ–M and Γ–M waveguides filter

incident light is mainly parallel to the $\Gamma-M$ direction, the PC will behave like an isotropic medium with a negative index of refraction in that particular frequency range.

5.1. Negative refraction

Based on the above analysis, we designed and fabricated an air-bridged PC structure that exhibited negative refraction of infrared light [46]. The structure is schematically shown in Fig. 16(a). These structures are directly drilled by FIB technique. The input infrared signal channel is a silicon wire waveguide, which is inclined with respect to the surface normal by 10°. The lattice constant *a* of the square array is 460 nm and the diameter of the air hole is 220 nm. We first use 3D FDTD method to simulate the electromagnetic field intensity distribution at wavelength 1503 nm. The result is displayed in Fig. 16(b). We find strong reflection and scattering at the interface between the input waveguide and the PC structure. This is induced by the serious impedance mismatch at the interface, although the high index contrast air-bridged structure can achieve good optical confinement. To surpass this obstacle, we use a tapered air-holes connection layer at the input surface of PC structure to reduce the reflection and scattering losses. As shown in Fig. 16(b), a large fraction of light power from the input waveguide is coupled into the PC structure and negative refraction of light beam within the PC structure is clearly seen. Besides, the reflection or scattering of light at the input interface of the PC is very much reduced. This clearly indicates that the designed tapered interface can reduce the interface impedance mismatch remarkably. The calculated value of negative refraction angle is –45°.

In our measurement, TE-polarized light from a tunable semiconductor laser (1500–1640 nm) was first launched into a tapered single mode fiber, and then coupled to the silicon wire waveguide, and finally incident on the PC structure. The ordinary way to see the light propagation behavior is to directly observe the pattern of the radiated light from the top of the sample using a conventional microscopy objective and an infrared CCD camera. The result is shown in Fig. 16(c). The light spot at the middle bottom part of the pattern is the radiated light from the input silicon wire waveguide. The big light spot at the center represents the scattered light at the interface between the input wire waveguide and the PC due to impedance mismatch. There is also a small bright spot at the top right corner of the pattern, and it is

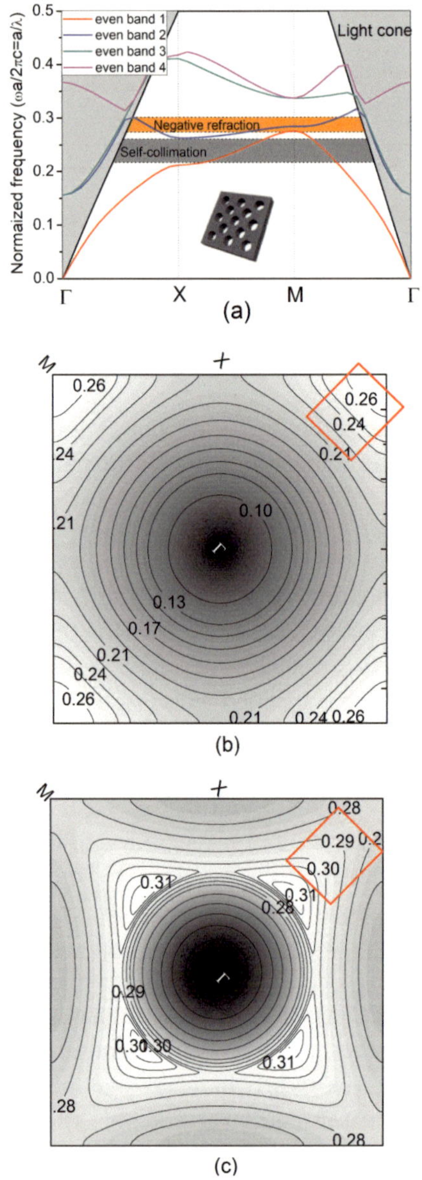

Figure 15. a) Photonic band structures of TE-like bands for an air-holes square-lattice PC slab; (b) EFS contours of the TE-like first band for the same PC show that self-collimation can occur in the direction around the Γ–M direction; (c) EFS contours of the TE-like second band show that negative refraction can occur in the direction around the Γ–M direction [46].

recognized to result from the radiated light when the negative refraction beam hits the end facet of the PC structure. Because the TE-like modes are strongly confined guided mode on the silicon slab and the surface fields are nonradiative and evanescent with respect to the vertical direction of the PC slab, the far-field pattern observed and recorded by the ordinary optical microscopy is not able to reveal the detailed process about how the negative refraction beam propagates inside the PC structure unless the scattering of light by roughness and irregularity is sufficiently strong on the beam propagation path. This is indeed the case for Fig. 16(c). In fact, the small bright spot could not appear without the intentional introduction of the air slot at the far end of the PC structure. It would not be possible to tell which way the infrared beam would refract if without the aid of this scattering light spot.

In order to observe clearly and tell unambiguously the ray trace of the negative refraction beam in the PC structure, we used the SNOM technology (SNOM-100 Nanonics, Israel). A probe scans in the vicinity of the surface of the PC structure and records the near-field intensity distribution. The tip has a resolution of about 100 nm, i.e., 1/15 of the wavelength. The signal is recorded by an infrared single-photon detector, which allows us to capture very weak infrared signals. The probed near field information directly reflects light propagation properties of the TE-like modes for the PC and enables one to visualize the ray trace of the negative refraction light beam because the near field at the surface is an integral part of the modal profile of the confined guided modes that exponentially decay away from the surface of the slab. In the SNOM picture [Fig. 16(d)], a bright spot also appears at the front interface of the PC structure, but it is much smaller than the one in Fig. 16(c). The ray trace of the incident light beam along the silicon wire waveguide and its propagation along the negative refraction direction inside the PC structure can be clearly seen. The negative refraction angle is about −45°, which is in good agreement with the FDTD simulation presented in Fig. 16(b). The SNOM detection unambiguously discloses the negative refraction property of the designed PC.

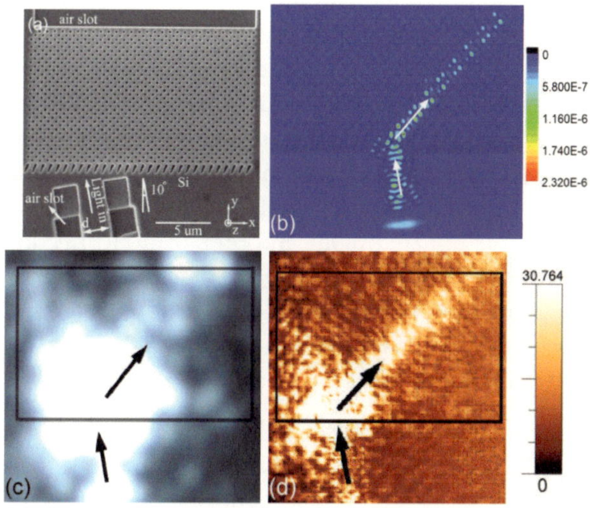

Figure 16. a) SEM picture of the PC structure and an input waveguide. The width of the waveguide *d* is 2 μm; (b) Light intensity distribution of TE-like modes for PC with deliberately designed tapered air-holes interface; (c) Directly observed pattern of the radiated light of $\lambda = 1500$ nm from the top using an objective lens; (d) SNOM picture of the negative refraction of the same wavelength. In each picture, the boundary of the PC structure is superimposed as solid lines [46].

On the other hand, ordinary positive refraction only occurs for TM-like confined modes, so the designed PC structure can behave as an efficient beam splitter in an integrated optical circuit. The high-resolution SNOM technology can greatly help one to directly visualize the ray trace and acquire deeper understanding on various anomalous wave propagation behaviors, such as super-prism, superlensing, self-collimation, and slow light in deliberately designed 2D PC slab structures in the optical wavelengths. This in turn can help to explore a wider regime of controlling light behaviors on the nanoscale for future basic science and high technology applications.

5.2. Self–collimation effect

Self-collimation effect is the propagation of light without diffraction along the propagation direction. This phenomenon has been used to construct non-channel waveguides, beam splitters and beam combiners [47,48]. The behaviors of these devices are determined by the performance of the self-collimation effect. Recently we have designed and realized a simple structure composed by a square lattice array of elliptical air-holes where broadband large-angle self-collimation effect is observed for TE-like guided modes in infrared wavelength [49].

Figure 17(a) shows our PC structure formed by a square lattice of elliptical holes. The calculated TE mode photonic band diagram of the fourth, fifth and sixth bands are shown in Fig. 17(b). The self-collimation effect can be observed at the gray regions within a broad normalized frequency range 0.36−0.39 and 0.43−0.46. For simplicity, we only consider the EFS contours of the fifth TE band [Fig. 17(c)]. The contours are flat at the normalized frequency between 0.43 and 0.46 for all the values of k_y and not just in the vicinity of $k_y=0$. This feature indicates that our structure can support self-collimation for incident light beams with large incident angles. Then the FDTD simulation by using the MEEP package is performed to verify our prediction. A structure with the size of 30 $a\times$ 45 a is considered. A Gaussian beam with a width of 4a propagates into the surface at 0°, 20° and 60° incident angles (Fig. 18). For simplicity, we only consider the minimum (0.36) and the maximum normalized frequency (0.43). Figure 18 shows the electric field intensity distribution in the xy plane with 0°, 20° and 60° incident angles at normalized frequency 0.36 [Fig. 18(a)] and 0.46 [Fig. 18(b)]. From the simulation results, we find that the light beam is collimated along the propagation direction for each situation. However, the couple efficiency of the incident light becomes lower and lower with the increase of the incident angle. We do not show the field distributions for those incident angles that are larger than 60°. These six situations in Fig. 18 are sufficient to show the broadband large-angle characteristic of the self-collimation effect.

Following our simulation results, we fabricate our PC structures in SOI substrate by EBL and ICP etching process. Figure 19 shows the SEM pictures of the designed PC structures with 0°, 20° or 60° incident waveguides. Ray trace of light beam is observed using IR camera and a high numerical aperture (NA = 0.50) objective. Detailed images of the field intensity of the scattered light are recorded for 0°, 20° and 60° incident angles for different incident wavelengths. Here we only show the patterns of the minimum and maximum wavelengths for each incident angle. They demonstrate strong light confinement along the propagation direction for all the situations. The experimental results are in good agreement with FDTD

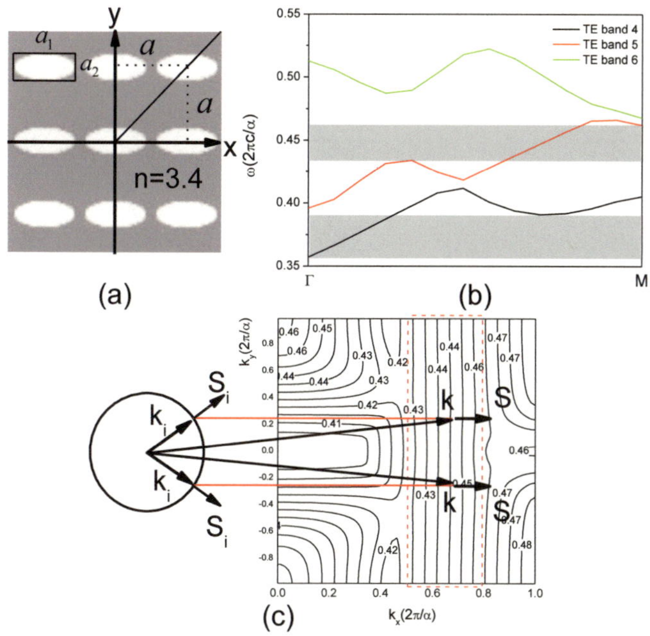

Figure 17. a) Schematic of the PC structure formed by a square lattice of elliptical holes; (b) band diagram of the fourth, fifth and sixth TE bands; (c) EFS contours of the fifth TE band [49].

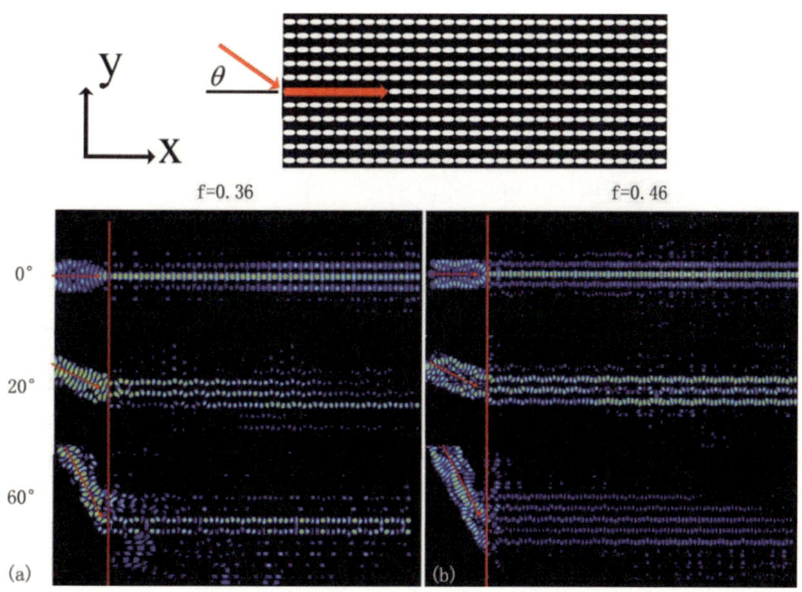

Figure 18. Electric field intensity distribution with 0°, 20° and 60° incident angles at the minimum normalized frequency 0.36 (a) and the maximum 0.46 (b). A FDTD method is used in the simulations [49].

simulations. We believe that this kind of structure may have potential applications in beam combiners and multiplexers.

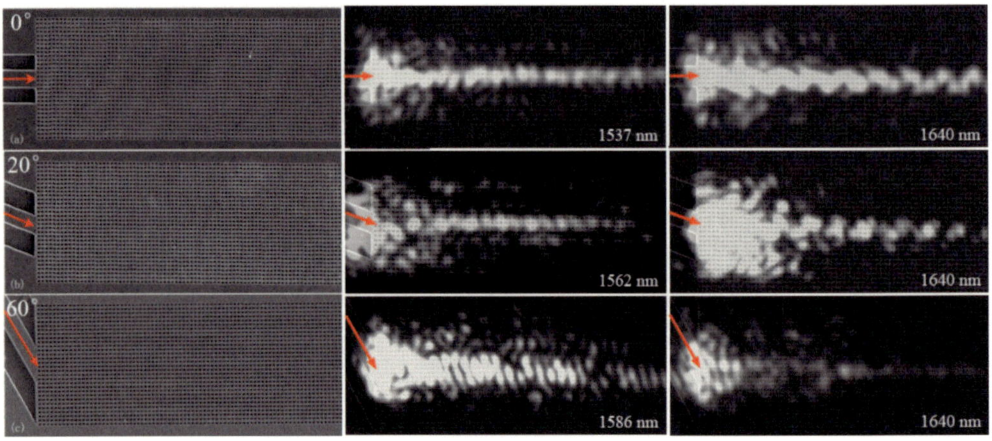

Figure 19. Left panels: SEM pictures of designed PC structures with 0° (a), 20° (b) and 60° (c) incident waveguide. Middle and right panels: Ray trace of light beam observed using IR camera and a high numerical aperture (NA = 0.50) objective. The patterns of the minimum and maximum wavelengths are shown for each incident angle [49].

6. On–chip wavelength–scale optical diode and isolator

Optical isolation is a long pursued object with fundamental difficulty in integrated photonics. The need to overcome this difficulty is becoming increasingly urgent with the emergence of silicon nano-photonics, which promises to create on-chip large-scale integrated optical systems. Motivated by the one-way effect, considerable effort has been dedicated to the study of unidirectional nonreciprocal transmission of electromagnetic waves, showing important promise in optical communications. Until now, on-chip integration of optical diode still stays in theory, particularly in silicon. These "optical diodes" include fluorescent dyes with a concentration gradient, absorbing multilayer systems, and second harmonic generators with a spatially varying wave vector mismatch. An electro-tunable optical isolator based on liquid-crystal heterojunctions, showing nonreciprocal transmission of circularly polarized light in photonic bandgap regions, has been reported. In another configuration using liquid crystals, linearly polarized light is used. In addition to many attempts on magneto-optical materials, optical isolators have also been fabricated using nonlinear optical processes and electro-absorption modulators.

An efficient routine to create optical diode is via time-reversal symmetry breaking or spatial inversion symmetry breaking [50], which could lead to the optical isolation in any device where the forward and backward transmissivity of light is very much different. We have reported a method for making unidirectional on-chip optical diodes based on the directional bandgap difference of two 2D square-lattice photonic crystals comprising a heterojunction structure and the break of the spatial inversion symmetry. Simulations confirm the existence of a clear isolation effect in the designed heterojunction structure. We fabricate these on-chip optical diodes in silicon and the near-infrared experiment results show high-performance optical isolation, in good agreement with the theoretical prediction [51]. This device may play the same basic role in photonic circuits as the electrical diode does in electronic circuits. It could further pave the way to achieve on-chip optical logical devices without nonlinearity or magnetism and bloom the photonic network integration.

Figure 20(a) shows the schematic configuration of the original diode structure under study, which consists of two PC slab domains (PC1 and PC2) with the same lattice constant a but different air hole radii (r_1 and r_2, respectively) comprising a heterojunction structure. These two PC regions stand at a silicon slab [grey area in Fig. 20(a)]. Each PC region has a square-lattice pattern of air holes [white holes in Fig. 20(a)], with the hetero-interface between PC1 and PC2 along the Γ-M direction. We set the two hole radii as a fixed ratio to the lattice constant a, which are r_1=0.24a and r_2=0.36a in order to simplify our discussion. These two composite PCs would comprise a pure PC region if r_1=r_2. The light source is placed symmetrically aside the structure with two 4a-wide ridge waveguides connecting the surface of the two PC regions. The whole area is surrounded by a perfectly matched layer.

We simulated the transmission spectra for a TE-like light signal transporting along the forward (from left to right) and backward (from right to left) direction. The refractive index of the dielectric slab was set to 3.4, corresponding to that of silicon at 1,550 nm. The slab thickness was h=0.5a. Figure 20(b) shows the calculated forward (black line) and backward (red line) transmission spectra. The frequency is normalized by a/λ. It is clearly seen that there exists an isolation band ranging from 0.2649 to 0.2958 (a/λ), where the forward transmission forms a peak with a transmissivity of about 6% while the backward transmissivity is down between 0.5% and 1%. The forward peak is located at 0.2793 (a/λ), just in the middle of the isolation band. We define $S=(T_F-T_B)/(T_F+T_B)$ as the signal contrast of the diode, where T_F and T_B denote the forward and backward transmissivity, respectively. The maximum S of this original diode equals 0.846 at the peak. Besides, there exists another isolation region from 0.2196 to 0.2649 (a/λ), where the backward transmissivity is higher than the forward transmissivity. This structure thus shows an extraordinary phenomenon of unidirectional transport property.

We calculated the band diagram of the TE-like modes of these two PC slabs using the 3D-FDTD method. Figures 21(a) and 21(b) show the calculation results. The first band (even mode) in bulk PC2 [Fig. 2(b)] is directional as the top mode frequency in the Γ-X direction (x-axis) is 0.2345 (a/λ) but that in the Γ-M direction (45°-direction) is 0.3087 (a/λ). Inside the region between 0.2345 (a/λ) and 0.3087 (a/λ), the all-directional transparent region of PC1 needs to be above 0.2633 (a/λ) [Fig. 21(a)] in order to match the bottom mode frequency in the Γ-X direction of the second band (odd mode). These two modes in PCs are the basic working mode of the diode structure. Here the even and odd modes are defined with respect to the off-slab mirror-reflection symmetry σ_z of the field component E_y. In the region between 0.2633 (a/λ) and 0.3087 (a/λ) PC1 is transparent in all directions, while PC2 is transparent along the Γ-M direction but opaque along the Γ-X direction. Compared with Fig. 20(b), the unidirectional transport region [0.2649 to 0.2958 (a/λ)] just coincides with the overlapped region between the directional bandgap of PC2 and the all-directional pass band of PC1. This simple picture indicates that

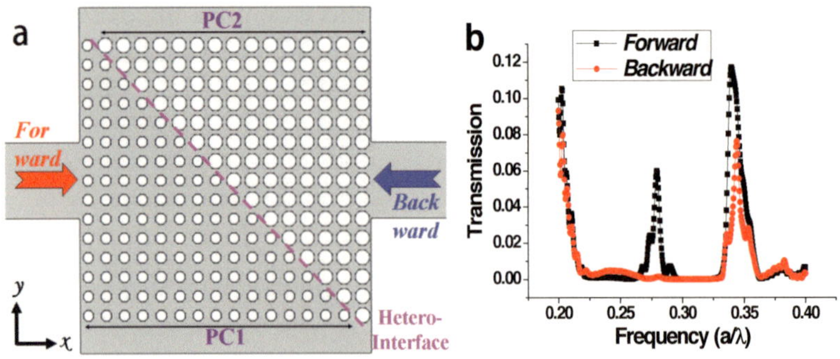

Figure 20. a) Schematic geometry of an original heterojunction optical diode formed by the interface (normal to the Γ-M direction) between two PC slabs (denoted as PC1 and PC2) with different hole radii (r_1 and r_2, respectively). (b) Simulated transmission spectra of the diode in the forward direction (the black line) and the backward direction (the red line). An input and output ridge waveguide has been used in the 3D-FDTD calculation. [51].

the current unidirectional transport effect involves two ingredients: (I) directional bandgap of PC2 and (II) all-directional pass band of PC1. Noting that the structure does not obey the spatial inversion symmetry alone the propagating direction, the principle of optical isolation can thus be summarized as follows:

1. Forward. When light goes across PC1 as the odd mode and reaches the hetero-junction along the Γ-X direction, it cannot stay in the Γ-X direction in PC2 further because of the Γ-X directional gap. But the hetero-junction is along the Γ-M direction, so light turns to the hetero-junction and diffracts as the even mode at any Γ-M direction into PC2, which passes through PC2 and eventually outputs.

2. Backward. When light goes directly into PC2 as the even mode, it turns to the two Γ-M direction paths which cannot convert to the odd mode of PC1 in the Γ-X direction and eventually leak out so that it does not output.

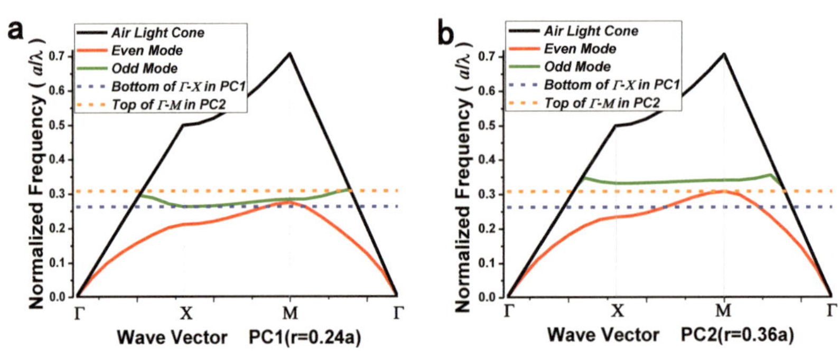

Figure 21. a) Calculated modal dispersion curve for PC1 (r=0.24*a*). (b) Calculated modal dispersion curve for PC2 (*r*=0.36*a*), in which the black line is the air light cone. The red curve denotes the first even mode, while the green curve denotes the second odd mode. The blue dashed line denotes the bottom frequency of the Γ-X directional odd mode [0.2633 (*a*/λ)] of PC1 and the orange dashed line denotes the top frequency of the Γ-M directional even mode of PC2 [0.3087 (*a*/λ)] [51].

http://dx.doi.org/10.5772/55041

Based on the numerical analysis of the optical diode, we have fabricated the original diode structure as well as a revised diode structure (with better performance) in silicon, whose SEM pictures are displayed in Fig. 22(a) and (c), respectively. Figures 22(b) and 22(d) show the theoretical and experimental results of the transmission spectra of the two diodes in the forward and backward directions. In Fig. 22(b) the theoretical forward peak of the original diode structure is at 1,575 nm [0.2793 (a/λ)] and the maximum transmissivity is 6%. The experimental forward peak is at 1,556 nm and the maximum transmissivity is 7%. In Fig. 22(d) the experimental forward peak is at 1,534 nm and the maximum transmissivity is 10% for the revised diode, whereas the theoretical forward peak is at 1,552 nm [0.2834 (a/λ)] with a transmissivity of 13%. The measured signal contrast S equals 0.718 (the original structure) and 0.831 (the revised structure) at the peak frequency. Both experimental peaks in Figs. 22(b) and 22(d) have a nearly 20 nm shift and 50 nm broadening against the theoretical simulations, which is probably due to the imperfections in fabrication. The experiment confirms the existence of the unidirectional transport effect in agreement with the theoretical prediction. Due to the arbitrariness of the lattice constant a, we can freely adjust the working frequency to anywhere as desired. This could be more convenient for the design of realistic photonic devices.

The principle for optical diode as analyzed in the above is robust as it is based on a simple directional bandgap mismatch effect of photonic crystal heterojunction. Yet, it should be noticed that in Figs. 22(b) and 22(d) the backward transmissions are fluctuating within 1% to 2% and both are higher than the simulation values, as a result, the signal contrast S degrades from 0.846 for the original structure and 0.92 for the revised structure in theory to 0.718 for the original structure and 0.831 for the revised structure in experiment. The performance improvement of the diode relies on how to maximize the peak of the forward transmissivity and minimize the backward transmissivity in experiment. Several means can help improve the forward transmissivity. First, one can change the air hole size of PC1 and PC2 and enlarge the directional bandgap. Calculations show that the forward peak transmissivity of the revised structure with $r_1 = 0.30a$ and $r_2 = 0.45a$ grows up dramatically to 29.4% while maintaining the same low level of backward transmissivity. Second, one can change the relative size of the input and output waveguides. Calculations show that by changing the input waveguide width to $2a$ and keeping the output waveguide width $4a$, the forward peak signal increases up to 20.8% in transmissivity.

To reduce the backward transmissivity, one can either enlarge the directional bandgap of PC2 to attenuate the backward signal more strongly, or eliminate the return of leak-out light from the outside of slab or the structure boundary by introducing the absorbing metal dots near the structure, or enlarge the heterojunction structure appropriately so that the leak-out light cannot enter the output waveguide. Following the above general ideas, we further optimize the optical diode structures as illustrated in Fig. 23. The structure has parameters of r_1=100 nm and r_2=160 nm, and the input and output waveguide width are $2a$ and $6a$, respectively. In experiment, we have got an optical diode with an maximum of 32.8% of the forward peak transmissivity and 0.885 of the signal contrast S at 1,557 nm.

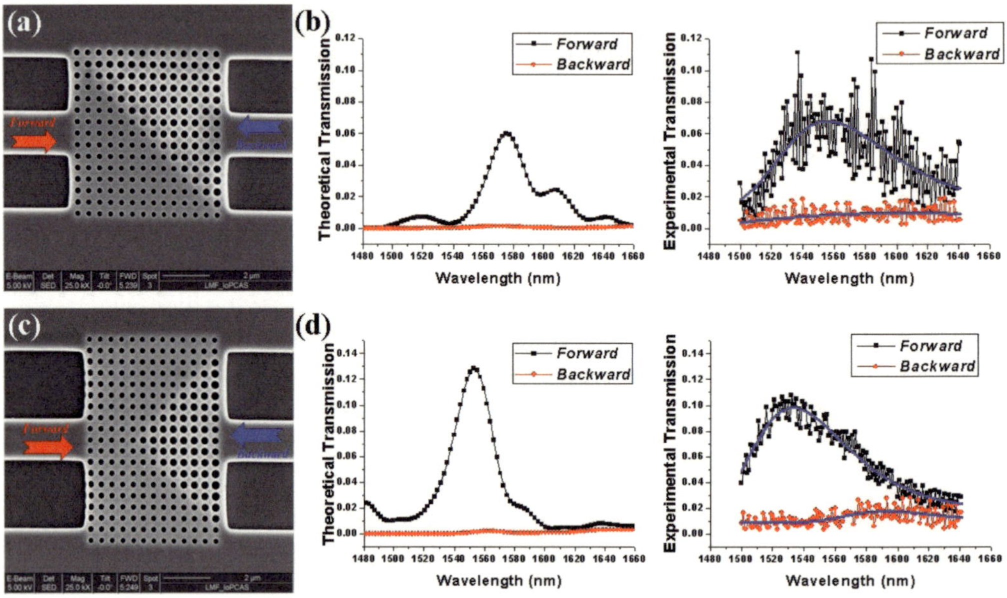

Figure 22. a) Scanning electron microscope images of the original optical diode structures. (b) Theoretical (left)and experimental (right) transmission spectra of the original diode structure. (c) Scanning electron microscope images of the revised optical diode structures. (d) Theoretical (left) and experimental (right) transmission spectra of the revised diode structure [51].

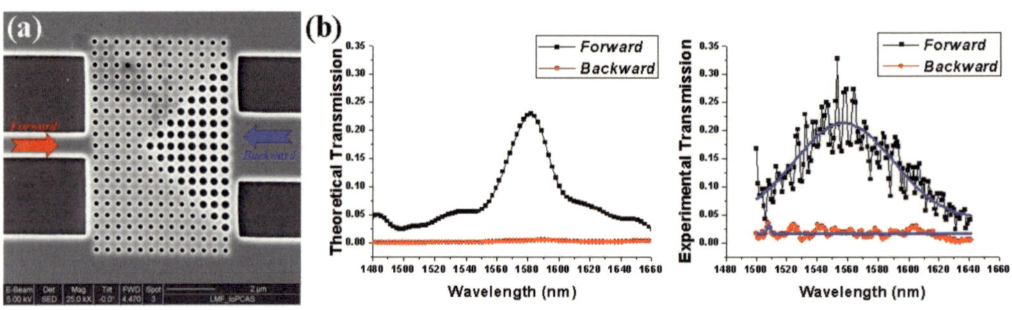

Figure 23. a) Scanning electron microscope images of the optimized optical diode system. (b) Theoretical (left) and experimental (right) transmission spectra of the optimized diode structure [51].

Our PC heterojunction diode has advantages of high signal contrast, wavelength-scale small sizes, and being all-dielectric, linear, and passive. Furthermore, it has a much smaller scale than those based on diffraction gratings and thus greatly facilitates large-scale integration. The high performance on-chip optical diode realized in silicon without nonlinearity or magnetism will stimulate the exploration of other more complex on-chip optical logical devices with ultra-high stability, integration and much less power consumption. Such an optical diode may play the same basic role in photonic circuits as the electrical diodes do in electronic circuits, which have significantly revolutionized fundamental science and advanced technology in various aspects of our routine life due to their capability of rectification of current flux. Furthermore, its large-scale fabrication could be readily achieved by the well-developed CMOS techniques. The realization of high-performance on-chip optical diodes may open up a road toward photonic computers.

Strictly speaking, the existence of unidirectional transport effect of light does not mean automatic achievement of optical isolation. Recently there have appeared hot controversies upon whether isolation of light can be realized via linear and passive photonic structures. Several schemes to realize unidirectional transport of light through linear and passive photonic structures have been proposed, which are essentially based on the principle of spatial-inversion symmetry breaking. Feng *et al.* reports a passive silicon optical diode based on one-way guided mode conversion [52]. However, whether or not nonreciprocal transport of light can happen in the structure has raised hot controversies [53,54]. Fan *et al.* made a scattering matrix analysis for relevant forward and backward modes of the structure and argued that the structure is essentially reciprocal and cannot enable optical isolation because it possesses a symmetric coupling scattering matrix. In their response, Feng *et al.* acknowledge that their structure, as a one-way mode converter with asymmetric mode conversion, is Lorentz reciprocal, which states that the relationship between an oscillating current and the resulting electric field is unchanged if one interchanges the points where the current is placed and where the field is measured, and on its own cannot be used as the basis of an optical isolator. The controversies have thus raised a fundamental question: Can one construct an optical isolator by using a linear and time-independent optical system? The answer to this question by the authors of Ref. [53,54] obviously is no.

But our theoretical and experimental study on the optical isolation performance of our PC heterojunction diode leads to a totally different answer to the above question, namely, the spatial inversion symmetry breaking diode can construct an optical isolator in no conflict with any reciprocal principle [55]. To see whether there is a good isolation effect of the silicon diode, we implement a direct method that is originated from the conventional magneto-optical isolator that has been popularly used in laser devices [Fig. 24(a)]. One places a total reflection mirror after the output port in the forward direction of the isolator device and monitor the reflection signal from the input port. This reflection signal well describes and measures the round-trip transmissivity of light across the isolator device. If the reflection signal is the same as or is comparable with the forward signal, then the structure does not have the desired isolation property. In contrast, if the reflection signal is much smaller than the forward signal, then a good isolation property is implied.

An equivalent way to investigate the optical isolation performance of the diode structure is to adopt a doubled-diode structure with a mirror-symmetrical plane at the forward direction output port of the diode, as depicted in Fig. 25. Obviously this method has set all the forward output signals as the backward input signal of the diode and thus can directly test the isolation property of the diode structure. By implementing this method, we calculate simultaneously the forward transmissivity and the round-trip transmissivity of the diode structure. Comparison of these two quantities would directly measure their isolation properties. Figure 25(a) is the schematic geometry of the doubled-diode structure corresponding to the diode depicted in Fig. 24(b). The parameters of the diode are the same as in Fig. 23. The width of the input and

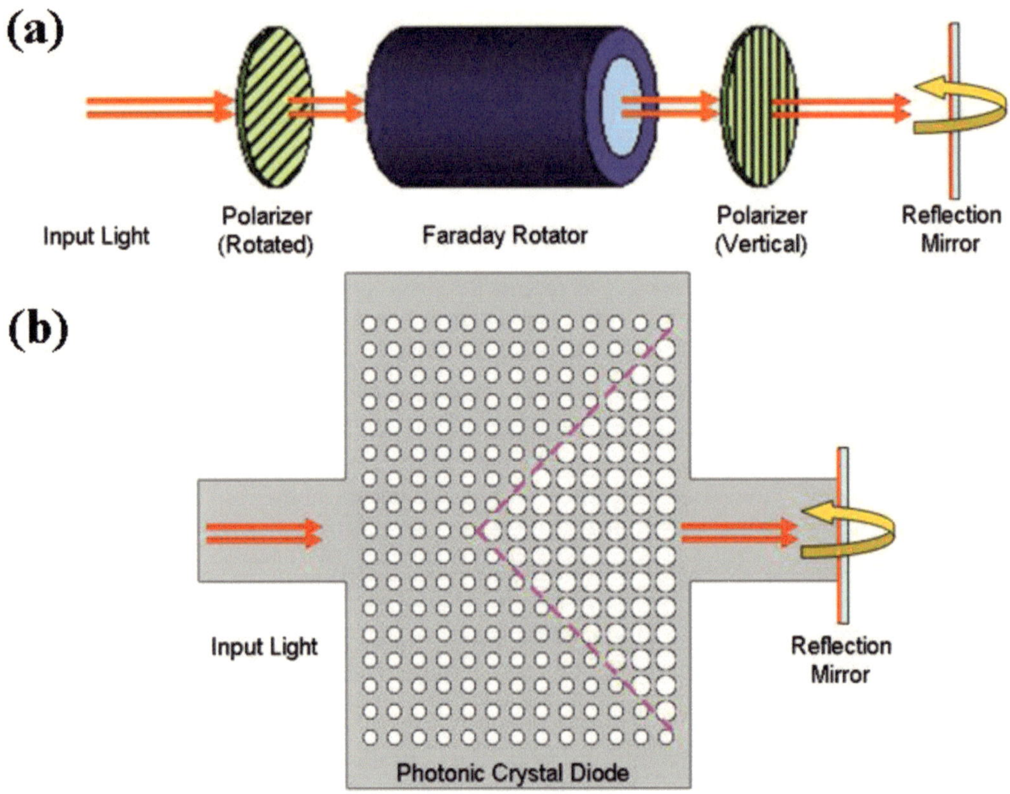

Figure 24. a) Traditional magnetic-optical isolator with reflection mirror at the output port. (b) On-chip optical diode system with reflection mirror at the output port. The absence of the reflection signal at the input port can indicate the isolation property of the system [55].

round-trip output waveguides is the same $2a$ (a=440 nm) and the center connection waveguide is $6a$. The length of the center connection waveguide, which measures the distance of the total-reflection mirror away from the output port of the diode, is $10a$. The spectra [Fig. 25(b)] show that the round-trip reflection peak is located at 1,582 nm and has a quantity of only 0.3%, which is almost two orders of magnitude smaller than the forward peak [with a maximum transmissivity of 22.9%]. The result indicates that the diode has a significant optical isolation property.

Based on the above numerical analysis, the double-diode structure was fabricated in silicon. Figure 25(c) shows the SEM images of the fabricated double-diode structures along the light path. The lattice constant a was 440 nm, and the radii r_1 and r_2 of air holes in the two photonic crystals of the heterojunction diode were approximately 110 nm and 160 nm. The length of the output waveguide is 4 μm (≈$10a$). The slab thickness was 220 nm. Figure 25(d) shows the experimentally measured forward and backward transmission spectra of the diode, as well as the round-trip transmission spectrum of the doubled-diode structure. The forward, backward and round-trip transmissions are optimized and the input/output loss has been removed. The spectra show that the maximum round-trip transmissivity, located at 1,553 nm, is only 0.4%, almost two orders of magnitude smaller than the forward peak [with a maximum transmissivity of 32.9%]. The experiment confirms the existence of a significant isolation effect in agreement with the theoretical prediction. We change the length of the center connection waveguide of the double-diode structure from 4 μm to 3 μm and 5 μm. The measured round-trip transmission spectra for the three structures are displayed in Fig. 25(e). The results show that the round-trip transmission signal decreases remarkably along with the increasing length of the center connection waveguide, and already reaches an extremely low level (below 0.05%) in the whole spectrum range when the output waveguide length increases to 5 μm. This clearly indicates that the mode dispersion in the output waveguide of the diode has no influence to the isolation property of the diode. Due to the arbitrariness of the lattice constant *a*, we can freely adjust the isolation frequency to anywhere as desired. This could be very convenient for the design of realistic photonic devices.

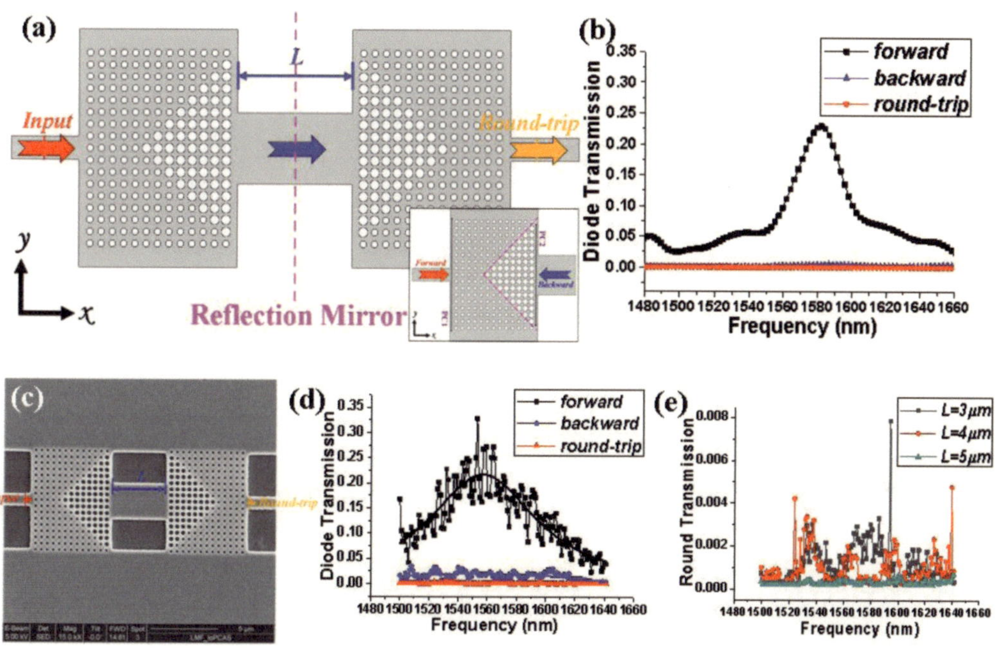

Figure 25. a) Schematic geometry of a doubled-diode structure with the total reflection mirror modeling the round-trip transmission of an isolator, and the corresponding single-diode structure under forward and backward transmissions. (b) Calculated forward (black line), backward (blue line), and round-trip (red line) transmission spectra of the diode. (c) Scanning electron microscope images of doubled-diode structure. (d) Experimental transmission spectra of forward transmission (black line), backward transmission (blue line), and the round-trip reflection (red line). (e) Experimental spectra of the round-trip reflection with changed center waveguide length L [55].

To better understand the underlying physics, we further make a detailed analyses based on the scattering matrix theory adopted in Ref. [53,54], and find that the above numerical results of optical isolation are in no conflict with the reciprocity theorem involved in our linear and passive silicon optical diode structure. Our diode basically consists of two in-plane information channels (A and B, the input and output waveguide channels for infrared signal, which can be either single mode or multimode channels.) as well as many in-plane side-way and off-plane scattering channels (denoted as C as a whole, which causes dissipation of information away the signal channels). At the two ends of the diode device the fields are written as follows:

$$\begin{bmatrix} A_{out} \\ B_{out} \\ C_{out} \end{bmatrix} = S \begin{bmatrix} A_{in} \\ B_{in} \\ C_{in} \end{bmatrix}, \tag{3}$$

in which A_{in} corresponds to the input signal from port A, A_{out} to the output signal from port A, B_{in} to the input signal from port B, B_{out} to the output signal from port B, C_{in} to the input signal from port C, and C_{out} to the output signal from port C. The scattering matrix S transforms the input state of all the channels [the column vector in the right hand of Eq. (3)] into the output state of all the channels [the column vector in the left hand of Eq. (3)]. The scattering equation of the forward transmission is written as:

$$\begin{bmatrix} a_{out1} \\ b_{out1} \\ c_{out1} \end{bmatrix} = S \begin{bmatrix} a_{in0} \\ 0 \\ 0 \end{bmatrix}. \tag{4}$$

As the silicon diode structure is linear and passive, the system as a whole is reciprocal in regard to time-reversal symmetry following the Lorentz reciprocity theorem. As a result, the scattering matrix S is symmetric with $S = S^T$ and further satisfies:

$$S^* = S^{-1}. \tag{5}$$

Suppose all the output signals are reversed and come back into the system, then the input at port B for the system is now exactly the same as $[a_{out1}^* \; b_{out1}^* \; c_{out1}^*]^T$. The scattering equation is then

$$S \begin{bmatrix} a_{out1}^* \\ b_{out1}^* \\ c_{out1}^* \end{bmatrix} = (S^* \begin{bmatrix} a_{out1} \\ b_{out1} \\ c_{out1} \end{bmatrix})^* = (S^{-1} \begin{bmatrix} a_{out1} \\ b_{out1} \\ c_{out1} \end{bmatrix})^* = \begin{bmatrix} a_{in0}^* \\ 0 \\ 0 \end{bmatrix}, \tag{6}$$

which is exactly the same as the initial input from port A. This clearly indicates that there is no isolation behavior in the structure if all information is reversed back into the system, consistent with the reciprocity theorem for a time-reversal symmetric system.

However, the story can be very different when the in-plane signal transport is concerned, as is always the case for 2D silicon PC slab structures. In our structure the information and energy involved in C channels are dissipated permanently against the in-plane channel A and B due to scattering loss (both in-plane and off-plane), and they cannot be reversed back totally and input again into the structure, so in practice, C_{in} in Eq. (3) can be assumed to be zero. As a result, Eq. (6) should be modified as:

$$S\begin{bmatrix}0\\ b_{out1}^{*}\\ 0\end{bmatrix}=\begin{bmatrix}a_{out2}\\ b_{out2}\\ c_{out2}\end{bmatrix}. \tag{7}$$

In general, Eq. (7) looks very different from Eq. (6), which indicates that the reciprocal transport of light in regard to the signal channel A and B has been broken. It shows that even if the same forward transmission signal of port B is reversed back and input into the diode, the output signal of port A can be much different from the initial input signal $a_{\text{in}0}$ of port A because no signal is reversed and input back into the channel C. Therefore, the considerable unidirectional transmission behavior can take place for the in-plane signal with no conflict with the reciprocal principle. In other words, the $a_{out2}(=S_{12}\bullet b_{out1}^{*})$ could be much different from $a_{in0}^{*}(=S_{11}\bullet a_{out1}^{*}+S_{12}\bullet b_{out1}^{*}+S_{13}\bullet c_{out1}^{*})$ when $S_{11}\bullet a_{out1}^{*}+S_{13}\bullet c_{out1}^{*}\neq 0$. This justifies the occurrence of a good isolation effect in the silicon optical diode. In ideal structures, both of them are zero, and Eq. (7) becomes

$$S\begin{bmatrix}0\\ b_{out1}^{*}\\ 0\end{bmatrix}=\begin{bmatrix}0\\ 0\\ c_{out2}\end{bmatrix}, \tag{8}$$

which implies a 100% signal contrast of the isolator.

It is worth saying a few more words here for better drawing a clear picture about the physics discussed in the above. In nature, as time always flows forward and cannot be reversed, one usually uses the term of reciprocal or nonreciprocal transport of light to describe a model system of back transport of light, in many cases to describe the reflection of light back into the considered structure. In this regard, simply consider a point source radiating an outgoing spherical wave front. If time can be reversed, the outgoing spherical wave front is contracted into an ingoing spherical wave front, eventually to a point. This is a very good picture to describe reciprocal transport of light in a linear system. However, to realize in real world such a concept, one needs to place a perfect spherical mirror concentric with the point source of light, which reflects back all information carried by the outgoing expanding spherical wave into the ingoing contracting spherical wave. If, however, one has only a small planar mirror placed at some distance and with a limited solid angle with respect to the source, the reflected signal can never return to the initial state of a point source when it reaches the position where the light source is located. The conventional magneto-optical isolator also works in this category of physical picture. It is used to block down the back-reflection signal of the transmission light, and the underlying physics can be well described by the model of time-reversal symmetry breaking. The same physics picture applies equally well to our optical diode. The fact that there exists information dissipation from the signal channels to other channels in a spatial-inversion symmetry breaking structure is sufficient to induce an optical isolation in regard to the signal.

The above numerical calculations and experimental results have shown that our silicon PC slab heterojunction diode exhibits promising performance of optical isolation, with a round-trip transmissivity two orders of magnitude smaller than the forward transmissivity for in-plane infrared light across the structure. Our scattering matrix analysis indicates that the considerable unidirectional transport of in-plane signal light can be attributed to the information dissipation and selective modal conversion in the multiple-channel spatial-inversion symmetry breaking structure and has no conflict with reciprocal principle for a time-reversal symmetric structure. It is expected that optimized connection interfaces between the input and output waveguides with the heterojunction diode can yield better impedance mismatch and bring higher forward transmission efficiency. That optical isolation can occur in a linear, passive, and time-independent optical structure would stimulate more thinkings and insights on the general transport theory of light in the fundamental side and open up a road towards photonic logics in silicon integrated optical devices and circuits in the application side.

7. Parallel–hetero photonic crystal structures

Photonic crystal heterostructures (PCHs) have attracted increasing interest in optical integrated circuits and cavity quantum electrodynamics (cavity QED) due to their useful photonic band-gap structure and the ability to provide nanocavities with ultra-high quality factor (Q factor). The properties of PCHs have been investigated both in theory and experiment over the past several years. In previous works, the transmission and reflection characteristics of PCHs were revealed only across the hetero-interface between two photonic crystals with different lattice constants. The basic character of PCHs was the shift of the band edge, which results in a transmission gap with approximately 100% efficiency [56]. In comparison, recently we have reported a method for making a parallel-hetero perturbation inside the waveguide and analyze the optical properties of the photonic crystal parallel-hetero perturbation (PHP) structure. It is expected that this new type of PCHs not only contains the band-edge shifting property but also has an additional transmission gap which can be easily regulated in the middle of the transmission band. Simulations and experiments confirm the existence of the additional transmission gap [57]. Our work can help to enlarge the usage of PCHs in the design and fabrication of novel cavities and filters via localized modulation of structural dispersion, which are key components in a photonic network. Based on the PCH structures, we have further proposed and realized a new scheme of cavity, an interface heterostructure cavity without any confinement barrier to confine light. Interestingly, the localized resonant mode lies in the pass band of the waveguide, in comparison of those cavities whose localized modes are always located within the band gap of the structures.

Figure 26(a) shows the geometry of the designed cavity, which consists of two identical parallel-hetero perturbation (PHP) waveguides. Each PHP waveguide consists of two semi-infinite PC (PC1 and PC2) slabs having slightly different lattice constants (a_1 and a_2, respectively). We assume a_2 is 5% larger than a_1. Each PC region has a triangular-lattice pattern of air holes, with the W1 waveguide formed by a line of missing holes along the Γ-K direction. Without periodicity along any direction, the PHP is essentially a kind of incommensurate PC superlattice along the waveguide direction. The transmission spectra of PHP waveguide structure exhibits an additional gap [0.29-0.30 (a_1/λ)] in the middle of the pass band of the two individual W1 waveguides besides the usual band edge shifting [0.26-0.27 (a_1/λ)] [56]. Figure 26(b) shows the calculated resonance spectrum of the cavity by using 3D FDTD method in association with the Pade approximation for the TE-like modes of the PC slab [58]. The resonance spectrum shows that a resonant mode surprisingly appears in the pass band region of the two PHP waveguides at 0.2855 (a_1/λ). In addition, this resonant mode (called band-pass mode) has a rather high value of Q factor reaching 5,340.

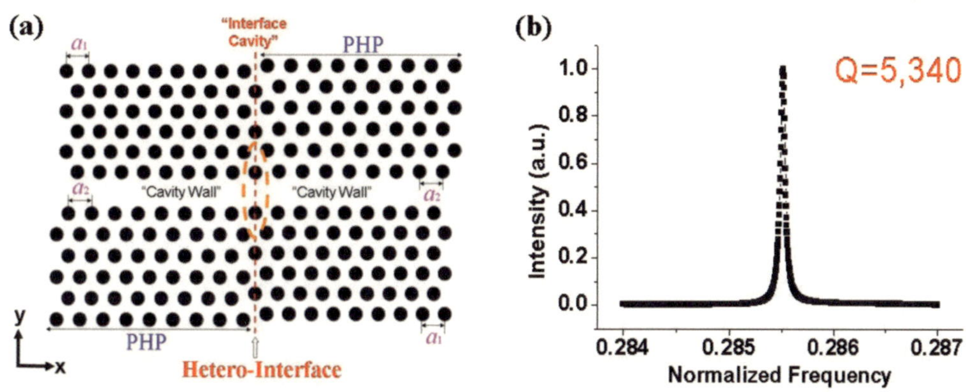

Figure 26. a) Schematic geometry of an anti-symmetric parallel-hetero cavity structure formed by only two PHP waveguides. (b) Calculated resonant mode distribution of the interface PHC [58].

We calculated the E_y field distribution of TE-like mode transporting along the waveguide at the resonance frequency 0.2855 (a_1/λ). It is surprising that the interface PHC has no influence on the propagation of waveguide mode [Fig. 27(a)]. To further confirm that the resonant mode really exists, we plot in Fig. 27(b) the calculated Ey field distribution of the resonant mode at the interface PHC at 0.2855 (a_1/λ) using a point source located within the interface cavity region. The mode does oscillate as a quadrupole form without barriers along the propagation direction even after the light source is turned off. This indicates that the resonant mode does exist and is not a calculation fault.

Based on the numerical analysis, an interface PHC structure was fabricated [Fig. 28(a)].The experimental results [Fig. 28(b)] confirms the theoretical prediction with the Q factor of the

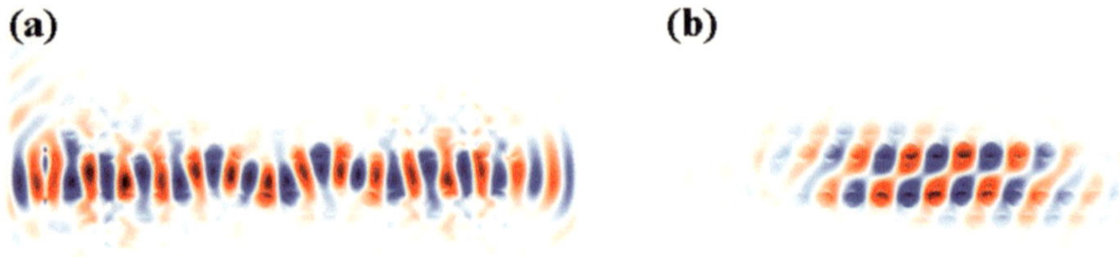

Figure 27. a) Calculated TE-like mode Ey field distribution of light transporting along the waveguide at the resonance frequency 0.2855 (a_1/λ). (b) Calculated TE-like mode Ey field distribution of the interface-cavity resonance at the resonance frequency 0.2855 (a_1/λ) [58].

cavity resonant mode decreases to 1,517 (centered at 1,570.2 nm) [Fig. 28(c)]. We then use an infrared CCD camera to monitor the resonant properties on top of the interface PHC region. When the laser wavelength is tuned to 1,570 nm, we can readily see a bright light spot located at the exact interface position [red circle in Fig. 28(d)] besides the other two light spots induced by the coupling between PHP waveguide and ridge waveguide. When the laser wavelength is tuned to 1,560 nm, the light spot in the red circle is gone while the other two light spots remain, indicating that the cavity is off resonance. This confirms the existence of the band-pass resonant mode.

The physics of PHC can be understood from the point of view of slow light effect. We have used a narrow light pulse centered at 0.2855 (a_1/λ) to pass through the interface PHC and a pure, same-length PHP waveguide, respectively, and recorded the output time of the pulse. The interface PHC is 599.25 time units, while the pure PHP waveguide is 606.3 time units, slightly longer than the interface PHC [Fig. 29(a)]. This indicates that the interface still contains one PC period length, so the whole interface PHC can be seen as a three-part structure, a combination of "slow light"–"fast light"–"slow light" region [Fig. 29(b)]. The central fast region in the PHC, which is only one period long, could also confine energy between two slow light regions and behaves as a cavity.

The physics underlying the interface PHC can also be analyzed by focusing on the phase shift in the waveguide of the interface PHC during propagation due to the fact that this band-pass mode is near the edge of the additional-gap region. The current PHC, which is made from an incommensurate superlattice structure, is similar to a periodic structure with dislocations. As a result, multiple scattering of light around the dislocation will occur and result in pinning effect of light (energy shifting across the dislocation line). Previous study shows that the PHP structure provides the asymmetric phase condition during the light propagation along the waveguide. The interface connects two same anti-symmetric PHPs and can be seen as a vertical edge dislocation line (phase reversal line), so that the phase condition is reversed to the opposite aside the interface. In material science the dislocation could cause the charge accumulation. Similarly our phase reversal dislocation gives an abrupt phase shift across the interface and may cause the photon accumulation around the phase reversal line. This phase

http://dx.doi.org/10.5772/55041

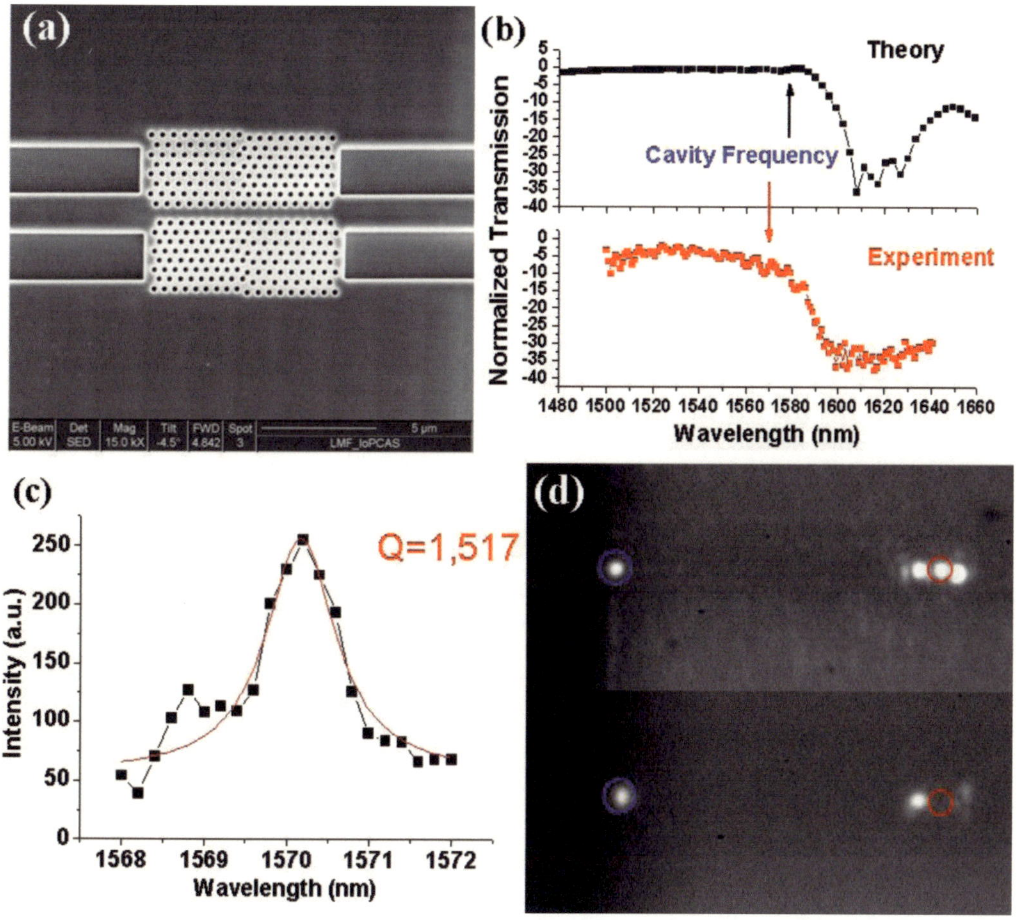

Figure 28. a) SEM image of the fabricated interface PHC composed of two PHP waveguides. (b) Experimental transmission spectrum of the interface PHC compared with the simulation results. (c) Resonance spectrum of the light spot obtained by its gray class analysis. (d) Infrared CCD images of the interface PHC structure on resonance (1,570 nm) and off resonance (1,560 nm) with output ridge waveguide [58].

reversal dislocation plays as a defect and forces the light energy distribution to become skew and swinging while crossing the interface [Fig. 27(a)]. The local energy oscillation acts as the cavity resonance and leads to the photon accumulation at the interface.

The PHC has several distinct properties. First, unlike the conventional PC cavities which are formed either by removing one or several air holes or by waveguide-like hetero-cavities, the current PHC does not have any confinement barrier to confine light. Second, the resonant mode is located within the pass band of waveguides, so the localized resonant mode and the continuous waveguide mode can easily co-exist in the same frequency and space regime. Third, the coupling efficiency between waveguide and cavity modes is much higher than the conventional cavities, which strongly depend on the transverse evanescent field profile overlap of the cavity mode and the waveguide mode. As a result, this PHC has nearly no

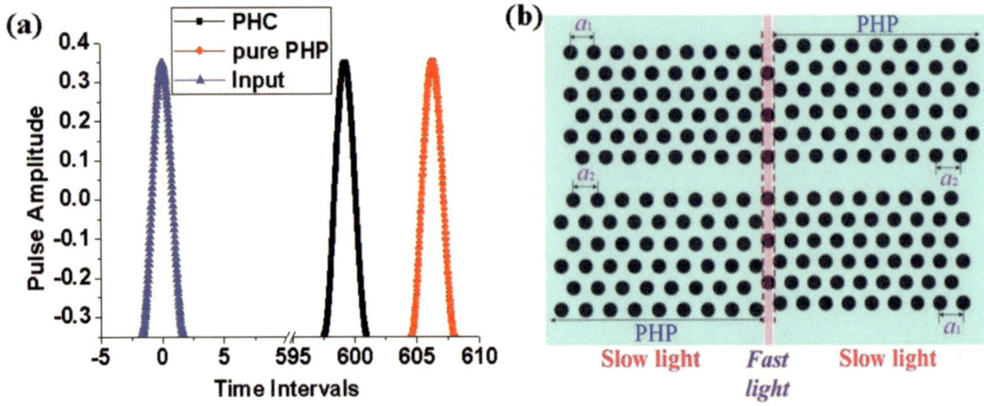

Figure 29. a) Time diagram of a narrow light pulse at 0.2855 (a_1/λ) passing through the interface PHC and a pure, same-length PHP waveguide. (b) Schematic slow light diagram of interface PHC [58].

influence on the propagation of waveguide mode despite the presence of localized cavity mode. High coupling efficiency between waveguide and cavity mode can help to reduce the operating power for PC lasers, and make it possible to integrate multiple lasers, photodetectors, and switches on a single chip.

8. Conclusions

In summary, we have presented recent progresses on infrared 2D air-bridged silicon PC slab devices that were made in our group in the past several years. 2D air-bridged silicon PC slab structures can confine light by the high index contrast in the vertical off-plane direction and manipulate light by photonic band and band gaps in the lateral in-plane direction. In addition, silicon is transparent and has a large refractive index in the infrared wavelengths. Therefore, the air-bridged silicon PC structures become one of the most important elements in integrated optics.

In this review, we have discussed several integrated optical elements and devices in regard to their design, fabrication, and characterization. These devices are based on either the PBG or photonic band structure engineering. To bring these devices into reality, we have made extensive efforts to construct high-efficiency numerical simulation tools for solution of photonic band structures, transmission spectra, light propagation dynamics, light wave patterns, and many others. These simulations allow for design of optimized PC structures for a specific application. The fabrication of these devices strongly depends on state-of-the-art nanofabrication technologies including EBL, FIB, ICP etching, and other wet etching techniques and procedures. We have constructed a convenient experimental setup to measure the transmission spectra and monitor the propagation path of infrared signals within PC structures simultaneously. The setup makes the optical characterization of PC devices accurate, user-friendly, fast, and convenient.

We have proposed Γ−M waveguides made in 2D triangular-lattice PC slabs. After a series of optimization and improvement, we find out the optimized geometries for high performance wide band Γ−M waveguides. As the Γ−M waveguide is perpendicular to the usual Γ−K waveguide, we combine the Γ−K waveguide and Γ−M waveguide together to form 90° waveguide bends and channel drop filters. The combination of Γ−M waveguide and Γ−K waveguide can offer a flexible way to interconnect and couple between different devices. In addition, we have shown design of two other kinds of channel drop filters. They achieve fine tuning of the resonant wavelengths by changing the size of the cavities or the shape of the air holes around the cavities. We have also designed and realized PC waveguide based on the coupled-cavity scheme. There are a lot of geometric parameters to fine tune the light transport properties in the structure. Moreover, we have successfully fabricated high-*Q* silicon PC microcavities with a *Q*-factor up to 70000 and this paves the way to experimentally explore light-matter interaction within the strong-coupling regime in the all-solid platform.

We have investigated optical devices that work on band structure engineering. We have explored PC structures that exhibit interesting and useful dispersion and refraction properties, such as negative refraction and self-collimation effects. We have designed and fabricated a kind of PC structure with negative refraction effect and use SNOM technology to observe the negative refraction ray trace of infrared light beam. In addition, we obtain broadband large-angle self-collimation effect for TE-like guided modes in infrared wavelength.

We have demonstrated the design, fabrication, and characterization of on-chip wavelength-scale optical diodes that are made from the heterojunction between two different silicon 2D square-lattice PC slabs with directional bandgap mismatch and different mode transitions. The measured transmission spectra show considerable unidirectional transmission behavior, in good agreement with numerical simulations. The experimental realization of on-chip optical diodes with wavelength-scale size using all-dielectric, passive, and linear silicon photonic crystal structures may help to construct on-chip optical logical devices without nonlinearity or magnetism, and would open up a road towards photonic computers.

We have demonstrated optical isolation of our diode structure. Both numerical simulations and experimental measurements show that the round-trip transmissivity of our diode could be two orders of magnitudes smaller than the forward transmissivity, indicating good performance of optical isolation. The occurrence of in-plane light isolation is attributed to the information dissipation due to off-plane and side-way scattering and selective modal conversion in the multiple-channel structure and has no conflict with the reciprocal principle. That optical isolation can occur in a linear, passive, and time-independent optical structure would stimulate more thinkings and insights on the general transport theory of light in the fundamental side and open up a road towards photonic logics in silicon integrated optical devices and circuits in the application side.

We have designed and fabricated cavities without confinement barrier by combining two incommensurate PC superlattice waveguides. A resonant mode with a high quality factor shows up in the pass band of waveguides. It has nearly no influence on the propagation of waveguide mode and can be directly coupled with the waveguide mode. The experimental measurement confirms the theoretical prediction of extraordinary coexistence of localized cavity mode and continuous waveguide mode with high coupling efficiency in the same frequency and space regime. The novel type of cavity has a number of unique properties that are advantageous to on-chip information transport and processing. The discovery of cavity without confinement barrier might attract interest in fundamental physics and optical engineering communities.

All of these results show that 2D air-bridged silicon PC structures can control light propagation in many flexible ways and have many potential applications in all-optical integrated circuits and other fields. The efforts that we have made and the experiences that we have accumulated in the past several years will allow us to design and realize PC devices with novel functionalities.

Acknowledgements

This work was supported by the National Key Basic Research Special Foundation of China (No. 2011CB922002), the National Natural Science Foundation of China, and National Center for Nanoscience and Technology of China. The authors would like to thank the previous members of our group, Cheng Ren, Haihua Tao, Yazhao Liu, and Changzhu Zhou for their contributions to the works presented in this review.

Author details

Zhi-Yuan Li*, Chen Wang and Lin Gan

*Address all correspondence to: lizy@aphy.iphy.ac.cn

Laboratory of Optical Physics, Institute of Physics, Chinese Academy of Sciences, Beijing, China

References

[1] Yablonovitch, E. Inhibited spontaneous emission in solid-state physics and electronics. Physical Review Letters, (1987).

[2] Joannopoulos, J D, Johnson, S G, Winn, J N, & Meade, R D. Photonic Crystals: Molding the Flow of Light. 2nd ed. Princeton: Princeton University Press, (2008).

[3] Valentine, J, Zhang, S, Zentgraf, T, Ulin-avila, E, Genov, D A, Bartal, G, & Zhang, X. Three-dimensional optical metamaterial with a negative refractive index. Nature, (2008).

[4] Yao, J, Liu, Z W, Liu, Y M, Wang, Y, Sun, C, Bartal, G, Stacy, A M, & Zhang, X. Optical negative refraction in bulk metamaterials of nanowires. Science, (2008).

[5] Johnson, S G, Fan, S H, Villeneuve, P R, Joannopoulos, J D, & Kolodziejski, L A. Guided modes in photonic crystal slabs. Physical Review B: Condensed Matter and Materials Physics, (1999).

[6] Painter, O, Lee, R K, Scherer, A, Yariv, A, Brien, O, Dapkus, J D, Kim, P D, & , I. Two-dimensional photonic band-gap defect mode laser. Science, (1999).

[7] Mcnab, S J, Moll, N, & Vlasov, Y A. Ultra-low loss photonic integrated circuit with membrane-type photonic crystal waveguides. Optics Express, (2003).

[8] Luo, C Y, Johnson, S G, Joannopoulos, J D, & Pendry, J B. All-angle negative refraction without negative effective index. Physical Review B: Condensed Matter and Materials Physics, (2002).

[9] Kosaka, H, Kawashima, T, Tomita, A, Notomi, M, Tamamura, T, Sato, T, & Kawakami, S. Self-collimating phenomena in photonic crystals. Applied Physics Letters, (1999).

[10] Yu, X F, & Fan, S H. Bends and splitters for self-collimated beams in photonic crystals. Applied Physics Letters, (2003).

[11] Kosaka, H, Kawashima, T, Tomita, A, Notomi, M, Tamamura, T, Sato, T, & Kawakami, S. Superprism phenomena in photonic crystals: toward microscale lightwave circuits. Journal of Lightwave Technology, (1999).

[12] Baba, T, Matsumoto, T, & Echizen, M. Finite difference time domain study of high efficiency photonic crystal superprisms. Optics Express, (2004).

[13] Berrier, A, Mulot, M, Swillo, M, Qiu, M, Thylén, L, Talneau, A, & Anand, S. Negative refraction at infrared wavelengths in a two-dimensional photonic crystal. Physical Review Letters, (2004).

[14] Born, M, Wolf, E, & Bhatia, A B. Principles of Optics: Electromagnetic Theory of Propagation, Interference and Diffraction of Light. New York : Cambridge University Press, (1999).

[15] Ho, K M, Chan, C T, & Soukoulis, C M. Existence of a photonic gap in periodic dielectric structures. Physical Review Letters, (1990).

[16] Johnson, S, & Joannopoulos, J. Block-iterative frequency-domain methods for Maxwell's equations in a planewave basis. Optics Express, (2001).

[17] Li, Z Y, Gu, B Y, & Yang, G Z. Large absolute band gap in 2D anisotropic photonic crystals. Physical Review Letters, (1998).

[18] Li, Z Y, Wang, J, & Gu, B Y. Creation of partial band gaps in anisotropic photonic-band-gap structures. Physical Review B: Condensed Matter and Materials Physics, (1998).

[19] Pendry, J B. Photonic Band Structures. Journal of Modern Optics, (1994).

[20] Chan, C T, Yu, Q L, & Ho, K M. Order-N spectral method for electromagnetic waves. Physical Review B: Condensed Matter and Materials Physics, (1995).

[21] Taflove, A. Computational Electrodynamics: the Finite-Difference Time-Domain Method. Boston : Artech House, (1995).

[22] Nicorovici, N A, Mcphedran, R C, & Botten, L C. Photonic band gaps for arrays of perfectly conducting cylinders. Physical Review E: Statistical Physics, Plasmas, Fluids, and Related Interdisciplinary Topics, (1995).

[23] Li, L M, & Zhang, Z Q. Muitiple-scattering approach to finite-sized photonic band-gap materials. Physical Review B: Condensed Matter and Materials Physics, (1998).

[24] Li, Z Y, & Lin, L L. Photonic band structures solved by a plane-wave-based transfer-matrix method. Physical Review E: Statistical, Nonlinear, and Soft Matter Physics, (2003).

[25] Li, Z Y, & Lin, L L. Evaluation of lensing in photonic crystal slabs exhibiting negative refraction. Physical Review B: Condensed Matter and Materials Physics, (2003).

[26] Lin, L L, Li, Z Y, & Ho, K M. Lattice symmetry applied in transfer-matrix methods for photonic crystals. Journal of Applied Physics, (2003).

[27] Li, Z Y, & Ho, K M. Light propagation in semi-infinite photonic crystals and related waveguide structures. Physical Review B: Condensed Matter and Materials Physics, (2003).

[28] Li, Z Y, Lin, L L, & Ho, K M. Light coupling with multimode photonic crystal waveguides. Applied Physics Letters, (2004).

[29] Che, M, & Li, Z Y. Analysis of photonic crystal waveguide bends by a plane-wave transfer-matrix method. Physical Review B: Condensed Matter and Materials Physics, (2008).

[30] Che, M, & Li, Z Y. Analysis of surface modes in photonic crystals by a plane-wave transfer-matrix method. Journal of the Optical Society of America a-Optics Image Science and Vision, (2008).

[31] Li, Z Y, & Ho, K M. Analytic modal solution to light propagation through layer-by-layer metallic photonic crystals. Physical Review B: Condensed Matter and Materials Physics, (2003).

[32] Li, J J, Li, Z Y, & Zhang, D Z. Second harmonic generation in one-dimensional nonlinear photonic crystals solved by the transfer matrix method. Physical Review E: Statistical, Nonlinear, and Soft Matter Physics, (2007).

[33] Li, Z Y, Li, J J, & Zhang, D Z. Nonlinear frequency conversion in two-dimensional nonlinear photonic crystals solved by a plane-wave-based transfer-matrix method. Physical Review B: Condensed Matter and Materials Physics, (2008).

[34] Oskooi, A F, Roundy, D, Ibanescu, M, Bermel, P, Joannopoulos, J D, & Johnson, S G. MEEP: A flexible free-software package for electromagnetic simulations by the FDTD method. Computer Physics Communications, (2010).

[35] Shinya, A, Mitsugi, S, Kuramochi, E, & Notomi, M. Ultrasmall multi-channel resonant-tunneling filter using mode gap of width-tuned photonic-crystal waveguide. Optics Express, (2005).

[36] Song, B S, Nagashima, T, Asano, T, & Noda, S. Resonant-wavelength tuning of a nanocavity by subnanometer control of a two-dimensional silicon-based photonic crystal slab structure. Applied Optics, (2009).

[37] Liu, Y Z, Liu, R J, Zhou, C Z, Zhang, D Z, & Li, Z Y. Gamma-Mu waveguides in two-dimensional triangular-lattice photonic crystal slabs. Optics Express, (2008).

[38] Zhou, C Z, Liu, Y Z, & Li, Z Y. Waveguide bend of 90° in two-dimensional triangular lattice silicon photonic crystal slabs. Chinese Physics Letters, (2010).

[39] Tao, H H, Ren, C, Liu, Y Z, Wang, Q K, Zhang, D Z, & Li, Z Y. Near-field observation of anomalous optical propagation in photonic crystal coupled-cavity waveguides. Optics Express, (2010).

[40] Hennessy, K, Badolato, A, Winger, M, Gerace, D, Atatüre, M, Gulde, S, Fält, S, Hu, E L, & Imamoglu, A. Quantum nature of a strongly coupled single quantum dot-cavity system. Nature, (2007).

[41] Akahane, Y, Asano, T, Song, B S, & Noda, S. High-Q photonic nanocavity in a two-dimensional photonic crystal. Nature, (2003).

[42] Zhou, C Z, Wang, C, & Li, Z Y. Fabrication and spectra-measurement of high Q photonic crystal cavity on silicon slab", Acta Physica Sinica, (2012).

[43] Ren, C, Tian, J, Feng, S, Tao, H H, Liu, Y Z, Ren, K, Li, Z Y, Cheng, B Y, Zhang, D Z, & Yang, H F. High resolution three-port filter in two dimensional photonic crystal slabs. Optics Express, (2006).

[44] Liu, Y Z, Feng, S A, Tian, J, Ren, C, Tao, H H, Li, Z Y, Cheng, B Y, Zhang, D Z, & Luo, Q. Multichannel filters with shape designing in two-dimensional photonic crystal slabs. Journal of Applied Physics, (2007).

[45] Liu, Y Z, Liu, R J, Feng, S A, Ren, C, Yang, H F, Zhang, D Z, & Li, Z Y. Multichannel filters via Γ-M and Γ-K waveguide coupling in two-dimensional triangular-lattice photonic crystal slabs. Applied Physics Letters, (2008).

[46] Gan, L, Liu, Y Z, Li, J Y, Zhang, Z B, Zhang, D Z, & Li, Z Y. Ray trace visualization of negative refraction of light in two-dimensional air-bridged silicon photonic crystal slabs at 1.55 μm. Optics Express, (2009).

[47] Nguyen, H M, Dundar, M A, Van Der Heijden, R W, Van Der Drift, E, Salemink, J M, Rogge, H W M, & Caro, S. J. Compact Mach-Zehnder interferometer based on self-collimation of light in a silicon photonic crystal. Optics Express, (2010).

[48] White, T P, De Sterke, C M, Mcphedran, R C, & Botten, L C. Highly efficient wide-angle transmission into uniform rod-type photonic crystals. Applied Physics Letters, (2005).

[49] Gan, L, Qin, F, & Li, Z Y. Broadband large-angle self-collimation in two-dimensional silicon photonic crystal", Optics Letters, (2012).

[50] Serebryannikov, A E. One-way diffraction effects in photonic crystal gratings made of isotropic materials. Physical Review B: Condensed Matter and Materials Physics. (2009).

[51] Wang, C, Zhou, C Z, & Li, Z Y. On-chip optical diode based on silicon photonic crystal heterojunctions. Optics Express, (2011).

[52] Feng, L, Ayache, M, Huang, J, Xu, Y L, Lu, M H, Chen, Y F, Fainman, Y, & Scherer, A. Nonreciprocal light propagation in a silicon photonic circuit. Science, (2011).

[53] Fan, S, Baets, R, Petrov, A, Yu, Z, Joannopoulos, J D, Freude, W, Melloni, A, Popovic, M, Vanwolleghem, M, Jalas, D, Eich, M, Krause, M, Renner, H, Brinkmeyer, E, & Doerr, C R. Comment on "Nonreciprocal light propagation in a silicon photonic circuit". Science, (2012). b.

[54] Feng, L, Ayache, M, Huang, J, Xu, Y L, Lu, M H, Chen, Y F, Fainman, Y, & Scherer, A. Response to comment on "Nonreciprocal light propagation in a silicon photonic circuit". Science, (2012). c.

[55] Wang, C, Zhong, X L, & Li, Z Y. Linear and passive silicon optical isolator. Scientific Reports, (2012).

[56] Song, B S, Asano, T, Akahane, Y, Tanaka, Y, & Noda, S. Transmission and reflection characteristics of in-plane hetero-photonic crystals. Applied Physics Letters (2004).

[57] Wang, C, Zhou, C Z, & Li, Z Y. Creation of stop band by introducing parallel-hetero perturbation in two-dimensional photonic crystal waveguides Journal of Optics, (2011).

[58] Wang, C, & Li, Z Y. Cavities without confinement barrier in incommensurate photonic crystal superlattices EPL, (2012).